Current Topics in Czech and Central European Geography Education

Petra Karvánková • Dagmar Popjaková
Michal Vančura • Jozef Mládek
Editors

Current Topics in Czech and Central European Geography Education

Springer

Editors
Petra Karvánková
Faculty of Education
Department of Geography
University of South Bohemia
in České Budějovice
České Budějovice, Czech Republic

Dagmar Popjaková
Faculty of Education
Department of Geography
University of South Bohemia
in České Budějovice
České Budějovice, Czech Republic

Michal Vančura
Faculty of Education
Department of Geography
University of South Bohemia
in České Budějovice
České Budějovice, Czech Republic

Jozef Mládek
Faculty of Education
Department of Geography
University of South Bohemia
in České Budějovice
České Budějovice, Czech Republic

ISBN 978-3-319-43613-5   ISBN 978-3-319-43614-2 (eBook)
DOI 10.1007/978-3-319-43614-2

Library of Congress Control Number: 2016954079

Printed on acid-free paper

This Springer imprint is published by Springer Nature
The registered company is Springer International Publishing AG Switzerland

Acknowledgments

Firstly, for the appearance of the publications, we thank Prof. RNDr. Jozef Mládek, DrSc., from the Department of Geography of the Faculty of Education of the University of South Bohemia in České Budějovice, who actually led, inspired, and encouraged all the team of editors and authors of the publication.

The publication has been financially supported by the Grant Agency of the University of South Bohemia in České Budějovice 162/2013/S *Geographical research and its implementation in inquiry-based teaching of Geography*. Thanks to all the students of the Department of Geography of the Faculty of Education of the University of South Bohemia in České Budějovice who participated in the formal preparation of the publication, mainly, Bc. Martina Kuřímska, Bc. Jiří Nerad, Bc. Jaroslava Tomková, and Bc. Barbora Rybaříková.

We thank the students of the Department of Geography of the Faculty of Education of Masaryk University in Brno Bc. Bohumír Veselý, Bc. Lucie Peřinová, and Bc. Pavlína Jedličková who were involved in the preparation of case study *A Powerful Way of Teaching Geography* within the project FRMU MUNI/FR/1315/2014. Under the project, a case study entitled *Analysis of the facilities and development of ATC Olšovec – Jedovnice* was also processed.

The publication was supported by the Czech Geographical Society and its departments for education and regional geography.

Contents

1 **Geographical Education from the Czech, Slovak and Polish Point of View** 1
Petra Karvánková, Miroslav Marada, Dana Řezníčková, Alena Madziková, and Danuta Piróg

Part I General Education System and Teaching Geography in the Czech Republic, Slovak Republic and Republic of Poland

2 **Geography Curriculum in Czechia: Challenging Opportunities** 15
Eduard Hofmann, Hana Svobodová, and Petr Knecht

3 **Current Geography Education in Slovakia: Conversions and Conditions** 33
Ladislav Tolmáči and Štefan Karolčík

4 **Teaching and Learning Geography in Secondary Education in Poland** 45
Wiktor Osuch

Part II Innovative Methods in Geography Teaching – Way to Mobilize the Students

5 **Inquiry-Based Education of Physical Geography** 63
Petra Karvánková, Dagmar Popjaková, Michal Vančura, and Šárka Nedvědová

6 **Global Development Education: Czech Approach in the Spread of Global Awareness** 87
Petra Karvánková, Dagmar Popjaková, Jana Kubelková, Petra Skalická, Pavel Horejš, Štěpán Klučka, Markéta Martínková, and Martin Kochánek

7 **Case Studies in Geography Education as a Powerful Way of Teaching Geography** 115
Eduard Hofmann and Hana Svobodová

8 **Project-Based Learning in Geography** 129
Petr Hynek

Part III Information Technology in Teaching Geography

9 **Use of Information and Communication Technology and Resources of the Internet in Education Natural Sciences** 147
Małgorzata Pietrzak

10 **Information Technologies in Teaching Geography from the Teacher's Point of View** 169
Martin Blažek, Martin Lána, Vojtěch Blažek, and Jan Dvořák

11 **Reading Satellite Images, Aerial Photos and Maps: Development of Cartographic and Visual Literacy** 187
Hana Svatoňová

Part IV Current Topics of Political, Regional and Social Geography and Their Didactic Applications

12 **The World's Major Regions as Part of Regional Geography Courses?** 211
Jiří Anděl, Ivan Bičík, and Kateřina Zavadská

13 **New Political Geography** 229
Barbara Baarová and Vladimír Baar

14 **Changes in Demographic Behaviour: Possible Use of Its Findings in Didactic Practice** 253
Jozef Mládek, Marcela Káčerová, Dagmar Popjaková, and Michal Vančura

Resumé 281

Streszczenie 285

List of the Authors

Jiří Anděl Faculty of Science, Department of Geography, University of J. E. Purkyně in Ústí nad Labem, Ústí nad Labem, Czech Republic

Vladimír Baar Faculty of Science, Department of Human Geography and Regional Development, University of Ostrava, Ostrava, Czech Republic

Barbara Baarová Faculty of Science, Department of Human Geography and Regional Development, University of Ostrava, Ostrava, Czech Republic

Ivan Bičík Faculty of Science, Department of Social Geography and Regional Development, Charles University in Prague, Prague, Czech Republic

Martin Blažek Faculty of Education, Department of Geography, University of South Bohemia in České Budějovice, České Budějovice, Czech Republic

UMR PASSAGES Université Montaigne-Bordeaux, Pessac, France

Vojtěch Blažek GIS administrator, Regioskop z.s.p.o., Písek; Faculty of Education, Department of Geography, University of South Bohemia in České Budějovice, České Budějovice, Czech Republic

Jan Dvořák Faculty of Education, Department of Geography, University of South Bohemia in České Budějovice, České Budějovice, Czech Republic

Grammar School Jírovcova, České Budějovice, České Budějovice, Czech Republic

Eduard Hofmann Faculty of Education, Department of Geography, Masaryk University in Brno, Brno, Czech Republic

Pavel Horejš Faculty of Education, Department of Geography, University of South Bohemia in České Budějovice, České Budějovice, Czech Republic

Petr Hynek Center for Innovation in Engineering Education, o.p.s., České Budějovice, Czech Republic

Marcela Káčerová Faculty of Natural Sciences, Department of Human Geography and Demography, Comenius University in Bratislava, Bratislava, Slovak Republic

Štefan Karolčík Faculty of Natural Sciences, Department of Didactics, Comenius University in Bratislava, Bratislava, Slovak Republic

Petra Karvánková Faculty of Education, Department of Geography, University of South Bohemia in České Budějovice, České Budějovice, Czech Republic

Štěpán Klučka Faculty of Education, Department of Geography, University of South Bohemia in České Budějovice, České Budějovice, Czech Republic

Petr Knecht Faculty of Education, Department of Geography, Masaryk University in Brno, Brno, Czech Republic

Martin Kochánek Faculty of Education, Department of Geography, University of South Bohemia in České Budějovice, České Budějovice, Czech Republic

Jana Kubelková People in need, o.p.s., Praha, Czech Republic

Martin Lána ZŠ a MŠ Dečín IV, Děčín, Czech Republic

Alena Madziková Faculty of Humanities and Natural Sciences, Department of Geography and Applied Geoinformatics, University of Prešov, Prešov, Slovak Republic

Miroslav Marada Faculty of Science, Department of Social Geography and Regional Development, Charles University in Prague, Prague, Czech Republic

Markéta Martínková Faculty of Education, Department of Geography, University of South Bohemia in České Budějovice, České Budějovice, Czech Republic

Jozef Mládek Faculty of Education, Department of Geography, University of South Bohemia in České Budějovice, České Budějovice, Czech Republic

Šárka Nedvědová Faculty of Education, Department of Geography, University of South Bohemia in České Budějovice, České Budějovice, Czech Republic

Wiktor Osuch Institute of Geography, Department of Geography Teaching, Pedagogical University of Cracow, Kraków, Poland

Małgorzata Pietrzak Institute of Geography and Spatial Management, Jagiellonian University in Kraków, Kraków, Poland

Danuta Piróg Institute of Geography, Department of Geography Teaching, Pedagogical University of Cracow, Kraków, Poland

Dagmar Popjaková Faculty of Education, Department of Geography, University of South Bohemia in České Budějovice, České Budějovice, Czech Republic

Dana Řezníčková Faculty of Science, Department of Social Geography and Regional Development, Charles University in Prague, Prague, Czech Republic

Petra Skalická People in need, o.p.s., Praha, Czech Republic

Hana Svatoňová Faculty of Education, Department of Geography, Masaryk University in Brno, Brno, Czech Republic

Hana Svobodová Faculty of Education, Department of Geography, Masaryk University in Brno, Brno, Czech Republic

Ladislav Tolmáči Faculty of Natural Science, Department of Geography and Geologie, Matej Bel University, Banská Bystrica, Slovak Republic

Michal Vančura Faculty of Education, Department of Geography, University of South Bohemia in České Budějovice, České Budějovice, Czech Republic

Kateřina Zavadská Faculty of Science, Department of Geography, University of J. E. Purkyně in Ústí nad Labem, Ústí nad Labem-město, Czech Republic

Chapter 1
Geographical Education from the Czech, Slovak and Polish Point of View

Petra Karvánková, Miroslav Marada, Dana Řezníčková, Alena Madziková, and Danuta Piróg

The world community of the first two decades of the twenty-first century represents a complicated system. Its functioning is greatly affected by information. The world is interconnected by information. People, including children, are overwhelmed, or even more attacked, with huge amounts of information and signals. Innovation is the driving force for development in all areas of life. Technological progress increases the technical literacy of the world's inhabitants. On the other hand, differences between advanced and less advanced areas are broadening. Global inequalities become the substrate for abuse of confidence, increase in tension, conflicts and terrorism. Regrettably, Democracy as a system of social governance has not worked universally. Political systems in many parts of the world where attempts to overthrow authoritative regimes have emerged are fragile and unstable. A great part of the world's inhabitants lives in poverty. People are more often than not confronted with diversity. Uncertainty is increasing; fear of insecurity is growing. Children and the younger generation are growing and maturing under these complex conditions.

P. Karvánková (✉)
Faculty of Education, Department of Geography, University of South Bohemia in České Budějovice, Jeronýmova 10, 371 15 České Budějovice, Czech Republic
e-mail: karvanko@pf.jcu.cz

M. Marada • D. Řezníčková
Faculty of Science, Department of Social Geography and Regional Development, Charles University in Prague, Albertov 6, 128 43 Prague 2, Czech Republic

A. Madziková
Faculty of Humanities and Natural Sciences, Department of Geography and Applied Geoinformatics, University of Prešov, 17. Novembra 1, 081 16 Prešov, Slovak Republic

D. Piróg
Institute of Geography, Department of Geography Teaching, Pedagogical University of Cracow, Podchorążych St. 2, 30-084 Kraków, Poland

P. Karvánková et al. (eds.), *Current Topics in Czech and Central European Geography Education*, DOI 10.1007/978-3-319-43614-2_1

These facts affect their development and quality of education has crucial importance.

The mission and commitment of states is the creation of conditions for children and young people to be provided with the most optimal education options. The state not only provides the education infrastructure, including teachers as mediators of education and their training, but should also assure an effective system of educational instruments and methods. Teaching methods used at schools change over time. They adapt to new social challenges and trends and new empiricism. The year 2000 became a turning point in school curricula reforms in Europe.

In the Central European area, in today's two independent states of the former Czechoslovakia, Czechia and Slovakia, and also in neighbouring Poland, the latest school reforms were affected by the fact that their preparation fell in a period of extensive social transformation related to the change of regime in these former socialist/communist states to pro-Western democratic states. Transformation in the former Czechoslovakia followed after the political changes of November 1989, known as the "Velvet Revolution".

It should be noted that the development of the territories of present-day countries such as the Czech Republic, the Slovak Republic and the Republic of Poland, just like the whole of Central Europe, was a fairly complex process not only at the end of the twentieth century but also far back in history. Central Europe represents a fairly narrow area between the Baltic Sea and the Mediterranean, and geopolitically, it was under pressure from its neighbouring powers, Germany in the west and Russia in the east. The Austro-Hungarian Empire exercised a major influence on this area until WWI. This was mostly shown in the development of Poland which, as a geomorphologically flat area, presented few obstacles for the interests of the powers. The powerful medieval kingdom of Poland ceased to exist at the end of the eighteenth century when Poland was divided into three parts. One part fell to Prussia, one part to the Austro-Hungarian Empire and one part to Russia. Historically and geographically, the Czechia, comprising Bohemia, Moravia and Czech Silesia, were protected by mountains and rivers from all four sides, and for many years, they belonged to the Austro-Hungarian Empire. Similarly, in the tenth century, Slovakia became a part of the Hungarian Empire and later of the Austro-Hungarian Empire. After World War One, in 1918, separate countries of the Republic of Poland and Czechoslovakia were established. These state formations were destroyed as a result of the Second World War and reinstated again in peace treaties after 1945. From the Second World War onwards, part of Central Europe was under the political influence of the Soviet Union. Poland and Czechoslovakia become socialist countries. The aforementioned political, social and economic changes of the 1990s that were caused by the "Velvet Revolution" resulted in the breakdown of the Soviet Bloc, the socialist system in Europe, and eventually also in the division of Czechoslovakia. A political agreement was made, and on 1 January 1993, two independent countries were established, the Czech Republic and the Slovak Republic (see more Chap. 13, Sect. 13.6).

Despite the fact that Czechia, Slovakia and Poland can be classed as relatively young countries, their educational system has a long tradition. Its beginnings date back to the end of the eighteenth century, to the times of the school reforms carried out by Maria Theresa. Also the educational system during so called the socialist era showed very good standards. We can say that the benchmark was the traditionally good level of school education in the preceding period. For example, the good results of the Czech education system are proven, inter alia, by an evaluation of the standards of pupils' knowledge in mathematics, TIMSS, of 1995, according to which Czech pupils, together with East Asian and Dutch pupils, ranked among the best, with a markedly higher level of excellence in comparison to other areas.

The purpose of the school reform after the political changes in Czechia, Slovakia and Poland in 1989 was to define new school curricula. Since the mid-1990s, the Czech and Slovak education systems have developed in two separate states. The reform of the curriculum culminated in 2004 in Czechia and in 2008 in Slovakia, where basic educational documents were innovated and accepted (so called Framework Educational Programme, Czech abbreviation RVP 2011; Kmeť 2009). In Poland, where efforts at the democratisation of the totalitarian society had appeared since the early 1980s, educational reform was commenced already before 2000 (in 1999). The processes of social transformation in the above-mentioned Central European countries are connected with certain economic tensions and limitations on their financial resources. This problem still strongly affects the whole education system today. Schools are allocated insufficient budget resources (state and private). This reflects in the low wages of teachers and school staff and partly in the under-dimensioning of financial resources for the material background of the educational process as such.

International tests of pupils' competencies such as PISA, TIMSS or PIRLS (Mullis et al. 2012; OECD 2014) suggest that the process to seek the optimum educational model is not yet accomplished in these countries. Despite the results of mapping the knowledge levels of Czech pupils, and in some indicators also Slovak and Polish pupils, which have again achieved above-average or at least average ranks, the school systems still appear to need considerable improvement. The improvement should above all change the surviving "traditional teaching" methods. Even though the disciplines of pedagogy and subject didactics (Stuchlíková and Janík 2015) achieve research and activity results comparable to the results of their foreign partners, and although there is a tradition of pedagogy research institutes and methodological centres asserting innovative teaching methods, the essence of Czech school education is still based on memorising encyclopaedic knowledge. This is the traditional teaching model based on active teacher and passive pupil. According to this model, the teacher transfers information to the pupils who memorise it. Evaluation of the pupils´ knowledge is then based on their ability and extent of reproduction of the memorised information.

Despite the reform efforts, Czech and Slovak school education has not grasped the basic development trend brought about by the changed social climate based on democratisation and the development of the information society since the 1990s. The current generations of children growing up against a background of the Internet, electronic communication and social networks are characterised by considerable technical skills and unlimited access to information. Pupils are confronted with great amounts of information in every school subject taught. Although official curicula have very framework nature, the teachers' curricula are too broad. The textbooks, considered by teachers as compulsory, include a lot of teaching material, too many descriptions and often just raw, superficial information about the studied theme. Teachers are under pressure to cover the textbook curricula and deliver the required volume of information in lessons in time. The broadly conceived curricula limit teachers' creativeness and didactic work. They become a barrier. Considering textbook content as compulsory curricula do not allow the teachers to apply teaching methods activating the pupils, evoking their interest and allowing them to go deep into the studied themes. Then, the lessons do not provide room for the application of innovative didactic methods. Pupils often learn with antipathy, superficially, without a deeper knowledge or understanding of the essence of the learned information. Many of them lose interest and motivation to absorb the provided information or are overloaded and exhausted with the quantity of information poured on them. In addition, they might lose orientation in the quantity of information and are unable to absorb the substantial and select it from the unsubstantial. The result is that the many pupils lack elementary knowledge. The knowledge level of new generations of children is often evaluated as lower than before. Children are said to be less capable and less comprehensive than the previous generations. But the problem of the lack of knowledge of current generations of pupils is not linked to decreasing mental capacity of children but with the capacity setting of the educational system at schools, especially elementary ones.

The educational system has made blunders in the area of the definition of a suitable teaching system in Czechia (and Slovakia) since the 1990s because (1) the system has been unable to respond to social changes and has failed to accept responsibility and take up the role of information selector and creator of a compendium of knowledge in individual subjects forming a limited but compulsory knowledge base for children which they would be obliged to master and (2) thus it has not made use of the opportunity provided by the certain limitation of the scope of information taught to create room for the application of methods developed by didactic experts and aimed at attracting the pupils' attention, provoking them to own activity and leading them to an active involvement in the teaching process, making them think, dive into the essence of problems, create, investigate, discover, look for logical relations and discuss deeper presented themes in individual subjects. The current crisis of elementary education in the Czech Republic could be resolved by the courage to appoint a team of experts in every school subject tasked to create the basic framework for the subject in the sense of the two basic points mentioned above and recommend alternative forms of teaching based on constructivist principles on the basis of this background.

Teachers of geography face the same problems as mentioned above. Czech experts in the didactics of geography are aware of them and discuss them at professional forums, often also with their Slovak and Polish colleagues. Generally, there is a strong tradition of professional meetings of Czech, Slovak and Polish geographers. That is why the Department of Geography of the Pedagogic Faculty of the South Bohemian University in České Budějovice came up with the idea to address

Czech experts in didactics of geography from Charles University in Prague, Masaryk University in Brno, J. E. Purkyně University in Ústí nad Labem, Ostrava University, the Institute of School Education Research in Brno as well as Comenius University in Bratislava, Matej Bel University in Banská Bystrica, the University of Prešov and last but not least the University of Education and the Jagiellonian University in Krakow. These experts were asked to write a publication together digestedly surveying the current issues, themes and methods of teaching consulted, recommended and used in the Central European area represented by them. The publication represents the Czech and partly the Central European approach and experience connected with application of the didactic methods worked out and for some time already successfully used in the Western world. The publication can be considered a concise insight into Czech Didactics of Geography. It was issued under the auspices of the Czech Geographical Society and its departments for education and regional geography.

The aims of the submitted publication are the current challenges related to teaching geography, mainly at the secondary schools and the higher education level (for children aged 12–16 years). It presents not only a platform of current issues in the teaching of geography, but mainly tries to propose possible changes, accept and involve modern trends into teaching and respond to impulses leading to an increase in interest among pupils and students in the subject of geography.

The authors approached execution of the publication via two methods. The first was a theoretical-methodological description of individual educational topics at the level of general educational theories. This is linked to some degree of specification of geography education methods and techniques and also their specific modification as a result of the actual conditions of education and upbringing in Central European countries, particularly under conditions in Czechia, Slovakia and Poland. The authors' second goal is the application of up-to-date methods and techniques in geography tuition in an attempt to contribute to the field. Several case studies, implementation and possibly exercises which contain instructions and procedures, and also the results of the application of the presented theoretical-methodological principles mainly in Czech Republic, are used to achieve this goal.

The publication applies selected modern methodical procedures in the teaching practice of geography. These include inquiry-based learning of geography, global development education and powerful teaching, which are supported by the high-quality use of information and communication technologies. Above all, the introduction of modern teaching methods and techniques which have already applied in developed European countries is there topical. These significant changes have brought mainly transformational experiences and changes in the educational systems in the observed countries. The application of innovative methods in teaching must be based on the creation of the historical development and political background of curricular changes which have taken place within the observed Central European "education" area. The area is represented mainly by the knowledge of the geography education in Czechia, Slovakia and Poland.

Therefore, the attention of *Part I* of the publication is focused on the state of the curricula in the area of interest. The conceptions as well as contents of geography education differ in every single country with respect to geographical knowledge, due to the country's history, language and culture. In other words, every curriculum is influenced by its national context. The creators of the *geography curriculum* will find lots of topics on how to realise the teaching designed to promote learning. Part I is divided into three main Chaps. 2, 3 and 4, devoted to the education of geography in *Czechia, Slovakia and Poland.* Each of these chapters can provide an understanding of the general education system in these countries, current issues, developments, problems, changes and trends in teaching geography. At the same time, attention is paid to the strengths and weaknesses of the current geography education. It includes the proposals of possible inspiration, challenges and changes which are brought about by school documents from developed countries.

The introduction of innovative methods and techniques in geography teaching, which arouse interest in pupils not only in the issue, but are also passed on in an entertaining and easily accepted modern form with the use of a wide range of information and communication technologies, is a prerequisite for successful teaching practice. The foundation as well as the goal of education is the ability to apply the acquired experiences and skills of pupils and students in their practical life. Some of the most current methodical attitudes, one which meets the above-mentioned criteria in direct practice and enabling their use on geography lessons, are presented in *Part II and Part III.*

Geography, with its traditional nature-based object of study, has an irreplaceable role among (natural) sciences. In addition to its multidisciplinarity, it has close links to all cross-sectional themes. Therefore, it is an appropriate subject for the application of *IBL (inquiry-based learning)* methodical procedures. IBL of geography arouses pupils' interest in their surroundings, nature and the landscape they live in. It develops their practical skills and abilities to apply theoretical knowledge in practice, thus such abilities and skills that the Czech pupils are still missing. Simultaneously, the thematic unit of Field Geographical Teaching, Practice and Application, which is a part of educational field geography in the Framework Educational Programme for Basic Education (Kol 2013), is realised in applying IBL of geography. Nowadays, among the main research questions in using IBL in practice is how IBL can be achieved in classroom. The submitted contribution tries to provide an answer to this question mainly from the Czech point of view. It familiarises readers with different approach, methods and options of using IBL in Czech secondary schools. At the same time, it presents examples of research tasks and case studies that can be appropriate inspiration for teachers. Case studies will mainly focus on examples of using IBL methods in teaching physical geography. The main reason is the low interest of Czech pupils in the natural sciences and low motivation to engage in natural science subjects. Chapter 5 proposes to provide inspirational ideas and ways to improve through IBL teaching physical geography and arouse the interest of pupils.

The anchoring of *global education* (GE or Czech term GDE – *global development education*) in Central European school practice and in school documents is still in its beginning. Distinctive social development of these countries is the main reason for slower implementation of the GE major themes into teaching. After the

Velvet Revolution in 1989, the transformation from a totalitarian regime to open democratic society took place. However, cultural, ethical and value changes have trailed behind the realised changes in political and economic areas. People have become apathetic to the perception of social differences, poverty, problems of marginal groups (Romani population, immigrants), environmental problems, etc. In such a social climate, it is therefore difficult to begin to teach children to think about events in a global scale. In school practice, teachers often face the question of how to grasp the global theme. Chapter 6 therefore first examines what global development education (GDE) is, what is its main aim and why it is necessary to use it in teaching practice. It will also mention specific development and application of methodology GDE in the Czech Republic. The contribution therefore introduces examples of good practice as well as case studies from Czechia, with the aim to show that it is not needed to be worried about current global events. On the contrary, arousing the interest of children in events in other parts of the world can be a very useful tool to support active learning and perception of sensitive topics in their surroundings.

A case study in Chap. 7 presents an appropriate form and method of providing students with a solution of real situations from the surroundings in which they live. This is called powerful teaching, and it is designed to help pupils and students to be able to cope with the rigours of everyday life through geography education. This method is not so well known and used in the Czech Republic as abroad, where it is known under the name "powerful knowledge" or "powerful teaching". For this reason, the introductory part of this chapter devotes enough space to understand "powerful learning" and noted how it differs from IBL, project-based, problem-based, student-centred and constructivist approaches to learning. Knowledge from the Czech geography education is in our case used for solving a case study in a decisive process in which students solve options and consequences of the construction of a ski resort in Brno (in Czechia). They submit their conclusions to the municipal council for assessment.

Project-based learning (PBL) is a philosophy of education which has been used from time immemorial. Very often it is confused with project teaching. However, the attractiveness of PBL is not given by project work, as many people believe, but by stimulating the pupil's internal interest. While the majority of teaching methods try to motivate a pupil more, PBL is based on the natural learning process and works with the own internal motivation of a pupil and their hobbies. The basic feature of PBL is a partnership between a teacher and a pupil. In PBL everything revolves around a student, and all choices and decisions are related to their needs. It is about student-centred learning. Applying principles of PBL in teaching geography leads to a significantly more effective study, to easier acquisition of appropriate knowledge or skills, to saving time for both the student and teacher but mainly to a return to the joy of learning by a student and to the joy of teaching by a teacher. Chapter 8 presents a PBL approach to a project as a tool for the acquisition of necessary knowledge and skills. The main goal of the chapter is to present the different approaches and strategies to realise PBL within Czech geographic education. Case studies will be given attention to common mistakes of teachers in implementing

PBL. Final summary will try to capture the main differences and approaches to the application of PBL in Czech geography education in the context of the examples and its application in an international (mainly the Central European) school practice.

Nowadays, we consider *information technologies in geography* teaching to be an essential part of material teaching tools, i.e. tools which the teacher and student use to achieve given educational goals. The *use of visual, auditory and audiovisual media* (Chap. 9) has become necessary in contemporary educational practice. This applies both in terms of modernisation of teaching practice and in terms of sustaining the interest of students who are more and more proficient in the use of these technologies.

Geography, dealing with a changing world, causes and impacts of incurred global, regional and local changes, whether in a natural or cultural environment, represents one of the main venues for using modern technologies within its teaching practice. Chapter 10 deals with the options of using *cloud tools* in teaching geography (e.g. Google, Disk, YouTube, forms, advanced search and easy publication of information, online cooperation in teaching, etc.). At the same time, it presents a practical example of using mobile technology, social networks and infographics in teaching geography. It introduces also a tool called ArcGIS Online for creating maps and map applications, which is a great help in the creation of interactive learning. Geography has a wealth of experience in this area – maps, cartograms and thematic maps.

The presentation of options of field education, which is a part of educational field geography, can be also found in this chapter. Field education belongs, not only in the Czech Republic but also in many other countries, among powerful learning strategies. However, its implementation is in many respects difficult. In the context of teaching geography, it brings us to the solution of so-called case studies. In practice, it means solving various real situations which happen or can happen in an environment, and pupils and students search for ways and mainly for geographic instruments for their solutions. The final goal is to teach pupils think. It cannot happen without a change of attitude among teachers as well as pupils. The teacher's role has been changing from an omniscient provider of knowledge to a manager or an organiser, and the pupil's and student's role has been changing from a traditional consumer of information to an active learning person.

Geocaching is a worldwide game, based on boxes hidden at interesting places around the world for which we know the geographic coordinates. Using geocaching in geography classes, teachers find an interesting motivational tool. The game helps teach new facts, gain experience and improve students' skills. Students must search for the required information on the Internet; they must be able to find their way in the field and have good spatial orientation. Caches are located at various points of interest linked to historical or ethnographical events or at places that are interesting in terms of physics, geography or socio-economics.

According to the trends, *satellite and aerial images* are very useful tools which can be integrated in geography lessons (Chap. 11). The research brings a contribution to supporting the use of satellite and aerial images in education. Research participants (11-, 15- and 19-year-old students) were asked to solve spatial tasks in images and maps of various types. Differences in the efficiency of task solution regarding various types of source documents were analysed, and the generated scores were evaluated according to the participants' age and gender. Schoolchildren and students were asked to provide their subjective opinions on the difficulty of reading the various image types and their personal preference for either maps or images as a source for acquiring information.

Geography is a discipline which lies on the border between the natural and social sciences. With its traditional natural-based object of study, it also has an irreplaceable role among sciences in the development of teaching at schools. At the same time, with its scope and focus on the landscape as a place of life for man, it also has ambitions to create space in the educational field in the near future for teaching socio-economic issues in geography lessons, like in the final *Part IV*. In general, Part IV deals with the issue of teaching social, regional, political and cultural geography at schools. The first contribution debates about the issues connected with the substance and focus of *regional geography* which have been discussed in Czech and Slovak geography for several decades already. Chapter 12 offers viewpoints of world macro-regions from Czech teaching practice. It focuses on the division of the world into major geographical regions and related matters. In the last section, the contribution presents the results of surveys of mental maps of major geographical regions of students from three universities: Charles University in Prague (Czech Republic), Lomonosov University in Moscow (Russian Federation) and the University of Maribor (Republic of Slovenia).

Political structures at the global, regional and local level are distinguished by relatively rapid transformation, democratic or otherwise. This requires that teachers regularly update their political-geographic knowledge and its correct perception and interpretation during tuition. After the end of a bipolar world, political, ethnic, religious, social and economic relates have entered into a new phase of development in an effort to discover new forms of balance. The political structures of the world are undergoing changes in different spatial dimensions from global through macro-regional to local ones. We are witnesses of processes leading to democratisation of political systems, but also of the opposite trends. These political processes are very strongly influenced by ethical and religious conflicts, nationalist tendencies, from positive emancipation to negative xenophobia. In terms of education, the disadvantage of reflecting changes of political structures is their relatively high speed, which cannot be reflected in textbooks and which requires constant updating from teachers. It is important to find an optimal level of politico-geographical knowledge for each school grade and to avoid the mere memorising of facts. The relation to political structures is most affected by subjective opinions of every teacher who passes them on to their students. Therefore, a strong emphasis should be placed on the training of pre-service teachers in the context of modern education of *political and cultural geography*. In Chap. 13, the political and economic changes in some Central European countries and their perception by the children of secondary schools will be reflected.

Changes of demographic behaviour have received a lot of attention from several scientific disciplines, and they are also a strong focus of social practices. Since the 1970s of the twentieth century, the *changes in demographic behaviour of population* have been reflected in the countries of Western and Northern Europe; later they were identified in Central, Eastern and Southern Europe. They are strongly manifested in three areas: (1) In the area of reproductive behaviour, we see a decrease of birth rate intensity, fertility and reproduction; (2) at the same time, the process of population ageing takes place, i.e. the number and proportion of seniors is growing; and (3) big changes take place in the area of family behaviour. The contribution of Chap. 14 contains an analysis of changes of demographic behaviour of European populations, with a special focus on the Czech, Slovak and other Central European population. This theme might be interesting and attractive already for children of primary schools. The contribution intends to help teachers to make the teaching of appropriate topics on geography lessons more attractive. It, namely, intends to guide pupils to actively participate in research and to guide them to investigate, recognise, evaluate and present demographic knowledge.

The approaches to and selected methods of teaching current geography topics presented in the submitted publication demonstrate how it is possible to guide students to contemplate, research, think critically and develop their independence. In contemporary Czech teaching practice, the weekly time allocation of classes for the educational subject of geography at the 2nd level of elementary schools and lower levels of secondary schools is approximately 2 h of tuition per week. This also demonstrates the low importance of geography compared to other taught subjects and its relegated position today within the terms of educational subjects. The application of suitably chosen methodological procedures of selected modern methods of teaching geography offers the opportunity to increase the attractiveness and effectiveness of classes by bringing tuition more in line with the practical lives of students.

Most of the mentioned modern methods of teaching awaken students' interest not only in geography but also primarily in their surroundings, the nature and landscape they live in, at a local, regional or global level. During tuition, students learn much more knowledge and more standpoints and values and develop their practical skills and abilities in applying theoretical knowledge in practice and understanding context. Geography is consequently a more than suitable subject for the application of all the aforementioned modern methodological procedures because of its multidisciplinary aspect, which provides close links to a cross section of topics within the terms of "Framework Education Programme". In this way, geography gains the possibility of increasing the attractiveness of teaching lessons. The successful resolution of major educational challenges in geography depends on many factors, both objective and subjective. The symmetrical balance between (geographical) scientific knowledge and teaching methods can be considered as one of the decisive factors.

The submitted publication tries to present the approaches in teaching geography based on experiences in Czechia, Slovakia and Poland too, in the background of selected Central European connections. Its ambition is to support the aims of the IGU Commission on Geographical Education (IGU CGE), which are contained in the International Charter on Geographical Education and in the International

Declaration on Geographical Education for Cultural Diversity. The publication is primarily intended for professional geography teachers and students of geography and also other people interested in geography education.

References

Kmeť, P. (2009). Reforma školského kurikula na Slovensku [Reform of the school curriculum in Slovakia]. In *Rozvoj a perspektívy pedagogiky a vzdelávania učiteľov*. Zborník z medzinárodnej vedeckej konferencie, pp. 189–192. Prešov: FHPV PU a MC Prešov.

Kol. (2013). *Rámcový vzdělávací program pro základní školy RVP ZV* [Framework Education Programme for Basic Education FEP BE]. Praha: VÚP.

Mullis, I., et al. (2012). *TIMSS 2011 International results in mathematics*. International Association for the Evaluation of Educational Achievement (IEA), TIMSS & PIRLS International Study Center. http://timssandpirls.bc.edu/timss2011/downloads/T11_IR_Mathematics_FullBook.pdf. Accessed 15 Sept 2015.

OECD (2014). PISA 2012. Results in focus what 15-year-olds know and what they can do with what they know. http://www.oecd.org/pisa/keyfindings/pisa-2012-results-overview.pdf. Accessed 15 Sept 2015.

RVP – Metodický portal. (2011). Kurikulární reforma [*Curriculum reform*]. http://wiki.rvp.cz/Knihovna/1.Pedagogicky_lexikon/K/Kurikul%C3%A1rn%C3%AD_reforma. Accessed 15 Sept 2015.

Stuchlíková, I., Janík, T., et al. (2015). Oborové didaktiky: vývoj – stav – perspektivy [Field didactics: developments – situation – perspectives]. Masarykova univerzita, Brno. http://www.ped.muni.cz/didacticaviva/data_pdf/knihy/oborove-didaktiky_online.pdf. Accessed 14 Sept 2015.

Part I
General Education System and Teaching Geography in the Czech Republic, Slovak Republic and Republic of Poland

Chapter 2
Geography Curriculum in Czechia: Challenging Opportunities

Eduard Hofmann, Hana Svobodová, and Petr Knecht

2.1 Brief Outline of the Development of the Geography Curriculum

In the Czech education system, the teaching of geography dates back to 1809, first in grammar schools and vocational schools. In 1869, with the issue of the Basic Imperial Education Act (Exner-Bonitz reform), geography was introduced to all years of primary and municipal schools. Gradually, geography also began being taught at most secondary schools (Knecht and Hofmann 2013). The teaching of geography during these early years and the concept of the curriculum were mostly static. School geography consisted of summaries and lists of countries, mountains, rivers, towns, important towns and cities, important places and battles (Janka 1970, page 32). Gradually, the first critics of this concept began to appear, who pointed to the low educational value of this kind of teaching. This is when the methodology and later the didactics of geography first appeared. However, this publication does not set out to provide a detailed historical development of the teaching of geography in Czechia. The dates of the introduction of the teaching of geography show that geography education underwent a long historical development which can be divided into several periods.

One of the most progressive periods is the years from the establishment of an independent Czechoslovakia to the start of World War II, which saw the rise of geography education. The shaping of schools was greatly influenced by the ideas of reform pedagogy. Formalism in education was suppressed, and the individuality of the pupil was promoted. Independent work increased, and in the area of geography, there was also teaching in the form of practical fieldwork and the like. After World

E. Hofmann (✉) • H. Svobodová • P. Knecht
Faculty of Education, Department of Geography, Masaryk University in Brno, Poříčí 7, 639 00 Brno, Czech Republic
e-mail: hofmann@ped.muni.cz

P. Karvánková et al. (eds.), *Current Topics in Czech and Central European Geography Education*, DOI 10.1007/978-3-319-43614-2_2

War II, the education system in Czechoslovakia underwent many reforms which were controlled for many years by the ideology of the system. However, the entire education system was marked by one common feature: the "single education system". This meant that teaching was based on a single curriculum. Up until the 1990s, there was no full alternative education system or even alternative teaching materials.

Hofmann et al. (2014) state that when looking back at the history of the teaching of geography in schools in Czechia, we discover that the creation of its contents has not changed in more than 80 years. In the 1920s, new curricula were introduced on the territory of Czechia, and intensive work was carried out on them from the mid-1920s onwards. The contents of these curricula were established on the "regional principle" based on learning about the place of residence, native region and republic where knowledge was presented of physical geography and the impacts of the position of the Earth in the universe. Despite learning about Europe, the curricula returned to the Czech Republic where their position was assessed within the broader community and concluded with the world and astronomy. In this period, the Pedagogy Department of the Central Association of Teachers' Unions was also active and elaborated these curricula into the form of detailed timetables, paying particular attention to links between subjects. It is interesting how the tables devoted to the detailed timetables resemble "today's efforts" to create the school educational programmes.

Although in the following years, the concept of education was more or less periodically reformed, the scheme of teaching geography still remained the same, if we ignore the ideological colouring of the entire education system from the 1950s to the 1980s. This period had an unfortunate effect on some geographic disciplines, especially political and regional geography. Traditionally, in geography the position of physical geography remained strong on which the unilateral ideological view did not have such an effect as on its socio-economic part. This classification of geography is also projected into the present period, when geography is still being classified in educational documents as a natural science (Hofmann et al. 2014).

As was stated above, the breakup of the single education system occurred in the 1990s. The introduction of plurality into the education system was related to the overall liberation of society, and besides alternative teaching programmes and educational materials led to a series of problems and unresolved issues. This saw the start of the search for an optimal system of teaching geography in Czech schools, which finally culminated in the discussion of curriculum reform at the turn of the millennium. The result of this discussion was a reform intended to implement innovative approaches, methods and forms in teaching that would activate pupils in the process of learning and prepare them to apply the knowledge they acquired in school in practical life and shape their personal attitudes and views. A significant feature of the "new" education strategy was the concept of lifelong learning, which was supported by the introduction of key competencies. Key competencies should be developed as a set of knowledge and abilities important for the personal development of an individual's role in society (Rychen and Slagnik 2001). Here attention was also drawn to the inadequate encyclopaedic concept of geography based on component approaches.

2.2 Geography and the Education System in the Czech Republic

In recent decades, there have been sweeping changes in the Czech education system. Moreover, these changes are ongoing, and the entire system is continuing to develop. According to the Constitution of Czechia, all citizens have an equal right to education. The Czech education system stems from the Education Act, which since 2004 sets a multilevel system of educational programmes that are created at the state and school level.

The main document which defines the principles of curriculum policy is the National Education Development Programme in Czechia, known as the White paper (Kol. 2001). This is a system project shaping the starting point of ideas, general plans and development programmes which are to be decisive for the development of the education system in the medium-term horizon (Kol. 2001). In the National Education Development Programme, educational objectives have been elaborated as set by the Education Act No. 561/2004 Coll., which defines the main areas of education, the contents of education and the means necessary to achieve these objectives (Kol. 2013a). The state level further consists of Framework Education Programmes (hereinafter "FEPs"), created by the Research Institute of Pedagogy and approved by the Ministry of Education, Youth and Sports. However, the Research Institute of Pedagogy was dissolved, and its duties are now carried out by the NIE – National Institute for Education. The FEPs are created for individual fields of education and define the compulsory contents, scope and conditions of education.

A two-level system of curriculum documents distinguishing between the state level and the school level was introduced. While the state-level curriculum was represented by the FEPs, the school-level curriculum was formulated in the school education programmes (SEPs). The implementation of the state curriculum at the school level is represented by the development of SEPs. In general, SEPs stand for educational autonomy of schools as well as teachers' professional responsibility for the outcomes of the educational process. The point of SEP lies in motivating teachers to elaborate on their own ideas regarding education in schools (Kol. 2013a). This resulted in the relaxation of the considerably centralised and directive system of basic and secondary education at a lower level of management, such as regions, municipalities and individual schools, and the powers and obligations of school headmasters were significantly increased.

The education process takes place in educational facilities, which, according to the International Standard Classification of education, is classified under the following levels:

International Standard Classification of Education (ISCED 1997)

0 – Pre-basic education (nursery school)
1 – Basic education (level 1 of basic school)
2 – Lower secondary education (grade 6–9)
3 – Upper secondary education (grade 10–13 of secondary school)
4 – Schools of follow-up education
5 – Upper vocational schools, bachelor's and master's university degrees
6 – Doctoral study programmes

2.2.1 Basic Education

Basic or primary education is provided at basic schools for a period of 9 years of compulsory school attendance, which was introduced in 1774. This is the only educational stage in Czechia in which every child participates. For many children, this is the only chance to come into contact with the teaching of geography. Pupils with all levels of abilities and from all sections of the population acquire unique experience here of social relations. Peers in classes can influence each other in a naturally heterogeneous collective and build irreplaceable social capital which serves as a precondition for future life. In Czechia, there are more than 3.6 thousand primary schools.

The Framework Education Programme for Basic Education (Kol. 2013a) defines the educational contents which are divided into nine educational areas. All subjects in educational areas are compulsory at primary schools. Each area consists of interlinked subjects. The areas marked in bold are those areas containing geography. The areas marked in bold and italics are those areas where geography has a significant overlap.

Educational Area in the Framework Education Programme for Basic Education

1. Language and Language Communication (Czech language and literature, foreign language)
2. Mathematics and its Application (Mathematics and its Application)
3. Information and Communications Technology (Information and Communications Technology)
4. **Man and his World (Man and his World)**
5. ***Man and Society (History, Citizenship Education)***
6. **Man and Nature (Physics, Chemistry, Natural History, Geography)**
7. Art and Culture (Music Education, Art Education)
8. Man and Health (Health Education, Physical Education)
9. Man and the World of Work (Man and the World of Work)

A compulsory part of the FEP is also cross-section subjects, whose objective is to fill in and strengthen links between subjects.

Cross-Section Subjects

- **Environmental education**
- Media Education
- ***Multicultural Education***
- Education of the Democratic Citizen
- ***Education in Thinking in the European and Global Context***
- Personality Education and Social Skills Education

The FEP for basic education (FEP BE) is drawn up in such a way, so pupils can attain key competencies or a summary of knowledge, skills, abilities, attitudes and values important for their personal development and use (Kol. 2013a).

Key Competencies

- Competencies for learning
- Competencies for addressing problems
- Communicative competencies
- Social and interpersonal competencies
- Civil competencies
- Working competencies

Basic education is divided into level 1 and level 2. Level 1 lasts 5 years (grades 1–5), attended by pupils at basic schools. This is the period when pupils are slowly prepared for systematic compulsory education. It is one of the most crucial periods because it brings great changes in the child's daily regimen. Care is taken at level 1 to provide comprehensive knowledge of the individual needs and capabilities of each pupil. The objective is to create the preconditions in pupils for lifelong learning.

Level 1 is directly followed by level 2, which completes compulsory school attendance. It lasts for 4 years (grades 6–9) and can take place at basic school, grammar school or a conservatory. The most common form of teaching is on a daily basis of study. The pupil, parent or legal guardian can also request an individual plan making education possible without daily school attendance. The period of study at level 2 is challenging, because pupils undergo significant hormonal, physical and personality changes. These changes are described under the term puberty. During this time, individual differences begin to appear between pupils, who form their own opinions, experience new feelings and often reject tested truths. The objective of level 2 is to provide pupils with the best foundation of general education.

Preschool education can precede basic education, which so far is not compulsory in the Czech Republic. This is provided in nursery schools normally attended by children ages three to six. Preschool education has a positive effect on a child's development, supports personality development, prepares the child for further education and complements family upbringing. Given its importance, the objective of Czech preschool education is to improve its quality and availability. Currently, there are intensive discussions concerning compulsory preschool education.

2.2.1.1 Anchoring of Geography in the Framework Education Programme for Basic Education (FEP BE)

In Framework Education Programme for Basic Education (Kol. 2013a), geography education is divided into two areas: level 1, "Man and his World", and level 2, "Man and Nature".

"Man and his World" is the only educational area of the FEP BE conceived only for level 1 of basic education. This comprehensive area defines the educational contents applying to man, family, society, homeland, nature, culture, technology, health, safety and other subjects. It uses a view of history and the present and guides towards skills for practical life. The broadly conceived synthetic (integrated) contents cocreate compulsory basic education at level 1. The educational content of "Man and his World" is divided into five thematic circles (while basic geographic knowledge, skills and attitudes are taught in the first thematic circle):

1. **Place where we live**
2. People around us
3. People and time
4. Diversity of nature
5. Man and his health

The level 2 of basic school is the educational area of "Man and Nature", which includes questions associated with the investigation of nature. It provides pupils with the means and methods for a deeper understanding of natural facts and laws. It also gives them the necessary foundation for a better understanding and use of current technology and helps them to find their way in ordinary life.

The fields of education in "Man and Nature" are physics, chemistry, natural history and geography, whose research nature gives pupils a deeper understanding of the laws of natural processes and teaches them about the usefulness of scientific findings and their application in practical life. In studying nature, pupils also acquire important skills.

Education in this area is directed at shaping and developing key competencies guiding the pupil towards:

- Investigation of natural facts and their context with the use of various empiric learning methods (observation, measurement, experiment) and methods of rational thinking
- The need to ask questions about the course and causes of various natural processes that have an impact on the protection of health, life, environment and property and to correctly word these questions and look for adequate answers
- A method of thinking that requires the testing of expressed assumptions about natural facts in more independent ways
- Assessment of the importance, reliability and accuracy of obtained scientific data to confirm or disprove expressed hypotheses or conclusions
- Involvement in activities friendly to natural systems, their health and the health of other people

- An understanding of the context between human activity and the state of the natural and living environment
- Considerations and actions that prefer the most effective use of sources of energy in practice, including the broadest use of their renewable sources, especially solar radiation, wind, water and biomasses
- Creation of skills suitable for contact with objects or situations potentially or currently harmful to human life, health, property or environment

The subject of geography is divided at level 2 into seven thematic circles:

1. Geographic source of data, cartography and topography
2. Natural picture of Earth
3. Regions of the world
4. Social and economic environment
5. The environment
6. Czech Republic
7. Geography fieldwork, practice and application

The thematic circles contain expected outputs and summarised education content. These elaborated thematic circles form the basic framework of education at level 1 and level 2 of basic school, which pupils are expected to have attained upon graduation. In a certain sense, the expected outputs take on the form of the intended standard (see further).

2.2.2 Secondary Education

The Czech education system is characterised by a high rate of completion of secondary education. Pupils can attend secondary school after completing compulsory primary school and successfully passing the entrance exam for the chosen school. After completing the educational programme, pupils will attain a secondary education with a vocational certificate or secondary education with a secondary school-leaving exam (in Czech, *maturita*). Pupils attain these educational levels at grammar schools, secondary vocational apprentice training centres, follow-up study schools or at conservatories.

Grammar schools provide a full general secondary education. The objective is to prepare students for a more challenging qualified profession or university study. It is completed with a secondary school-leaving exam and the acquisition of a secondary school-leaving exam certificate. The period of study may last 4 years for students after having completed the 9th year of basic school or 6 and 8 years. Secondary vocational schools work in the same way as grammar schools. Study at these schools lasts 4 years and is completed with a secondary school-leaving exam. These schools specialise in vocational fields as well as preparing students for university. Secondary vocational apprentice training centres offer study in fields which aim to prepare students for qualified activities mostly involving manual work. This is completed by a final exam after 2–3 years of study and the acquisition of a final examination certificate and vocational certificate. If students with a vocational certificate decide to also take the secondary school-leaving exam, they can also attend a 2-year follow-up study. Secondary education can also take place at a conservatory. This is a specific type of education preparing students for challenging artistic or teaching activities in music, dance, singing and dramatic art. This lasts 6–8 years and is completed with a secondary school-leaving exam or certificate.

2.2.2.1 Anchoring of Geography in the FEP for Secondary Education (FEP GS)

Geography education is most comprehensively represented at grammar schools (the following paragraph will therefore be devoted only to grammar schools) and should have direct continuity in geography education that pupils undergo in basic education. The educational areas in the framework educational programme for grammar schools (Kol. 2007) are the same as for basic education and differ in the order and some formulations. The same applies to cross-section subjects and key competencies when the working competence is replaced with the competence for enterprise. This compliance should guarantee direct continuity in the lower educational level. In Czechia, there are more than 350 grammar schools.

Educational Area in the Framework Education Programme for Secondary Education

1. Language and Language Communication (Czech Language and Literature, Foreign Language, Second Foreign Language)
2. Mathematics and Its Application (Mathematics and Its Application)
3. Man and Nature (Physics, Chemistry, Biology, Geography, Geology)
4. Man and Society (Basics of Civics and Social Sciences, History, Geography)
5. Man and the World of Work (Man and the World of Work)
6. Arts and Culture (Music, Fine Arts)
7. Man and Health (Health Education, Physical Education)
8. Information Science and Information and Communication Technologies (Information Science and Information and Communication Technologies)

As conceived, the Framework Curriculum Timetable:

- Determines only the basic parameters of education organisation, through which it provides schools with ample scope for variability in preparing curriculum timetables for their school education programmes
- Reduces the number of normative elements in higher forms and thus makes it possible for schools to implement their educational objectives much more effectively and react flexibly to the educational needs and interests of the pupils
- Defines the overall minimum time allotment for individual educational areas (fields) in the 4-year grammar-school education programme and in the upper stage of 6- or 8-year grammar schools
- Enables integration of educational content

The first and second forms are conceived as periods during which pupils jointly acquire the functional basics of grammar-school education with a school-leaving examination, which may be further developed, reinforced and complemented variously in subsequent forms. For this reason, education in the first and second forms is conceived as compulsory in the great majority of educational areas. *The third and fourth forms* allow considerable variability on the level of school education programmes in the educational offer, which makes it possible for schools to meet the educational needs and interests of their pupils most effectively. Except for the educational areas, Language and Language Communication, Mathematics and Its Application (in the third form) and the educational field Physical Education, no other educational area or field is conceived as compulsory. Whether at all and in what forms the other educational areas (fields) will be included, what time allotment they will be assigned and how they will be implemented are determined by the school education programme. Optional educational activities must be offered to pupils by schools in at least the third and fourth forms (Kol. 2007, p. 82).

2.2.3 Upper Vocational Education

This is an education that deepens the knowledge and skills acquired in secondary education preparing students for challenging professions that do not require a university diploma. To be accepted by an upper vocational school, students must have the secondary school-leaving exam certificate and pass the entrance exam. The study lasts 3 years and includes vocational practice and is completed with a certificate. Graduates gain the title DiS (diploma specialist). Geography education here is represented only if it has direct continuity in the study field and applies only to individual geographic disciplines (e.g. tourism geography). Currently, this is more of a marginal form of education. A number of upper vocational schools have been transformed into universities only offering bachelor degrees.

2.2.4 University Education

University education in the Czech Republic has a long tradition. Charles University was founded here in 1348 as the oldest university in Central Europe. Study at the university is offered at the bachelor's and master's levels. The bachelor study programme lasts 3 or 4 years, and the student can begin to study after passing the secondary school-leaving exam and successfully completing the entrance procedure. The programme provides preparation for a profession and for study in a follow-up master's programme. The study ends with a state final exam, a part of which involves the defence of a bachelor's thesis, and the graduate gains the title Bc. The follow-up master study programme lasts 1–3 years and can also exist as a separate study programme, in which case it can be 4–6 years. It aims at deepening theoretical knowledge and its application in practice. It is completed by a state final exam and defence of diploma thesis. The graduate gains an academic title such as Mgr., Ing. or MgA. Graduates of the master's programme can also go on to study a doctoral study programme. The entire system is shown in the diagram in Fig. 2.1. "specialised" geography, and its individual fields and geography for teachers are studied at universities in Czechia.

2.3 Selected Problems of the Current Curriculum

Under the "single curricula", stress was placed on the contents of education. The curriculum was then divided into basic (sometimes "core") and additional. Textbooks formed the basis for school teaching and were the biggest source of information. Under the latest reform, stress was placed on the development of thinking and above all on the development of competence. The development of new communications technology and related amount of available information is forcing school reforms even in a number of other developed countries. The function of what pupils and students should know in the plural system of education began to be met by educational standards. The problem of educational standards not only from geography education is our first stop.

2.3.1 Creation of Geographic Education Standards in Czechia

In the case of Czechia, the creation of geography education standards is somewhat controversially conceived immediately in several respects:

1. Although Czechia has had the Basic Education Standard since 1995 (Kol. 2013b), this was not a standard which could meet the new curriculum reform launched at the turn of the millennium. The wording of this standard corresponded with the contents of basic school curricula.
2. The Ministry of Education, Youth and Sports of the Czech Republic did not decide to create new standards until the turn of 2011. The team that prepared the geography education standards was set up almost at "random". The general public, such as from the members of the Czech Geographic Society, was not included in the creation of the standards.
3. The time for drawing up geography education standards was shorter than half a year, whereas in other countries, it is a long-term process lasting several years.
4. After submitting these standards, a new group was formed in the next half year which resubmitted the standards again under time pressure. They did not acquire their final form until the third elaborated version.

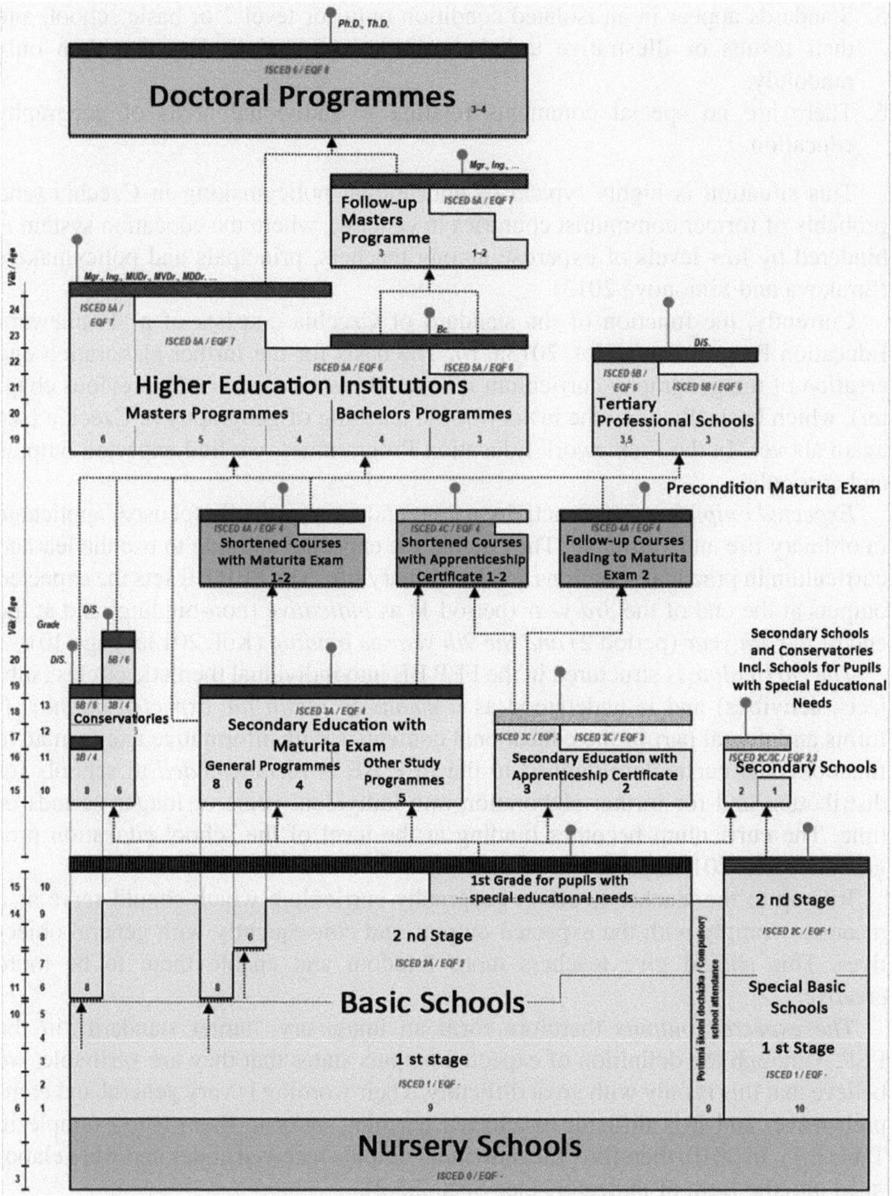

Fig. 2.1 The education system of the Czech Republic (Source: MŠMT ČR (2012))

5. Standards appear in an isolated condition only for level 2 of basic school, and their results or illustrative tasks are not tested, and if they are, then only randomly.
6. There are no special comments relating to individual areas of geography education.

This situation is highly typical of educational policymaking in Czechia (and probably of former communist countries in general), where the education system is hindered by low levels of expertise among teachers, principals and policymakers (Straková and Simonová 2013).

Currently, the function of the standard of Czechia consists of a "Framework Education Programme" (Kol. 2013a, b). The basis for the further elaboration and creation of the geography curriculum is seven thematic circles (see previous chapter), which basically copy the history of the teaching of geography in Czechia (see again above). In the Framework Education Programmes, we find expected outputs and curricula.

Expected outputs are of an activity nature and are practically focused, applicable in ordinary life and verifiable. They define the expected aptitude to use the learned curriculum in practical situations and in ordinary life. The FEP BE sets the expected outputs at the end of the *3rd year* (period 1) as *indicative* (non-binding) and at the end of the *5th year* (period 2) *and the 9th year* as *binding* (Kol. 2013a, page 10).

The curriculum is structured in the FEP BE into individual thematic circles (subjects, activities) and is understood as *a means to attain the expected outputs.* It forms an integral part of the educational contents for its informative and formative function. The curriculum defined in the FEP BE is *recommended* to schools for distribution and for further elaboration into individual years or longer periods of time. The curriculum becomes binding at the level of the school education programme (Kol. 2013a, page 10).

It is up to the teacher to select geography curriculum which should serve as a means to comply with the expected outputs and consequently with general objectives. This should give teachers more freedom and enable them to be more creative.

The expected outputs therefore form an imaginary "target standard" in the FEP. Although the definition of expected outputs states that they are verifiable, we believe that this is only with great difficulty. Their wording is very general and comprehensive, and it is difficult to allocate teaching tasks to them (see example in Table 2.1). In 2010, therefore, the individual outputs received codes and were elaborated into the form of indicators (see Sect. 2.3.2).

We have already stated at the start of Sect. 2.2 that geography education standards appear in Czechia in an isolated condition for individual school levels. If we compare the FEP for level 1 of basic schools, for level 2 of basic schools and for grammar schools, we can expect differences in the wording and number of thematic circles (Table 2.2).

Based on a comparison of the columns in Table 2.2, it can be briefly stated that all the complexity of elaborating the FEP for various school levels disappears,

Table 2.1 Expected outputs for the thematic circle of geographic information, sources of data, cartography and topography

Geographic information, sources of data, cartography and topography
Expected outputs
Pupil
Organises and reasonably assesses geographic information and sources of data from available cartographic products and papers, graphs, diagrams, statistical and other information sources
Uses with understanding basic geographic, topographic and cartographic terminology
Reasonably assesses geographic objects; phenomena and processes in the landscape sphere; their certain regularity, legitimacy and diversity; and their interrelation and conditionality, and distinguishes border (barriers) between fundamental spatial components in the landscape
Creates and uses personal mental diagrams and mental maps for orientation in specific regions; for spatial perception and assessment of places, objects, phenomena and processes in them; and for creating attitudes to the surrounding world

Table 2.2 Comparison of thematic circles in the FEP for level 1 of basic schools, level 2 of basic schools and for grammar schools

Thematic circles for level 1 of basic schools	Thematic circles for level 2 of basic schools	Thematic circles for grammar schools
Place where we live	Geographic sources of data, cartography and topography	Geographic information and fieldwork teaching
People around us	Natural picture of Earth	Natural environment
People and time	Social and economic environment	Social environment
Diversity of nature	Regions of the world	Regions
Man and his health	Czech Republic	
	Living environment	Living environment
	Geography fieldwork teaching, practice and application	Geographic information and fieldwork teaching

because the thematic circles do not follow on one from the other. The expected outputs for the thematic unit of the Czech Republic in the FEP for grammar schools are apparently missing, because they are in the "Regions" unit. The links from the cartography and topography unit that are found in the FEP BE in the first circle are concealed in the FEP for grammar schools in the geographic information and fieldwork teaching circle. The assessment of the actual sequence in the deepening of individual thematic units in both documents is for separate comparison, and *it is more convenient to compare the expected outputs than just the circles* (Table 2.3).

At first glance, these facts can appear unsubstantiated, but the opposite is the truth. If we compare the creation of these documents in the geography education environment of NGS (National Geography Standards; see Sect. 2.3.2) USA, we find that it is difficult to expect the conceptual development of geographic knowledge, skills and attitudes which logically follow one from the other in the education system in Czechia. For the creation of the expected outputs and subsequent elaboration, we can for the above-mentioned reasons describe this method of conceiving the

Table 2.3 Comparison of expected outputs in a circle that applies to physical geography

Level 1 BS – *Place where we live*	Level 2 BS – *Natural picture of Earth*	Gramm. – *Natural environment*
Expected outputs, pupil:	Expected outputs, pupil:	Expected outputs, pupil:
Will distinguish natural and artificial elements in the surrounding landscape and express various ways of their aesthetic value and diversity (period 1)	*Will assess the position of Earth in the universe and compare the fundamental properties of Earth with the other bodies of the solar system*	*Compares the position of Earth in the universe and fundamental properties of Earth with the other bodies of the solar system*
Will find typical regional peculiarities of nature, settlements, economy and culture in a simple way and will assess their significance in natural, historical, political, administrative and proprietary terms (period 2)	*Will compare the action of internal and external processes in the natural sphere and their effect on nature and on human society*	*Compares using examples the mechanisms of the action of endogenic (including plate tectonics) and exogenic processes and their impact on the formation of the Earth's surface and on human life*
–	*Will demonstrate using specific examples the shape of the planet Earth and assess the consequences of the motions of Earth on the life of humans and organisms*	*Explains the mechanisms of global circulation of the atmosphere and its consequences for the formation of climate zones*
–	*Distinguishes and compares components and elements of the natural sphere, their interrelation and conditionality, distinguishes, names and classifies the shapes of the Earth's surface*	*Will distinguish the components and elements of the physical geographic sphere and will recognise the relationships between them*
–	–	*Will explain big and small circulation of water and distinguish individual components of the hydrosphere and their function in the landscape*
–	–	*Assesses the waters and pedosphere of the Earth as the basis for life and sources of the development of society*
–	–	*Will distinguish the main biomes of the world*

geography curriculum as problematic. As was already stated in the previous text, the wording of the expected outputs is too general, and sometimes it is difficult to imagine what they mean, what precisely is to be developed and what in their sense is to be achieved. Therefore, in 2010–2013 attempts were made to elaborate them into the form of *indicators*.

Table 2.4 Basic thematic circles of the standards of geographic education (level 2 of basic school)

	USA	Czech Republic	Slovak Republic
1.	World in a spatial context	Geographic sources of data, cartography and topography	Planet Earth
2.	Places and regions	Natural picture of Earth	Earth models
3.	Physical geography systems	Regions of the world	Travelling around the Earth
4.	Human systems	Social and economic environment	Geographic excursions and outings
5.	The living environment and society	Living environment	Australia and Oceania
6.	Application level of geography	Czech Republic	Polar regions
7.	–	Geography fieldwork teaching, practice and application	America – New World
8.	–	–	Africa, Asia
9.	–	–	Europe
10.	–	–	Slovakia

When conceiving the first version of elaborated standards, the group of authors attempted to include all their experience in this process under *time pressure*. It began systematically with a discussion concerning the wording of the expected outputs, which it first explained and then adjusted according to the revised Bloom taxonomy of educational objectives. This step was not accepted in the final version, and the wording of expected outputs remained the same as proposed in the original proposal of the FEP. The authors also attempted to present the circles using basic theses to explain how to proceed further in the creation of the indicators. Once the proposal for these indicators was presented, the working group no longer consisted of the same members, and besides discussions on the web portal of the National Institute for Education, it was not even tested.

Moreover, Czechia, with no national assessment (neither high stakes nor low stakes) of learning in basic and lower secondary schools, has become a unique case in Europe. Two attempts to introduce compulsory tests in the last grade of both basic and lower secondary schools (grade 5 and grade 9) were discontinued within the past 10 years. Schools can use tests developed and sold by several commercial providers. Since 2014, the national tests for system-level monitoring based on sample surveys shall be run every year (Dvořák et al. 2014).

2.3.2 Comparison of Basic Geographic Circles of Standards of the USA, Czech Republic and Slovak Republic

At first glance, it was obvious from Table 2.4 that the wording was already different from the basic circles discussed as part of geography education. However, this does not primarily mean that their contents also differed.

For example, the absence of the regional geography of the USA or fieldwork teaching does not mean that attention is not devoted to them in the later drafts. The regional geography of the USA is found in all the circles. Data collected from various sources, i.e. even by means of fieldwork teaching, is also carried out as a cross-section on suitable examples. The circles of standards of Slovakia have the most differences, especially in their specific form. The contents of physical geography and socio-economic geography in Slovakia are, for example, concealed under the title to the circle travelling around Earth. Meanwhile, the wording of the circles of the standards of the USA and the Czech Republic makes it possible to structure the curriculum into individual years in various ways, while circles in the draft standards of Slovakia are set for individual years. Sometimes, there is a tendency for teachers to no longer consider the circles that were discussed, even though in this fixed model, the model of a spiral curriculum structure is respected, which is discussed in other subjects in upper years. Table 2.5 breaks down the thematic units of NGS USA into a total of 18 standards.

2.4 Conclusion

At this point, it is very difficult to compare the philosophy of the creation of a geography curriculum in Czechia, Slovakia or Poland with the creation of the same curriculum in other countries (in this case, the USA). The purpose of the first chapter was to briefly introduce the creation of the geography curriculum in Czechia and some problems associated with this topic. This very brief comparison is not comprehensive but only illustrative. We are aware of the fact that the concept and contents of geography education in individual countries differ from one another in the context arising from geographic knowledge, which may be due to its development, language and culture. In other words, each curriculum is influenced by its national character. Of course, this does not mean that we must follow only the knowledge at the national level, especially in a country like the Czech Republic, which does not have a broad base of experts in geography education or geography teaching. We can be greatly inspired by comparing the curriculum with a different concept. The challenges of the Geography Education Section at the IGU can also be inspirational for us.

Table 2.5 Thematic units of NGS USA

Page no.	A geographically educated person will learn:
	World in a spatial context
1	How to use maps and further geographic models, aids, technology and geospatial technology of spatial thinking to gain, process and submit information about the world from a spatial perspective
2	How to use mental maps to structure information about people, places and the environment in a spatial context
3	How to analyse the spatial structure of information about people, places and the environment in a spatial context
	Places and regions
4	Physical geographic and socio-economic characteristics of places
5	Methods of creating regions to facilitate the interpretation of the complexity of the world
6	How culture and experience of people affect the perception of places and regions
	Physical geography systems
7	Physical geographic processes that shape the structure of the Earth's surface
8	Characteristics and the spatial layout of ecosystems and biomes on the Earth's surface
	Human systems
9	Characteristics, distribution and migration of people on Earth
10	Characteristics, distribution and diversity (complexity) of the world cultural mosaic
11	Structures and networks of mutual economic links of the world
12	Processes, structures and functions of human settlements
13	How cooperation and conflicts between people affect the division of the Earth's surface and control over it
	The living environment and society
14	How human activity modifies the physical geographic environment
15	How physical geographic systems affect socio-economic systems
16	Understanding changes that are linked to the problem of the assessment of the significance, use and distribution and importance of natural resources.
	Application level of geography
17	How to use geography to interpret the past
18	How to use geography to interpret the present and plan the future

Source: Heffron and Downs (2012), Matoušek (1997)

References

Dvořák, D., Urbánek, P., & Starý, K. (2014). High autonomy and low accountability: Case study of five Czech schools. *Pedagogická orientace, 24*(6), 919–940.

Heffron, S. G., & Downs, R. M. (Eds.). (2012). *Geography for life: National geography standard.* Washington, DC: National Council for Geographic Education.

Hofmann, E., et al. (2014). *Standardy geografického vzdělávání v Česku na Slovensku a v USA* [Geographical education standards in the Czech Republic, Slovakia and USA]. Materiály z projektu Educoland. Brno: Pdf MU. http://educoland.muni.cz/geografie/novinky-z-oboru/. Accessed 20 Aug 2015.

ISCED. (1997). *International standard classification of education.* http://www.uis.unesco.org/Library/Documents/isced97-en.pdf. Accessed 20 Aug 2015.

Janka, J. (1970). Vývoj a význam školského zeměpisu [Development and importance of school geography]. *Sborník Československé společnosti zeměpisné, 75*(1), 32–39.

Knecht, P., & Hofmann, E. (2013). K problému řazení geografického učiva ve školních vzdělávacích programech [Geography curriculum scope and sequence in Czech lower secondary schools]. Informace České geografické společnosti. *Česká geografická společnost, 32*(2), 13–25.

Kol. (2001). Národní program vzdělávání ČR (Bílá kniha) [National education development programme in the Czech Republic (White Paper)]. http://www.vzdelavani2020.cz/narodni-program-vzdelavani-cr-bila-kniha.html. Accessed 20 Aug 2015.

Kol. (2007). *Rámcový vzdělávací program pro gymnázia RVP G* [Framework education programme for grammar schools FEP GS]. Praha: VÚP.

Kol. (2013a). *Rámcový vzdělávací program pro základní školy RVP ZV* [Framework education programme for basic education FEP BE]. Praha: VÚP.

Kol. (2013b). *Standardy pro základní vzdělávání zeměpis (geografie).* Národní ústav pro vzdělávání. http://clanky.rvp.cz/wp-content/upload/prilohy/17383/zemepis_geografie.pdf. Accessed 20 Aug 2015.

Matoušek, A. (1997). *Standardy geografického vzdělávání v USA* [Geographical education standards in the USA]. Studijní materiály. Brno: PdF MU.

MŠMT, Č. R. (2012). *The education system in the Czech Republic.* http://www.msmt.cz/file/27043/. Accessed 15 Aug 2015.

Rychen, D. S., & Slagnik, L. H. (2001). *Key competencies.* Göttingen: Hogrefe and Huber.

Straková, J., & Simonová, J. (2013). Assessment in the school systems of the Czech Republic. *Assessment in Education: Principles, Policy & Practice, 20*(4), 470–490.

Chapter 3
Current Geography Education in Slovakia: Conversions and Conditions

Ladislav Tolmáči and Štefan Karolčík

3.1 Brief Outline of the Development of Geography Teaching in Slovakia

Slovakia was under the influence of the geography teaching practices of the Hungarian educational system during the nineteenth century and the Czech and Czechoslovak educational system during the first 90 years of the twentieth century. Probably the best influence on geography teaching came with Czech professors during the period of the First Republic of Czechoslovakia from 1918 to 1939. The era of the communist dictatorship was tightly connected to communist propaganda even in geography teaching. This meant a complete disaster for creative and independent thinking. Geography and history were badly influenced by the distorting of facts and the whole content of these subjects.

As Hofmann et al. (2014) stated:

> the situation in Czechia and also Slovakia as far as introducing innovations and creativity into geography teaching during the years 1948 to 1989 was down to zero. The geography curriculum was closely tied to communist propaganda and distorted the way the world really looks. The content of the geography curriculum was based on regional geography and its aim was to compare good socialist countries to bad capitalist regions. Special attention was paid to the home country – Czechoslovakia or Slovakia in more details. (Zaťková et al. 1992; Zaťková 1993)

After 1993 when Czechoslovakia split into two independent countries, Slovakia stayed much more conservative, and no break-up of the single education system

L. Tolmáči (✉)
Faculty of Natural Science, Department of Geography and Geologie,
Matej Bel University, Tajovského 40, 974 01 Banská Bystrica, Slovak Republic
e-mail: ladislav.tolmaci@umb.sk

Š. Karolčík
Faculty of Natural Sciences, Department of Didactics, Comenius University in Bratislava,
Ilkovičova 6, 842 15 Bratislava, Slovak Republic

P. Karvánková et al. (eds.), *Current Topics in Czech and Central European Geography Education*, DOI 10.1007/978-3-319-43614-2_3

occurred there. The system was and still is very static. In geography teaching a strong tendency appeared to remove the propaganda and incorrect information in the curriculum, but the overall concept of teaching remained the same. Naturally, many formal declarations appeared, e.g. We teach for life, Geography is very practical, etc. (Likavský 2005), but the system and sadly even the society of geography teachers were very formalistic and not innovative. Modern approaches, methods and forms in teaching appeared very rarely and were not upheld (Likavský and Ružeková 2004a, b). New approaches had to appear paradoxically after changes to the whole educational system reducing the teaching of almost all subjects in Slovak schools according to the new state educational programme (SEP) in 2008. It incorporated the option for schools to develop their own school educational programmes (SCED). The idea itself was very good but society was not ready to take advantage of it. The teachers were not able to and sometimes also not willing to implement new ways of teaching not just geography but also other subjects. SEP and SCED are still in use, but after the experience of almost 10 years, there is a tendency to enforce SEP and to reduce SCEP.

3.2 Geography and the Education System in Slovakia

Slovakia shared the education system with the Czech Republic in former Czechoslovakia. After 1993, when these countries became independent states, their education systems went their own way but not markedly. In the last two decades, there have been changes in the Slovak education system, but not progressive enough. The Constitution of the Slovak Republic states that all citizens have an equal right to education. The Slovak education system is based on the concept of the SEP (state educational programme, Štátny vzdelávací programme) and the SCEP (school educational programme, Školský vzdelávací programme). This two-tier system of curriculum documents is very similar to the Czech model of education and is explained clearly in Chap. 2.

In Slovakia, as in the majority of EU countries, the education process takes place in educational facilities classified under the following levels:

International Standard Classification of Education – ISCED 1997 (ŠPÚ 2016a)

0 – Pre-basic education (Nursery School)
1 – Basic education (level 1 of primary school grades 1–4)
2 – Lower secondary education (grades 5–9)
3 – Upper secondary education (grades 10–13 of secondary school)
4 – Schools of follow-up education
5 – Upper vocational schools, bachelor's and master's university degrees
6 – Doctoral study programmes

3.2.1 Basic Education

Slovak basic education is provided for a duration of 9 years at basic (primary and secondary) schools. According to Slovak Law 245/2008, codex on education and training, as amended, compulsory education lasts 10 years and no longer than until the end of the school year in which the pupil reaches 16 years of age. Every child in Slovakia attends primary school. Geography as a subject of education from the age of 8 till 15 is attended by all children in Slovakia.

The SEP for primary education divides the education contents into eight educational areas. All of them are compulsory:

Educational Areas in the State Education Programme for Primary Education

1. *Man and values (ethics, religion)*
2. *Man and society (history, citizenship education, geography)*
3. *Man and the world of work (practical work)*
4. *Man and nature (biology, physics, chemistry, natural history)*
5. *Language and communication (Slovak language and literature, foreign languages)*
6. *Mathematics and information (mathematics and its application)*
7. *Art and culture (music education, art education)*
8. *Health and physical training (health education, physical education)* (CGE IGU 1992)

The SEP also has cross-sectional subjects. They are also compulsory:

Cross-Sectional Subjects

- Personality education and social development
- Education on marriage and parenthood
- Environmental education
- Media education
- Multicultural education
- Regional education and folk culture
- Education on road safety and protection of life and health (CGE IGU 2015)

Basic education in Slovakia is divided into two levels. The first level (grades 1–4) is attended by pupils at primary schools. They take geography in this stage as part of home country (vlastiveda) education. The second level (grades 5–9) till the age of 15 years can be taken at primary school, grammar school or a conservatory. Preschool education in Slovakia is quite common and has a very sophisticated curriculum. It is provided for children from 3 to 6 years of age. It is not compulsory.

Geography in the state education programme for basic education is included in the framework of the topics of "man and society" together with history. "Man and society" at the primary level (8–9 years old) is a comprehensive subject with content connected to the home country, family and life. Geography at level 2 of primary schools (10–14 years old) is devoted to elements of physical and human geography, but again it is mostly regional geography which fills the curriculum. Methods and approaches to discover them range from imaginary travelling to field work.

The SEP summarises key competencies of a broad scope:

- Search, compare and evaluate the veracity and assess the available information on regions from a variety of sources
- Present information about the country (region) in various forms (charts, tables, diagrams, photographs, films, etc.)
- Interpret maps of various kinds in digital and printed form
- Give reasons for the diversity of natural conditions on Earth and their impact on human life
- Take a stand for humanity's most serious problems and offer appropriate solutions
- Understand the complexity of the country and the strong interdependence of the natural and socio-economic components (Likavský 2006)

Geography at the second level of primary schools in Slovakia is divided into several elementary topics:

1. Maps and reading maps, geographical data
2. Physical geography and natural resources of the Earth
3. Regions of the world, based on traditional regionalisation
4. Human geography
5. Slovakia and its regions
6. Fieldwork, excursions, etc. (Nogová and Tolmáči 2008a; Karolčík et al. 2013)

3.2.2 Secondary Education

In Slovakia secondary education has an ever higher rate of completion every year. As with anywhere in the EU, students can enter secondary school after completing basic (primary) school education. The secondary school-leaving exam is necessary to enter universities or other types of higher education. Slovak grammar schools are intended to prepare students for higher education levels. Geography is closely tied to these schools, except for them geography is taught just in hotel academies and secondary business schools (Nogová and Tolmáči 2008b, c).

3.2.2.1 Geography in the SEP for Secondary Education

Educational Topics in the State Education Programme for Secondary Education (ŠPÚ 2016b, c)

1. *Man and nature (biology, physics, chemistry)*
2. *Man and values (ethics, religion)*
3. *Man and society (history, geography, citizenship education)*

4. *Language and communication (Slovak language and literature, foreign languages)*
5. *Mathematics and information (mathematics and its application)*
6. *Art and culture (music education, art education)*
7. *Health and physical training (health education, physical education)*

In the SEP the key competences to be achieved in secondary grammar schools from geography are:

- Use maps of various kinds in digital and printed form as a basic source of geographical information.
- Correctly and accurately interpret information about the country in various forms (charts, tables, diagrams, photographs, films and others).
- Seek comparisons and assess and evaluate the veracity of the information available about the country from a variety of sources.
- Give reasons for the diversity of natural conditions in different parts of the Earth and their impact on human life.
- Understand the principle of diversity of human society and its variable activities.
- Express a standpoint on the most serious question of the existence of mankind on earth and offer appropriate solutions.
- Interpret the complexity of the landscape and the strong interdependence of the natural and socio-economic components.
- Comprehensively assess the prospects of the development of individual regions (Nogová 2002; Nogová and Tolmáči 2008d).

3.2.3 University Education

University geography education in Slovakia was founded by the assistants of Czech geographers, e.g. Prof. Hromádka, between the two world wars. Comenius University in Bratislava today provides the best geography education in Slovakia. Geography programmes vary from seven bachelor's geography programmes, six master's geography programmes to three doctoral geography programmes. The bachelor's study programme usually lasts 3 years. The master's study programme usually lasts 2 years, and doctoral (Ph.D.) studies take 4 years to complete after finishing the master's degree (FPV UMB 2015).

3.3 Standardised Basics of Geography Teaching in Slovakia

More recently (since 2014), the fundamental transformation of geographic education in Slovakia has been anchored in determining performance – a first, and content – a second (ŠPÚ 2016a, b). This trend is extremely progressive, unfortunately, especially in connection with the freedom of conversion and overall conceptions, which are strictly given in Slovakia. This step in teaching geography in Slovakia is still pending. In the presented materials, we summarise the contemporary Slovak basics of teaching geography that provide outputs of the education content of geography. In Slovakia since 2015, there are fixed topics that will be taught each year (age of students). Their transfer is not possible. Performance and content are defined as the minimum to learn, and all the students have to achieve them. This is based on general didactic principles, the science curriculum and the International Charter on Geographic Education. They are divided into standards of geography for primary schools and secondary schools. Elementary geography and geography at universities are incorporated in specific materials and accreditation files.

In geography at secondary schools (stage 3 in England), there are two examples of standards (one problem-oriented topic which creates one-fourth of the whole curriculum and one regional-oriented topic – Africa, regional-oriented topics make up three-fourths of the geography curriculum in Slovakia). The basic geography teaching documents of Slovakia characterise the fundamental, structural elements of geography.

Geography (as a subject of learning) is meant to develop pupils' knowledge of the excellence and uniqueness of planet Earth. It helps them to understand the driving forces of phenomena and processes that take place on Earth and realise the principles on which the existence of life is based. Studying geography allows students to discover the country in all its complexity. It supports efforts to understand the relationships between the components and elements of the landscape and their interdependence.

Basic skills developed in geography are working with maps in analogical or digital format. Understanding how to find maps, read, work with them, analyse their content and interpret them and use them in an unfamiliar environment. They are the basics of geographical thinking and learning (Nogová and Hájek 1997).

Thorough knowledge of the Earth is a prerequisite for its protection. Each place on Earth is unique. One place differs from another through its special climate, plants, animals as well as inhabitants with their own language, culture and way of life. Each student is expected to be tolerant to otherness. They need to understand the fundamental causes of the variety and diversity of individual countries. This requires respect for the principles of democracy and civil liberties. In case they are not observed, it may lead to military conflict and global catastrophe. Geography in this respect plays an irreplaceable role. It emphasises reasonability and puts together in cases of not always clear interconnectedness the causes with their consequences. It shapes the personality of a young person by highlighting the similarities but also the peculiarities of regions. They are compared with Slovakia, pointing out its uniqueness in the context of Europe or the world (Likavský 2006).

In the next chapters, there are two examples of standardised curriculum outputs. The whole curriculum can be found in sources (ŠPÚ 2016a, b, c, Ušáková et al. 2009, 2010).

3.4 Education Standards in Geography: Stage 3 (11-Year-Old Students)

3.4.1 Planet Earth

Performance Standards

- Recognise the basic elements of the solar system in a simple sketch.
- Describe the apparent path of the sun and moon in the sky (drawings, sketches).
- Explain the causes of alternation of day and night on Earth.
- Determine according to the map of time zones where on the Earth there are more hours than in Slovakia and where there are less.
- State the causes of temperature zones on Earth.
- Explain the changes in climatic seasons.
- Recognise on the globe (maps) the continents and the oceans on the Earth.

Content Standards
Spherical shape of the Earth, angle of impact of sunlight, Earth rotation, inclination of the Earth's axis, Northern and Southern Hemispheres, Eastern and Western Hemispheres, orbit around the sun, inclination of the Northern and Southern Hemispheres, time zones, continents, oceans

3.4.2 Projection of the Earth

Performance Standards

- Distinguish the Earth's hemispheres on a globe.
- Determine a selected location on a map using geographic coordinates.
- Compare distance on maps of different scales.
- Identify basic objects in the country on the map ("read" the content of the map).
- Find specific locations on digital maps.

Content Standards
Globe, prime meridian, meridians, the equator, latitude, Eastern and Western Hemispheres, geographical coordinates, projection, map, map content (map symbols, graphic map scale), GPS

3.4.3 *Travelling on Earth*

Performance Standards

- Talk (according to maps and images) about travelling in selected mountains:
  - Explain (in your own words) the causes of the origin of mountains.
  - Appoint factors that contribute to the flattening of mountains.
  - Evaluate according to the map (globe) earthquake-risk areas on Earth.
  - Explain the arrangement of plants and animals according to altitude.
- Describe by maps and images travelling along a river from its source to its mouth:
  - Explain the formation of rivers and lakes.
  - Describe how valleys (canyons) and waterfalls were created.
- Talk about balloon travel from the equator to the polar countries:
  - Summarise the changes in the air with increasing altitude.
  - Explain why plants and animals on Earth are arranged into zones.
  - Assign to each latitudinal band two typical plant species and two species of animals.
  - Name two examples and evaluate the importance of protecting natural monuments registered in the list of natural and cultural heritage of UNESCO and show them on the map (globe).
- Speak about travelling in urban and rural areas:
  - Compare living conditions of people in urban and rural areas.
  - Assess the impact of natural conditions on population distribution.
  - Name two examples and evaluate the importance of protecting cultural sites registered in the UNESCO World Heritage List and show them on the map (Globe).

Content Standards
Mountains, plains, volcanoes, earthquakes, valleys (canyons), rivers, waterfalls, lakes, climatic zones (hot, mild, cold), air, clouds, types of landscapes (tropical rain forests, savannas, deserts, subtropical landscape, steppe, temperate deciduous forests, coniferous forests, tundra, polar landscape, mountain landscape), conservation, towns, rural settlements, natural and cultural monuments included in the UNESCO list

Performance Standards

- Design the route of a geographical excursion (walk), and describe its course.
- Estimate the distance to interesting points on the map (peaks, cultural-historical and natural monuments) and the difficulties in accessing them.
- Use a GPS device (mobile phone) to find points of interest.

Content Standards
Tourist facilities, tourist map, map orientation, compass, diversity of the terrain, altitude, points of Interest, peaks, view, tourist signs, geocaching, GPS

3.4.4 Africa

Performance Standards

- Define the location and description of Africa's coast on maps (bays, islands, peninsulas, seas, oceans) ("read" maps).
- Describe the trade wind's effect on the formation of deserts and semi-deserts in Africa.
- Explain the causes of distribution of climate zones in relation to the spread of flora and fauna in Africa.
- Compare the natural conditions in four climates of Africa.
- Give two examples of typical plant and animal species living in different landscape zones of Africa.
- Summarise the reasons for differences in population distribution around the major rivers of Africa.
- Evaluate the impact of natural conditions on the uneven distribution of the population of Africa.
- Explain a frequent cause of ethnic and religious disputes and conflicts between nations in Africa.
- Explain the reasons for low proportion of the population living in African cities.
- Identify the four most densely inhabited areas of Africa on thematic maps.
- Evaluate the economic maturity of various regions of Africa.
- Name three major problems that lower the living standards of the population in different regions of Africa.
- Explain the inclusion of one of the sights of Africa in the UNESCO cultural and natural heritage and show it on the map.

Content Standards
Africa, Madagascar, the Horn of Africa, Gulf of Guinea, the Mediterranean Sea, Red Sea, Strait of Gibraltar, equator, tropics, basin, mountains, platforms, deserts, Sahara, Namibia, Atlas, Kilimanjaro, trade winds, Nile, Congo, Niger, vegetation zone, animals, Cairo, Lagos, Johannesburg, Kinshasa, Republic of South Africa, Kenya, Egypt, Nigeria, population distribution, population density, ethnic structure of Africa and cities with more than five million inhabitants ("reading" the map), economy and mineral resources, Africa's population problems, monuments included in the UNESCO list of natural and cultural heritage

3.5 Secondary Schools (Stage 4 in England): The Geographical Curriculum

In secondary schools (students aged 15–18), the ratio between problem-oriented themes and regional topics is better balanced. The theme of regions is still prevalent as an essential element. The ratio of problem-oriented issues to regional-oriented issues is one to two (in stage 3 at secondary schools the ratio of these themes is one to four).

As in the curriculum of secondary schools (stage 3) in first place is performance, which sets out the content. The structure of the national curricula of geography stage 4 is the same as the document for stage 3. The extent of teaching geography at secondary schools (stage 4) is not so different from the overall representation of teaching geography in stage 3. Geography is allocated 4 h over 4 years of secondary or high school. This is increased by the optional teaching of geography in the school-leaving year.

3.6 Conclusion

Teaching geography has seen over the past 10 years quite a significant retreat within the national education programme in Slovakia. In this process, the most involved is the transformation of the entire education system in Slovakia. This means the increasing priority given to the language teaching block and optional subjects. A move towards teaching geography in the sense of a nonsystemic approach has not occurred. One positive trend is the particularly clearly defined set of minimum requirements (performance, content) of education. However, much more can be expected from a clearly defined deviation from too deeply regionally oriented geography teaching. Strengthening of the material-technical foundations of teaching is evident from resources of information and communications technology for a wide range of activity books and atlases. Slovak geography is still awaiting to offer schools more than one state-guaranteed textbook for each grade. This can enable a breakthrough to alternative approaches to teaching but at set standards that guarantee stability and a high level of geographic education.

References

CGE IGU Commission on Geographical Education of the International Geographical Education. (1992). *International charter on geography education*. http://www.igu-cge.org/charters_1.htm. Accessed 6 Oct 2015.

CGE IGU Commission on Geographical Education of the International Geographical Union. (2015). *The 2016 international charter on geography education*. http://www.igu-cge.org/Charters-pdf/Draft%20Charter%20IGU-CGE%20June%2011th%202015.pdf. Accessed 6 Oct 2015.

FPV UMB Fakulta prírodných vied Univerzity Mateja Bela. (2015). *Didaktika geografie* [Didactics in geography]. http://www.fpv.umb.sk/o-fakulte/akreditacia/phd/didaktika-geografie.html. Accessed 6 Oct 2015.

Hofmann, E., et al. (2014). *Standardy geografického vzdělávání v Česku a Slovensku a v USA* [Geographical education standards in the Czech Republic, Slovakia and USA]. Materiály z projektu Educoland. Pdf MU Brno. http://educoland.muni.cz/geografie/novinky-z-oboru/. Accessed 6 Oct 2015.

Karolčík, Š., Likavský, P., & Tolmáči, L. (2013). Profil absolventa učiteľského štúdia geografie a rámcový návrh nového študijného programu [The graduate profile the teaching of geography and a framework proposal for a new study program]. *Geografia, 21*(1), 21–23.

Likavský, P. (2005). Odpovedajú poznatky žiakov 8. ročníka základných škôl požiadavkám vzdelávacieho štandardu? [Does the knowledge of pupils of 8th classes of elementary schools correspond to the national educational standard?]. *Geografia, 13*(3), 116–120.

Likavský, P. (2006). *Všeobecná didaktika geografie* [General didactics of geography]. Bratislava: Univerzita Komenského.

Likavský, P., & Ružeková, M. (2004a). Vedomostná úroveň žiakov 6. a 7. ročníka ZŠ vo vzťahu k vzdelávaciemu štandardu: 1. časť [Knowledge level of students of 6th and 7th grades in relation to the school curriculum standards: Part 1]. *Geografia, 12*(3), 128–131.

Likavský, P., & Ružeková, M. (2004b). Vedomostná úroveň žiakov 6. a 7. ročníka ZŠ vo vzťahu k vzdelávaciemu štandardu: 2. Časť [Knowledge level of students of 6th and 7th grades in relation to the school curriculum standards: Part 2]. *Geografia, 12*(4), 163–166.

Nogová, M. (2002). Vzdelávací štandard zo zemepisu a oblasti jeho aplikácie [Educational standards of geography and areas of application]. *Geografia, 10*(3), 125–129.

Nogová, M., & Hájek, T. (1997). Realizácia koncepcie školských zemepisných atlasov [Realization of school geographical atlases conception]. *Kartografické listy, 5*(7), 117–120.

Nogová, M., & Tolmáči, L. (2008a). Gymnázium: predmet Geografia [Grammar school: Subject geography]. *Geografia, 16*(2), 64–66.

Nogová, M., & Tolmáči, L. (2008b). Vzdelávacie štandardy: Geografia – gymnáziá [Educational standards: Geography – Grammar schools]. *Geografia, 16*(2), 67–69.

Nogová, M., & Tolmáči, L. (2008c). Vzdelávacie štandardy: Geografia – základné školy [Educational standards: Geography – Elementary schools]. *Geografia, 16*(2), 59–64.

Nogová, M., & Tolmáči, L. (2008d). Základná škola: predmet Geografia [Elementary school: Subject geography]. *Geografia, 16*(2), 56–59.

ŠPÚ Štátny pedagogický ústav. (2016a). Inovovaný ŠVP pre 2. stupeň ZŠ [Innovative ŠVP for 2nd grade of elementary school]. http://www.statpedu.sk/clanky/inovovany-statny-vzdelavaci-program-inovovany-svp-pre-2stupen-zs/clovek-spolocnost. Accessed 14 Oct 2016.

ŠPÚ Štátny pedagogický ústav. (2016b). Inovovaný ŠVP pre gymnáziá so štvorročným a päťročným vzdelávacím programom [Innovative ŠVP for grammar schools with four-year and five-year educational programme]. http://www.statpedu.sk/clanky/inovovany-statny-vzdelavaci-program-inovovany-svp-pre-gymnazia-so-stvorrocnym-patrocnym-1. Accessed 14 Oct 2016.

ŠPÚ Štátny pedagogický ústav. (2016c). Inovovaný ŠVP pre gymnáziá s osemročným vzdelávacím programom [Innovative ŠVP for grammar schools with eight-year educational programme].http://www.statpedu.sk/clanky/inovovany-statny-vzdelavaci-program-inovovany-svp-pre-gymnazia-s-osemrocnym-vzdelavacim-1. Accessed 14 Oct 2016

Ušáková, K., Gálová, T., & Čipková, E. (2009). Biologické vzdelávanie v Česku a v Maďarsku [Biological education in the Czech Republic and Hungary]. *Biológia, ekológia, chémia, 13*(3–4), 2–9.

Ušáková, K., Gálová, T., & Čipková, E. (2010). *Biológia v Štátnom vzdelávacom programe ISCED 3 na Slovensku a kurikulárne trendy v biologickom vzdelávaní vo vybraných krajinách EÚ a světa* [Biology at the state educational programme ISCED 3 in Slovakia and curriculum trends in biological education in selected EU countries and the world]. Bratislava: Univerzita Komenského.

Zaťková, M. (1993). Zmeny v učebných osnovách zemepisu na základnej škole [Changes in the curriculum of teaching geography at elementary school]. *Geografia, 1*(1), 10–11.
Zaťková, M., Čižmárová, K., & Šimerová, J. (1992). Návrh zmien obsahovej koncepcie zemepisu v 5. ročníku ZŠ [Proposal for changes to the content of the concept of geography in the fifth grade of elementary school.]. *Učitelské noviny, 42*, 19.

Chapter 4
Teaching and Learning Geography in Secondary Education in Poland

Wiktor Osuch

4.1 Selected Reports Regarding the Training of Geographers and Geography Teachers

In the wake of the transformation of Poland's education system after 1989, some extensive reports about the education of geographers and geography teachers were provided.

The report on the organisation and scope of teacher training of candidates for different subjects, including geography, was an initiative of the Polish Ministry of National Education, and a survey was conducted by the Central Methodological Centre of Teachers Studies (*Centralny Ośrodek Metodyczny Studiów Nauczycielskich*). The team dedicated to geography developed the first (and to this day probably the only so complete and thorough) analysis and evaluation of geographers and geography teacher training in Poland (Jelonek et al. 1996). The report shows that in 1995, geographical studies were conducted in ten universities and three pedagogical higher education schools. The report included an analysis of geographical training on different types of studies in these universities and students' preparation for the teaching profession, including an analysis of pedagogical practices. The potential of directional and psycho-pedagogical staff was also analysed (Jelonek 1996). Numerous demands and proposals relating to the further education of geography teachers were formulated, and the most important ones stipulate the correlation of programme content related to psychology, pedagogy and geography teaching (Jelonek et al. 1996). The detailed results of the report were presented by Piskorz (1996) at the anniversary conference "Different Paths of Education and Training of Geography Teachers" during the 50-year anniversary of the Institute of

W. Osuch (✉)
Institute of Geography, Department of Geography Teaching, Pedagogical University of Cracow, Podchorążych St. 2, 30-084 Kraków, Poland
e-mail: wiktor_osuch@up.krakow.pl

P. Karvánková et al. (eds.), *Current Topics in Czech and Central European Geography Education*, DOI 10.1007/978-3-319-43614-2_4

Geography at the Higher Pedagogical School of Cracow in 1996 (currently the Pedagogical University of Cracow).

This report showed that already in the 1990s, in spite of harmonised rules, there were large differences in the plans, programmes and organisational solutions of geographical studies in various centres in Poland (Osuch 2010).

Due to the implementation of the objectives of the Bologna Declaration, since 2007 there has been an alteration in plans and programmes of geographical studies in all geography institutes and chairs in Poland. These changes are the consequence of a broad education strategy adopted by almost all of Europe in order to implement the vision of a Europe of Knowledge. The changes introduced were the inspiration for developing the next report on the state of geography education in the new Polish education system (Groenwald et al. 2008). This report also contains comments or concerns regarding the quality of education in relation to core subjects and teaching blocks, as well as inappropriate selection of candidates for teachers' studies.

The author of this paper attempts to analyse and evaluate the process of training future geography teachers in Poland, trying to develop a newer report. This attempt was not welcomed by all universities because current time of change in geographers' education system (including changes in programmes and plans) does not promote any summaries or conclusions. The presented results of the survey were selective (Osuch 2010) and relate to an analysis and evaluation of solutions in selected academic centres.

Hibszer and Tracz (2011) developed a publication about organisational issues and geographical study plans after the introduction of the Bologna Declaration with regard to the implementation of contemporary standards of education, mainly on the basis of surveys and interviews. In practice, however, the studies were incomplete, and the results showed wide variation in the training of geographers and future geography teachers.

4.2 Training of Geographers and Geography Teachers in Poland

Currently, there are 15 geographical centres in Poland: IGiPZ at the Polish Academy of Sciences, University of Warsaw (UW), Jagiellonian University (UJ), University of Gdańsk (UG), Adam Mickiewicz University in Poznań (UAM), Pedagogical University of Cracow (UP), University of Silesia (UŚ), University of Łódź (UŁ), Nicolaus Copernicus University in Toruń (UMK), Kazimierz Wielki University in Bydgoszcz (UKW), University of Wrocław (UWr), University of Szczecin (USz), Jan Kochanowski University (UJK) in Kielce and Pomeranian University in Słupsk (AP). Kostrzewski et al. (2015) presented the results of the survey on the current state of Polish geography during the transformation of the education system.

Becoming a geography teacher is voluntary in most universities and depends on the choice of specialty (survey by Osuch 2010). The rapidly changing socio-economic situation in Poland has determined the enrichment of the educational offer with non-teaching specialties, which are promoted more frequently than teacher training. In an era of huge competitiveness on the education labour market, only proven field experience of students could convince potential employers – that is why pedagogical universities, as well as other types of higher education institutions, cannot ignore their graduates' possibility of future employment in the profession. According to Piróg (2012), 23 % of students at UP, 18 % of students at UMK and 16 % of students at UWr were clearly determined to work as geography teachers. Many graduates of geographical centres in Poland also see the school as the main place of future employment.

The Institute of Geography (UP) in Cracow offers the following teaching specialisations: geography with basics of entrepreneurship, geography with social sciences and geography with natural sciences, which is one of the most popular geography teaching specialisations in Poland. In principle there used to be an equal number of teaching and non-teaching specialisations, but this was abandoned. The current offer at the Pedagogical University includes many more non-teaching courses. Other academic centres in Poland present a similar situation, where the number of non-teaching specialisations and courses is greater than teaching specialisations.

The teaching specialisations were determined by standards for teacher training, according to which the teacher training course was set at 390 h (105 h of the main subject and 60 h of the second subject) (Osuch 2010). Usually, teacher training was provided at the first (bachelor's) degree level. Currently, due to the implementation of the learning outcomes for individual courses and specialisation effects, this situation has changed. Teacher training courses were divided according to stages of school education (Ministry of Science and Higher Education – Act of 17 January 2012). As natural sciences is taught in primary school, at the bachelor's level, students of "geography with natural sciences" pursue the didactics of specialisation in this course. At the master's level, students pursue didactics classes in the main subject – geography – which is taught in lower and upper secondary schools.[1] This newer solution is organisationally more difficult to achieve, so it is hard to determine if it is better. The vast majority of teachers' training classes and school practice in "geography with natural sciences" and "geography with social sciences" specialisation are carried out during master's degree studies. It may be a heavy load for a student to cope with a large number of these activities as well as other subjects and a thesis paper.

As already mentioned in the abstract, there are three levels of education in Polish educational system (beyond university level): the 6-year primary school (divided into two 3-year educational levels), the 3 years lower secondary school

[1] *Lower secondary school (grammar school) – 3-year comprehensive school, obligatory for 13–16-year-old students. Upper secondary school (lyceum) – a non-mandatory three-year comprehensive school with specialised classes, for 16–19-year-old students. It is leading to the award of the matriculation certificate (świadectwo maturalne) upon passing the external matriculation examination (egzamin maturalny) required to enter a university.*

(grammar school) and the 3-year upper secondary school (lyceum). In elementary school, geography is not offered as a stand-alone teaching subject, although selected geographic content is included in natural sciences taught in primary school from classes IV to VI. This is a result of the reform of the education system conducted in 1999 when geography and other natural sciences like biology, physics and chemistry were "derived" from primary to lower secondary schools (grammar school). That is why geography teachers from grammar schools are still interested in teaching natural sciences in primary schools. These opportunities will be explained in details in Sect. 4.5.

There are many ways to acquire the necessary teaching skills. Students can register at a teacher training college, which organises teacher training, including teaching practice or, like at UP, join the Studium Kształcenia Nauczycieli (SKN) designed for external students who did not decide to specialise in teaching, but want to acquire the necessary accreditation.

In recent years in Poland, as in other European countries, we pay special attention to the effects of academic education. Among the expected learning outcomes, we should especially consider useful competencies in the rapidly changing labour market and those that contribute to shaping skills, not just gathering information. According to Szkurłat (2011), we have to precisely define a new foundation of the academic process: the objectives to be achieved, the outline of geographical studies and a new curriculum. We also need to continuously diagnose the quality of educational effects and outcomes. All of these assumptions are already being implemented, although opinions among academics are very diverse (Osuch 2012).

Although the offers are very interesting for the prospective students, the lack of choice, e.g. biology or history as a second subject for students of geography, is striking. The correlation of the educational content at school between geography and biology and geography and history, as well as the relationship between these disciplines, tends to be quite obvious and has a long tradition. The main reason for this is probably organisational reasons that prevent the free choice of the second field of study and also the subject of education. Western Europe for years successfully used a wide choice of solutions for the second track of education, from a large range of subjects to correlate the student's interest and ability to find a job in school. In the near future, we will see whether individual universities educating geographers create a sufficiently wide range to choose another subject or whether the fear of major organisational challenges will reduce or severely limit the choice.

Piróg and Tracz (2013) in their deliberations on the concept of teacher training presented changes in the functioning of geography teachers' training studies for over 65 years of history of the Institute of Geography (Pedagogical University of Cracow). We clearly notice that demands on the education of geography teachers formulated decades ago by Prof. Jan Flis (1955, 1984) are still valid and should be taken into account in subsequent changes in plans and programmes of study. The authors draw attention to the vocational practices that are a very important element and an opportunity to prepare students to enter the labour market. They also advise further implementation of ambitious and interdisciplinary regional and field exercises (Piróg and Tracz 2013, p. 27). The author of this publication referred to the theory of the "pedagogisation of studies" proposed years ago, in light of the geographical concept of Prof. Jan Flis, confirming that many suggestions and solutions from distant years work today and resulted in a generation of well-educated and competent geography teachers (Osuch 2013).

The tradition of solid, 5-year master's degree studies is deeply rooted among employees of universities, and therefore many state universities in Poland are very reluctant to put the principles of the Bologna Declaration into practice. In addition, ongoing research in Austrian and German universities educating future geography teachers (Osuch 2004, 2006) has confirmed that geography teachers training at the bachelor level do not lead to appropriate didactical or factual preparation of students.

4.3 Geography in Grammar School and Lyceum: Analysis and Evaluation of the Implementation of the Programme Assumptions

Human geography in Western Europe is the dominant trend in geography, which Poland accepted only after strong resistance (during the transformation), evidenced by the inclusion of geographic content into "natural sciences" in elementary school and geography into the group of some environmental subjects in grammar school. In the twenty-first century, some interesting books in which socio-economic geography took its rightful place finally appeared, and thus Poland has shortened the distance in this regard to Western Europe. There was no good tradition in Poland in selecting educational content of socio-economic geography (human geography). It was classified in the group of natural sciences for many years and has lost ties with humanistic sciences, especially with history. The crisis of socio-economic geography related to all stages of education. According to Owczarz and Plit (2003), it was infected with the teaching of facts, boring textbooks and long series of numbers, e.g. the location of industrial facilities illustrating the power of the socialist economy. University entrance exams required statistical data, which was indeed the easiest to verify. It often happened that shortly before exams the new statistical yearbook appeared and new numbers were required on the exam. School and university studies leaked in time with an interpretation of the data, their critical analysis and evaluation explaining the relationships between phenomena (Owczarz and Plit 2003). Socio-economic geography content was undervalued and not popular among teachers – they preferred to carry out topics related to physical geography.

An interesting example of the use of field lessons in teaching-learning geography in the Polish lyceum were geographical workshop projects presented in the curriculum and textbooks prepared by one of the oldest and still popular Polish educational publishing houses Wydawnictwo Szkolne i Pedagogiczne SA (WSiP). They were concerned, for example, with improving the geographical environment in Poland and the region (municipality, district). Issues pointed out in the extended version of exam sheets were not well known among teachers. The new interests of geographers related to political geography: the geography of electoral activities of non-governmental organisations in the world, in Poland and in the local environment, and analysis of the causes and attempts to resolve political conflicts (Plit et al. 2003). Another textbook extended the relationship between man and the environment to concepts of sustainable development, extreme events, global threats to the environment and human health (quality and lifestyle, tropical diseases, AIDS – "the plague of the century") (Podgórski et al. 2004), which contributed effectively to an expansion of the geographical horizons (Osuch 2009).

Rodzoś et al. (2008) wrote about geography education changes raising the problem of the theoretical basis of geography, pluralism of textbooks, reform of the Polish education system and the "new school-leaving exam". The advantages and opportunities for geography included ease at attracting students' attention to the content, ability to take a critical view of key problems of the modern world – European integration, globalisation, sustainable development – and the usefulness of the course in shaping many competencies as well as the existence of many good textbooks. The main threats included information chaos in the textbooks resulting from trivial content, inadequate selection and arrangement of their content, eliminating geography by incorporating particular content into other subjects and the low level of the school-leaving exam.

4.3.1 Geography at Grammar School

Geography in compulsory grammar school is taught 4 h in a 3-year cycle. The same number of hours is dedicated to the other sciences: biology, physics and chemistry. Therefore, the rank of natural science school subjects at this stage of education is the same. Unfavourable demographic changes, resulting in a smaller number of classes at school, influence the number of geography hours, forcing geography teachers to complete their jobs in other educational institutions. Geography teachers are forced to acquire additional qualifications to teach a second subject, in order to stay in the profession. According to Tracz and Świętek (2014), this phenomenon is visible especially in small towns, where the educational offer is very limited.

Tracz and Świętek (2014) conducted recognition studies to check progress in the implementation of programme changes in grammar school geography education. Moreover, the difficulties faced by geography teachers in grammar schools were analysed and assessed. Based on research conducted in 50 schools, grammar school teachers pointed out an increase in behavioural problems in grammar school, reduced level of students' knowledge of the natural sciences and reduction of appropriate content in particular courses. The advantages were better learning outcomes demonstrated in PISA research, structural solutions similar to other European countries and schools better equipped with teaching aids (Tracz and Świętek 2014, p. 58–59). An important conclusion for analysing and evaluating the educational content under the assumptions of the curriculum is the fact that, in practice, the distribution of hours of geography is not compatible with the intentions of the authors of the curriculum. This implies that the choice of some textbooks, which provide the teaching of Polish geography in third class, is followed by the marginalisation of this content, as approximately 42 % of surveyed teachers implement 2 h of the subject in the second, not the third, class (Tracz and Świętek 2014). In many cases, moreover, physical geography issues implemented in the first class are too difficult for students (Tracz and Świętek 2012) and thus are continued at the beginning of the second class. This is also confirmed by previous studies by Tracz (2008) relating to the results of the competence test after grammar school.

Szczęsna (2011) researched the geography achievements of students before grammar school and has come to the conclusion that the integration of science content in the school contributed to a lowering in the level of natural sciences competencies among students, including geography. In addition, "natural sciences" poorly prepared students for further geography education in grammar school, due to insufficient training of cognitive and practical skills.

4.3.2 Geography at Lyceum

The reform of the education system in Poland, which started in 1999, has applied in upper secondary schools since the 2011/2012 school year. Geography is taught at the primary level (30 h in first class) and advanced level (240 h in second and third class). Only the primary level is compulsory for all students.

According to the assumptions of curriculum reform, general education on the third and fourth levels, realised in two different schools (grammar school and lyceum), is a consistent foundation of education (Marciniak 2011; Adamczewska 2014). The first class in lyceums focuses on issues of socio-economic geography, according to the adopted earlier assumption that at the primary level students should learn about the socio-economic and environmental problems of the modern world (Czerny 2011). According to Adamczewska (2014), geography teachers generally assess the content selection carried out in the first class of lyceums positively, believing that it is interesting and may attract young people, as it covers political geography, cultures, high-tech industries, demographics and tourism. One downside is the lack of Polish geography content, meaning that teachers cannot make references to specific examples. Also, the limited number of hours devoted to each issue seriously impedes the development of critical thinking and analytical skills in the geographical environment, which may cause stereotypes to be reproduced and reinforced (Adamczewska 2014).

A detailed analysis of educational content for the extended level of geography, conducted exclusively by willing students in the second and third class, has many flaws. Despite the 240 h allocated to the extended level, there are problems with the implementation of this level, because the new core curriculum differs only a little from the previous one, which in turn means that educational content must be implemented in a shorter cycle (only 2 years). Moreover, even a short weekly absence may cause a backlog of four to five topics in one subject, which can be extremely difficult for the student to catch up with. Issues implemented in the second class are markedly different from those of the first class, because they require specific knowledge, assimilation of geographical terminology, free use of the map, perceiving and explaining relationships and dependencies between individual elements of the geographical environment (Adamczewska 2014). During the implementation of physical geography content in the second class of lyceum, students should remember the issues discussed in grammar school, which can be an obstacle for them and for teachers.

In summary, the geography education in lyceums demotes geography as a subject, but also as a scientific discipline. A detailed SWOT analysis of geography education in lyceums (Adamczewska 2014) is not very positive. There are many more threats and vulnerabilities than opportunities and strengths. The weaknesses include: completion of compulsory geography education after the first year of lyceum, insufficient number of hours, the absence of physical geography and very limited role of Polish geography at the primary level. The threat is a decrease in the level of geographical knowledge in society and knowledge about the country and increasing anonymity of students. Profiled classes (and extended level of geography) are rarely chosen by students. Existence of these classes depends on the number of applicants – that is why the number of dismissed geography teachers is increasing. The following were considered as strengths and opportunities: dynamic and global approach to teaching geography, large number of hours for extended level of geography (profiled classes), opportunity to learn and understand the surrounding world, different cultures, shaping attitudes of openness and the possibility of solid and deliberate preparation for the school-leaving exam (Adamczewska 2014). It is impossible not to agree with the results of research conducted by M. Adamczewska in schools and the general experiences of geography teachers in this field.

4.4 School-Leaving Exam in Geography

After completing geography education at upper secondary school, students should have the knowledge and skills necessary to take the school-leaving exam in geography.

Since 2005, lyceum students could choose geography as the selected subject for the written part of the school-leaving exam. To date the school-leaving exam can be taken at the basic or advanced level. The choice of geography in the advanced level posed the necessity to acquire and deepen skills to solve more difficult geographical problems, which are sets of tasks that the students should analyse and learn. A popular way to prepare for the school-leaving exam were practice sheets developed by

Regional Examination Boards or task sheets compiled by educational publishing houses such as Wydawnictwo Szkolne i Pedagogiczne (Owczarz et al. 2005).

Wójcik (2013) analysed and evaluated school-leaving exam tasks in geography, especially the scope of content, their construction and technical and educational correctness. Tasks from 2005 to 2011 were analysed and the following conclusions were made:

- There was an increase of closed tasks, an increase of easy-to-solve tasks, especially at the primary level, and an increase of tasks that check knowledge with a decrease of tasks that check skills, which should be considered unfavourable.
- In school-leaving exam sheets, we observed slightly more tasks on world geography than Polish geography, and the overall concept of the sheet emphasised facts and drawing conclusions for accompanying attachments like texts, drawings, photographs, maps and models.
- In the school-leaving exam sheet (at the primary level), tasks connected with working with the map are highlighted, while explaining the processes and phenomena in the geographical environment is less important.
- There was a clear decrease of factual errors and inaccuracies in formulating commands (in comparison to previous sheets, e.g. 2015) (Wójcik 2013, p. 97–98).

As a consequence of the earlier choice of geography at the advanced level, currently (from 2015) lyceum students can pass geography only at the advanced level of the school-leaving exam. Polish language, foreign language and mathematics are compulsory for the school-leaving exam, and students must obtain at least 30 % to pass. In addition, they are obliged to choose at least one additional subject at the advanced level (maximum five), but without the selected threshold of completion.

Graduates now often choose geography as an additional subject. In 2015 in Poland, 45,634 students attempted to pass geography at the advanced level, which accounted for 22 % of the students who chose geography as an additional subject. For comparison, biology was chosen by 42,572 students, chemistry by 27,313 students, physics by 17,330 students, history by 17,328 students and social sciences by 28,027 students. The average exam result obtained in geography at the advanced level is 41 % (Preliminary information on the results of the matura exam, Central Examination Board 2015. www.cke.edu.pl). After the introduction of a compulsory exam in mathematics, the number of candidates for passing geography decreased. So far, a large proportion of students was interested and showed positive attitude to geography during school education. Also, average and poorly performing students willingly chose geography for the school-leaving exam because they were even weaker in other subjects and they thought they could quickly catch up with geography. Sometimes such thinking proved wrong. Since 2015, the chief motivation for choosing geography is higher education, where the school-leaving exam in geography is required or additionally rewarded.

Dzięcioł-Kurczoba (2014) presented interesting research on the reasons for the different results of the school-leaving examination, but they refer to previous versions of the school-leaving exam. It is necessary to repeat studies that show the impact of recent programme changes at lyceums on the current results of the school-leaving exam, introduced in 2015.

By 2014, the school-leaving exam tested mastery of knowledge and skills that were specified in the standards of examination requirements. Since 2015, learning objectives described in the core curriculum (in the form of general requirements which define the specific skills and learning content – specific requirements, including messages) replace the existing standards for educational requirements. The new school-leaving exam requires learning complex skills: understanding concepts, recognising geographical correctness, formulating causal relationships and analysing and processing information (Geografia Matura 2015).

4.5 Natural Sciences in Primary School and Upper Secondary School: New Opportunities for Geography Teachers

According to the reform of the Polish education system in 1999, primary schools introduced a new subject called "natural sciences", which is taught in classes IV–VI (3 h per week). The possibility of teaching a new subject in primary school appeared for geographers because geography and other natural sciences (biology, physics and chemistry) have been "pushed" into the newly created grammar school. Initially, natural sciences teachers, and therefore biology, geography, physics and chemistry teachers, could teach "natural sciences" without any additional courses or postgraduate studies. The opportunity to teach this subject was used quickly by teachers of other subjects from primary school, especially early school education, who completed postgraduate studies giving them permission to teach "natural sciences". In subsequent years, natural sciences teachers were obliged to hold an appropriate three-semester postgraduate degree to obtain formal permission to teach "natural sciences" in primary schools. Currently postgraduate studies in "natural sciences" last two semesters for active teachers. Geography students can obtain permission to teach "natural sciences" by choosing it as their second specialisation subject in the *Geography with natural sciences* track. For a few years now, it has been a very popular specialisation in geography teaching studies.

Due to the consistent implementation of education system reform for upper secondary schools, especially in general upper secondary schools, new opportunities for teachers appeared, including geography teachers.

"Natural sciences" was introduced as a supplementary subject upper secondary school (lyceum) (Regulation of the Minister of National Education of 27 August 2012. – Dz. U. of 2012 No. 0 pos. 977) in September 2013 and is aimed at students who have not chosen any of the natural sciences subjects (biology, chemistry, physics, geography) at the advanced level. "Natural sciences" is taught from class II, after finishing primary level natural sciences subjects in class I. At present, this subject can be taught in lyceums by biology, chemistry, geography or physics teachers, and the final decision in this matter belongs to the headmaster. Usually all of these teachers teach "natural sciences", and, depending on the implementation of specific educational content, headmasters choose those teachers who feel competent enough. There are also cases of top-down allocation of hours for teachers who need to deal with their workload. Ministerial assumptions indicate that teachers' competencies should influence decisions about teaching "natural sciences", not organisational issues. Nevertheless, it seems possible that in the future its teaching in lyceums will be sanctioned by additional postgraduate studies.

According to the framework curriculum for upper secondary schools, 120 h are dedicated to "natural sciences". In addition to developing appropriate curricula and textbooks, innovative solutions like educational projects were also permitted. For example, the "Kształcenie Pełne Wyobraźni (KPW)" project implemented by the Tischner European University curriculum (Ćwiro et al. 2012) presents detailed educational content with the educational objectives in four colours representing assignment to a specific subject, biology, geography, physics or chemistry, which will help the teacher to prepare the lesson concepts and specialised class literature. It should be noted that the implemented project does not allow the use of any manual, because that is not needed nor was planned. Students benefit from the materials of the Moodle platform and the multimedia presentations prepared and provided therein, as well as source texts, glossaries, sets of test tasks and control questions. In addition, the project developed lesson plans with attachments for teachers.

An important feature of the programme is to present the issues in an interdisciplinary way, so that students perceive the presented topics as well as their practical use and the topicality of the issues raised in a comprehensive manner. This innovative concept of teaching allows students to develop interests, understand and use modern research methods used in the natural sciences. Many of the lesson topics are very interesting and compel students to seek original, custom solutions (Osuch 2014).

The project is currently being tested in four general upper secondary schools. After the full 2-year cycle has been implemented, we will be able to judge the success of this exemplary educational project.

Dzięcioł-Kurczoba (2014) wrote specifically about natural sciences training for teachers in colleges and universities in Poland, demonstrating the legal possibilities of natural sciences studies at different levels and forms of education, and the list of courses implemented within the framework of studies at four selected departments in Poland (p. 84). The author is concerned about the great diversity of programmes at the postgraduate level, as well as the limited substantive foundation for future natural sciences teachers.

Piróg and Jania (2013) wrote about the improvement and further training of geography teachers. They showed in detail the degree of professional activity, the reasons for activity as well as the preferences of geography teachers. Research shows that younger geography teachers (with up to 5 years of teaching experience) prefer postgraduate studies and qualification courses to gain additional qualifications (confirmed by a certificate or diploma), while more senior teachers willingly participate in training seminars, conferences and consulting methodology (Piróg and Jania 2013). Unfortunately, there are serious barriers to making improvements, such as high training costs and the inability to reconcile deadlines for training and work.

References

Adamczewska, M. (2014). *Geografia w zreformowanym liceum – doświadczenia nauczycieli z Łodzi i województwa łódzkiego* [Geography in reformed Lyceum – Experience of teachers from Lodz and Lodz province]. Annales Universitatis Peadagogiscae Cracoviensis. Studia Geographica VI, 162 (pp. 207–216). Krakow: Scientific Publishing House of Pedagogical University of Krakow.

Ćwiro, E., Fryt, Z., Słowiak, P., & Ślósarz, J. (2012). *Program nauczania. Przedmiot uzupełniający: Przyroda. IV etap edukacyjny* [Curriculum. Supplementary course: Nature. IV stage of education]. Krakow: Tischner European University.

Czerny, M. (2011). Komentarz do podstawy programowej przedmiotu geografia [Comments to geography core curricilum]. In *Podstawa programowa z komentarzami. Tom 5. Edukacja przyrodnicza w szkole podstawowej, gimnazjum i liceum* Warsaw: Ministry of National Education. *Natural Science Core Curriculum, 5*, 177–194.

Dzięcioł-Kurczoba, B. (2014). *Uczelnie wyższe w Polsce a kształcenie przyrodnicze* [Natural sciences education on Polish universities]. Annales Universitatis Peadagogiscae Cracoviensis. Studia Geographica VI, 162, (pp. 80–87). Krakow: Scientific Publishing House of Pedagogical University of Krakow.

Flis, J. (1955). Wakacyjne prace polowe w programie studiów geograficznych [Holiday field work in geography studies curriculum]. Krakow. *Scientifical and Educational Yearbook of Higher Pedagogical School – Geographical and Biological Sciences, 4*, 3–28.

Flis, J. (1984). Pedagogizacja nauczycielskiego kierunku studiów geograficznych [Pedagogization of teaching faculty at geographical studies]. In *Dydaktyki przedmiotowe na kierunkach nauczycielskich w uniwersytetach i wyższych szkołach pedagogicznych* (pp. 51–56). Krakow: Materials and Reports of the Teachers Studies Methodological Centre of Higher Pedagogical School.

Geografia Matura 2015 – najważniejsze zmiany. (2015). *Przewodnik dla nauczycieli szkół ponadgimnazjalnych* [Geography. Matura exam 2015 – The most important changes. Guide for teachers of higher secondary schools]. Warsaw: Nowa Era. ISBN 978-83-267-1867-0.

Groenwald, M., Plit, F., Rodzoś, J., & Tracz, M. (2008). Raport o stanie geografii szkolnej w nowym systemie oświaty w Polsce. In *Geografia we współczesnym systemie kształcenia* [The report on the state of geography education in the new education system in Poland] (Geographical Documentation, 38, pp. 5–17). Warsaw: Institute of Geography and Spatial Organization Polish Academy of Sciences.

Hibszer, A., & Tracz, M. (2011). Studia geograficzne w Polsce po wprowadzeniu Deklaracji Bolońskiej [Geographical studies in Poland after the introduction of the Bologna declaration]. In M. Tracz & E. Szkurłat (Eds.), *Efekty kształcenia geograficznego na różnych poziomach edukacji* [The effects of geographical education on different levels] (Work of Educational Comission of Polish Geographical Society, 1, pp. 137–151). Lodz: Publishing House of University of Lodz.

Jelonek, A. (1996). Kształcenie geografów w wyższych uczelniach w Polsce, w tym kształcenie nauczycieli geografii [Education of geographers at Universities in Poland, including the training of teachers of geography]. In J. Jarowiecki & S. Piskorz (Eds.), *Różne drogi kształcenia i dokształcania nauczycieli geografii Scientific conference 23–24.04.1996 materials* (pp. 53–65). Krakow.

Jelonek, A., Kucharska, M., Kusiński, W., & Piskorz, S. (1996). *Raport "Kształcenie geografów w wyższych uczelniach w Polsce, w tym nauczycieli geografii"* [Report "Geographers education at Universities in Poland, including geography teachers"]. Krakow: Bulletin of the Teachers Studies Methodological Centre of Higher Pedagogical School of Krakow.

Kostrzewski, A., Roo-Zielińska, E., Krzemień, K., & Lisowski, A. (2015). Geografia w okresie transformacji system nauki w Polsce – aktualny stan, perspektywy rozwoju [Geography during the transformation of education in Poland – Current status, prospects for development]. Warsaw-Lublin. *Geographical Journal, 86*(1–2), 23–47.

Marciniak, Z. (2011). O potrzebie reformy programowej kształcenia ogólnego [The need for general educational program reform]. In *Podstawa programowa z komentarzami. Tom 5. Edukacja przyrodnicza w szkole podstawowej, gimnazjum i liceum* (Natural Science Core Curriculum, 5, pp. 7–21). Warsaw: Ministry of National Education.

Ministry of National Education. (2012). Act of 27 August 2012 on core curriculum for preschool education and general education in particular types of schools (Journal of Laws from 2012, No. 0, item 977). Accessed 15 Mar 2016.

Ministry of Science and Higher Education. (2012). Act of 17 January 2012 on educational standards in preparation for the teaching profession. www.procesbolonski.uw.edu.pl/dane/standardy_ksztalcenia_dla_nauczycieli.pdf. Accessed 30 July 2015.

Osuch, W. (2004). *Forming Subject Matter and Didactic Competence by Geography Students – Candidates for Teacher Profession in Poland and Selected European Countries.* Geografie a promeny poznani geograficke reality (pp. 532–540). Ostrava: Ostravska Univerzita.

Osuch, W. (2006). *Formation geography professional competences – Candidates for teachers in selected universities in Poland, Germany, Slovakia and Czech Republic* (Geographic review, 2/2, pp. 558–565). Banska Bystrica: Department of Geography Faculty of Natura Science Matej Bel University Banska Bystrica.

Osuch, W. (2009). *Selected aspects of teaching of geographers in Poland against geographical education at school.* Geodays Liberec 2008. Book of Proceedings (pp. 242–249). Liberec: Technical University of Liberec.

Osuch, W. (2010). *Kompetencje nauczycieli geografii oraz studentów geografii – kandydatów na nauczycieli* [The competences of teachers of geography and geography students – Candidates for teachers] (Monographic works, 570). Krakow: Scientific Publishing House of Pedagogical University of Krakow. ISBN 978-83-7271-620-0.

Osuch, W. (2012). *Changes in university education of geography students in Poland against the competences of geography teachers.* Europäische Kooperationen. Europäische Perspektiven 4. Pädagogische Hochschule Wien (Hg.) (pp. 71–85). Wien: LIT VERLAG GmbH & Co. KG Wien.

Osuch, W. (2013). Wybrane aspekty pedagogizacji nauczycielskiego kierunku studiów geograficznych [Selected aspects of pedagogization of teaching faculty at geographical studies]. Annales Universitatis Peadagogiscae Cracoviensis. Studia Geographica IV, 148 (pp. 59–71). Krakow: Scientific Publishing House of Pedagogical University of Krakow.

Osuch, W. (2014). Przykłady innowacji w liceum – wstępna ocena programu nauczania przyrody w ramach projektu "Kształcenie Pełne Wyobraźni – KPW" [Examples of innovation in Lyceum – A preliminary assessment of the natural sciences curriculum within the framework of the project "Education full of imagination"]. Annales Universitatis Peadagogiscae Cracoviensis. Studia Geographica VI, 162 (pp. 230–241). Krakow: Scientific Publishing House of Pedagogical University of Krakow.

Owczarz, M., & Plit, F. (2003). *Geografia część 2. Człowiek i jego działalność. Poradnik dla nauczyciela w liceum ogólnokształcącym* [Geography part 2. Human and his activities. Textbook for Lyceum]. Warsaw: WSiP SA. ISBN 83-02-08524-3.

Owczarz, M., Szewczyk, I., Karcz, W., & Osuch, W. (2005). *Maturalnie, że zdasz. Geografia-zakresy podstawowy i rozszerzony: pytania, zadania, testy, arkusze egzaminacyjne* [N(M)aturally, you will pass. Geography – Basic and extended range: Questions, tasks, tests, examination papers]. Warsaw: WSiP SA. ISBN 83-02-08603-7.

Piróg, D. (2012). Poziom zainteresowania pracą w zawodzie nauczyciela wśród studentów geografii [Level of interest in teaching profession among students of geography]. Lodz: Torun. *Work of Educational Commission of Polish Geographical Society, 2*, 194–210.

Piróg, D., & Jania, R. (2013). *Dokształcanie i doskonalenie zawodowe nauczycieli geografii jako instrument dostosowania się aktualnych wyzwań rynku pracy* [Training and professional development of teachers of geography as an instrument to adapt to current challenges in the labor market]. Annales Universitatis Peadagogiscae Cracoviensis. Studia Geographica IV, 148, (pp. 72–83). Krakow: Scientific Publishing House of Pedagogical University of Krakow.

Piróg, D., & Tracz, M. (2013). Koncepcja nauczycielskich studiów geograficznych według Jana Flisa [The concept of geographical studies for teachers, according to Jan Flis]. Annales Universitatis Peadagogiscae Cracoviensis. Studia Geographica IV, 148 (pp. 17–29). Krakow: Scientific Publishing House of Pedagogical University of Krakow.

Piskorz, S. (1996). Rola i zadania polskiego nauczyciela geografii na przełomie XX i XXI wieku. In J. Jarowiecki & S. Piskorz (Eds.), *Różne drogi kształcenia i dokształcania nauczycieli geografii* [The role and tasks of Polish geography teacher at the turn of XX and XXI century]. Scientific conference 23–24 April 1996 materials (pp. 144–152), Krakow.

Plit, F., Osuch, W., Sielatycki, M., & Wrona, J. (2003). *Geografia część 2- Człowiek i jego działalność. Kształcenie w zakresie rozszerzonym. Podręcznik do liceum ogólnokształcącego* [Geography part 2. Human and his activities. Textbook for Lyceum]. Warsaw: WSiP SA. ISBN 83-02-08524-3.

Podgórski, Z., Marszelewski, W., & Becker, K. (2004). *Geografia część 3. Człowiek i środowisko. Podręcznik dla liceum ogólnokształcącego, liceum profilowanego i technikum* [Geography part 3. Human and environment. Textbook for higher secondary schools]. Warsaw: WSiP SA. ISBN 83-02-08855-2.

Preliminary Information on the Results of the Matura Exam Conducted in May 2015. (2015). Website of Central Examination Board (Centralna Komisja Egzaminacyjna) www.cke.edu.pl/index.php/egzamin-maturalny-left. Accessed 9 Sept 2015.

Rodzoś, J., Szczęsna, J., & Wojtanowicz, P. (2008). Transformacje geografii szkolnej [Transformations of School Geography]. In A. Hibszer (Ed.), Polska dydaktyka geografii – idee- tradycje – wyzwania. Sosmowiec. *Work of Faculty of Earth Sciences of University of Silesia, 47*, 99–108.

Szczęsna, J. (2011). Osiągnięcia uczniów z zakresu geografii i progu gimnazjum [Students' achievements in the field of geography and the threshold of lower secondary school]. In M. Tracz & E. Szkurłat (Eds.), *Efekty kształcenia geograficznego na różnych poziomach edukacji. Work of Educational Commission of Polish Geographical Society*, 1 (pp. 67–79). Lodz: Publishing House of University of Lodz.

Szkurłat, E. (2011). Problemy jakości kształcenia geograficznego na tle zmian w kształceniu akademickim w Europie i w Polsce [The problems of quality of geographical education on the background of changes in European and polish universities]. In M. Tracz & E. Szkurłat (Eds.), *Efekty kształcenia geograficznego na różnych poziomach edukacji*. Work of educational commission of Polish geographical society, 1 (pp. 29–40). Lodz: Publishing House of University of Lodz.

Tracz, M. (2008). *Znaczenie geografii jako przedmiotu ogólnokształcącego na przełomie XIX i XX wieku – studium przypadku* [The importance of geography as a general school subject in the late nineteenth and early twentieth century – A case study]. Geographical documentation, 38 (pp. 72–79). Warsaw: Institute of Geography and Spatial Organization Polish Academy of Sciences.

Tracz, M., & Świętek, A. (2012). Nauczyciele o nauczaniu przyrody w szkole podstawowej [Teachers voice about teaching geography in primary school]. *Geography at school, 64*(4), 15–19.

Tracz, M., & Świętek, A. (2014). *Zmiany programowe nauczania geografii w rzeczywistości szkolnej na przykładzie gimnazjum* [Program changes in teaching geography in school reality, based on lower secondary school]. Annales Universitatis Peadagogiscae Cracoviensis. Studia Geographica VI, 162 (pp. 53–66). Krakow: Scientific Publishing House of Pedagogical University of Krakow.

Wójcik, J. (2013). *Geografia na maturze – zróżnicowanie merytoryczne i ocena zadań z arkuszy egzaminacyjnych w latach 2005–2011* [Geography at Matura Exam – Substantive differentiation and evaluation of tasks of the examination papers in 2005–2011]. Annales Universitatis Peadagogiscae Cracoviensis. Studia Geographica IV, 148 (pp. 84–100). Krakow: Scientific Publishing House of Pedagogical University of Krakow.

Part II
Innovative Methods in Geography Teaching – Way to Mobilize the Students

Chapter 5
Inquiry-Based Education of Physical Geography

Petra Karvánková, Dagmar Popjaková, Michal Vančura, and Šárka Nedvědová

5.1 Introduction

The methodology concept of inquiry-based education (IBE) (or inquiry-based science education (IBSE), inquiry-based learning (IBL)) has the same primary goal as other innovative methods, i.e. to make the learning process more attractive for learners and to activate learners. On the other hand, this method is different as it simulates research and applies scientific methods, which are adjusted to the age of learners, to make the learning process more interesting. The development of IBE was initiated by social demand for the development of science, which led to the "Renaissance" of scientific literacy and natural sciences. Natural sciences were seen as "subject fields in crisis", since there was a dramatic drop in interest in these subjects by the young generation. A clearly defined request for a research approach in learning was stated in the National Standards for Natural Sciences Education in the USA in 1996 (Stuchlíková 2010, s. 150). In Europe, it was the European Commission and its experts who started promoting and supporting the new learning methods, in the spirit of "Europe needs more scientists?", in order to support children's interest in natural sciences and to spread inquiry-based education (EC 2004, 2007, 2013; Rochard et al. 2007; Osborne and Dillon 2008). Specialist didactics came into focus in connection with the development of IBE (Janík and Stuchlíková 2010). On the other hand, the paradigm of interdisciplinarity and multidisciplinarity of natural sciences education is emphasised (Papáček 2010a, s. 38). What is the role of geography and what can it offer, as a typical interdisciplinary science, during the implementation and spreading of IBE topics and methodology procedures? Which specific inquiry methods can be used during practical geography lessons in the field? These are the questions and objectives addressed in this chapter.

P. Karvánková (✉) • D. Popjaková • M. Vančura • Š. Nedvědová
Faculty of Education, Department of Geography, University of South Bohemia in České Budějovice, Jeronýmova 10, 371 15 České Budějovice, Czech Republic
e-mail: karvanko@pf.jcu.cz

P. Karvánková et al. (eds.), *Current Topics in Czech and Central European Geography Education*, DOI 10.1007/978-3-319-43614-2_5

5.2 Inquiry-Based Education in the Czech Republic

The change in education policy in Czechia at the beginning of the twenty-first century brought along a change in the objectives of geography education and its curriculum, in particular by building up and developing key competencies of learners, focusing on situations that are close to their lives and that are practical (for more information, see Chap. 2, Fričová 1982; Hájek 2003). In line with this change, the concept of IBE started to be promoted in the last years. The term IBE first appeared in the English-Czech Pedagogy Dictionary in the Czech Republic (Mareš and Gavora 1999 in Janík and Stuchlíková 2010, s. 21). Methodology teachers from the natural sciences education team of Dr Janoušková from the Charles University Environment Center in Prague and methodology teachers from the South Bohemian Didactic School, represented by Professor Stuchlíková and Professor Papáček from the Faculty of Pedagogy and Psychology, and the Faculty of Biology at the Faculty of Pedagogy of the University of South Bohemia in České Budějovice, greatly contributed to the development of the IBE concept in Czechia. The papers by Janoušková et al. (2008) and papers presented at the DiBi conference in 2010 were the first theoretical and empirical outputs of IBE (Papáček 2010b; Petr 2010; Stuchlíková 2010). A very detailed study of the prerequisites for IBE and its importance when educating a new generation of learners was prepared by M. Papáček in his article "Inquiry-Based Natural Sciences Education – a road for biology education for Y, Z and alpha generations?" (2010a).

The fact that IBE methodology processes are becoming a matter of interest in the Czech and Slovak Republics in the last 5 years is demonstrated by the number of projects which are focusing on making science and research more popular thanks to the incorporation of research processes into learning.

The Tereza Association was the first to start implementing projects and activities on the basis of IBE of natural sciences. In 2010, this association implemented a project 3 V – *Vědě a výzkumu vstříc! (3 V – Give a Go to Science and Research!)* with the main aim of contributing to the development of scientific forms of work at secondary schools (SS) in Czechia and to encourage interest in natural sciences at SS and to enhance cooperation between SS and universities (U) (Sdružení Tereza 2010). The main vision of the project was to increase the attractiveness of scientific careers for the young generation. A similar vision of popularisation of natural sciences, such as biology, chemistry, geography and geology, was presented by the communication project of the Faculty of Natural Sciences at Charles University in Prague, named *Přírodověci.cz (Natural Scientists.cz)* in 2011 (PrF UK 2011). The main objective of this concept was to approach primary and secondary school students, their teachers, the media, professionals in the field and the public and to increase their interest in natural sciences. Subsequently, a similar project named *Badatelé.cz (Researchers.cz)* started in Czechia in 2012 as a result of the cooperation of the Tereza Association and a team of teachers from Czech primary schools (PS) and guarantors from universities (U) (Přírodovědecká fakulta UK v Praze, Pedagogická fakulta JU v Českých Budějovicích, Pedagogická fakulta MU v Brně). The aim of this 2-year project, supported from the EU programme investment into education, was to introduce inquiry-based education to primary schools as part of standard teaching practice (Sdružení Tereza 2012a). In addition, the Tereza Association helped to get involved in the GLOBE (Global Learning and Observations to Benefit the Environment) worldwide programme, which strongly supports the methods of inquiry-based education. Every year, more than 130 primary and secondary schools are involved in the GLOBE programme in the Czech Republic (GLOBE 2015). The University of South Bohemia in České Budějovice, as part of *scienceZOOM* projects (JU CB 2012), has been developing inquiry-based education of natural sciences and technical fields, and it has been working on the popularisation of these subjects since 2012. The main objective of these projects, financially supported from the European Social Fund, is to spread scientific and research findings amongst the widest circle of people, ranging from primary school students to scientific and research workers.

From 2010 to 2014, Slovakia, Czechia, Poland and other countries were involved in a European project, called ESTABLISH *(European Science and Technology in Action: Building Links with Industry, School and Home)* (ESTABLISH 2015), that was spreading IBE principles of natural sciences amongst students, ages 12–18, within the European Union. Slovakia, contrary to Czechia, together with other 11 EU countries, has already joined another of the European projects, named PRIMAS *(Promoting Inquiry in Mathematics and Science Education Across Europe)*, focusing on inquiry-based teaching of mathematics and natural sciences. The aim of this 3-year project (2010–2013) is to help teachers to integrate the features of "discovery" into teaching (PRIMAS 2014) and to create methodology support for teachers, amongst others.

The positives and advantages of inquiry-based education in natural sciences subjects, including geography, are clearly visible from the above-mentioned. The question remains if Czech teacher training institutions sufficiently prepare their students, i.e. future geography teachers, to teach geography in line with inquiry-based education. And consequently, if the graduates are able to teach geography using the principles and procedures of inquiry-based education so that the teaching is efficient, beneficial and useful, what knowledge, skills and experience do future geography teachers need to teach in line with inquiry-based education? These and other issues are being addressed by the authors of this chapter in their 3-year project run by the Grant Agency of the University of South Bohemia entitled: "Geographical Research and its Implementation in Inquiry-Based Geography Education" (2013–2015). The main objective of the project is to propose new topics and to verify methods suitable for the implementation of inquiry-based geography education. The project is primarily focusing on a municipal region, the town of České Budějovice, which represents a "living space" with significant dynamic changes that can be well monitored in terms of methodology, both in terms of socio-economic changes (e.g. migration flows, changes of industrial and urban structures, etc.) and in terms of physical geography (e.g. town climate, quality of environment, development of green areas in the town, etc.), that allow to understand their mutual correlation.

Despite all activities, it must be stated that the incorporation of methodology processes of inquiry-based education into a standard Czech learning process is still at the very beginning, compared to the world, in particular with regard to geography.

5.3 Geography and Inquiry-Based Education

The actual position of geography amidst other scientific fields is, as stated by, for example, Mičian (1988), Wahla (1983, 1991), Herink (2009) or Hynek (2002) and Vávra (2008), at the crossroads of natural, social and technical sciences (Fig. 5.1). Geography is not only an interdisciplinary but also a multidisciplinary scientific field, covering more research areas, which have, quite logically considering the genetic viewpoint, arisen from natural sciences (Matlovič 2006). As a unit it creates a separate category known as *natural and social sciences*.

Just like geography within the science system, geography as a school subject taught at primary and secondary schools has irreplaceable prerequisites of interlinking natural, social science, technical and other findings connected to the environment, place and time (Herink 2009). This interdisciplinary and wide multidisciplinary thematic scope ensures a major integration role within the school education system, and it is important in the development and formation of nearly all cross-sectional topics (Karvánková 2013; Karvánková et al. 2013, 2015).

The concept of IBE for geography didactics was presented at the Symposium of IGU Commission that took place in Helsinki on 6–10 August 2001. It is mentioned and analysed by, for example, Hindle (1993); Healey, Roberts (2004); Healey (2005) and others, and in Czechia by Hynek (2002) and later by Vávra (2009). The methodology of inquiry-based education has not yet been applied in geography lessons at Czech schools. Of course it is present as an activating method for problem-based learning. Many of the IBE methods and procedures are applied during project learning, focusing on a constructivist approach to learning (Kühnlová 1997; Hynek et al. 2011; Hynek and Svozil 2011), practical field learning (Hofmann 2005; Marada 2008) and geolaboratories (Kuldová 2008; Řezníčková 2003), etc. The didactic approach to inquiry-based teaching in geography and other natural sciences, such as chemistry and biology, was presented, using specific practical tasks under the title "task-oriented education" in a monograph by D. Řezníčková et al. (2013). The term "discovery learning" (Dudášová 2013; Lacinová Svoradová 2013;

Fig. 5.1 Position of geography in the science system (Source: Authors)

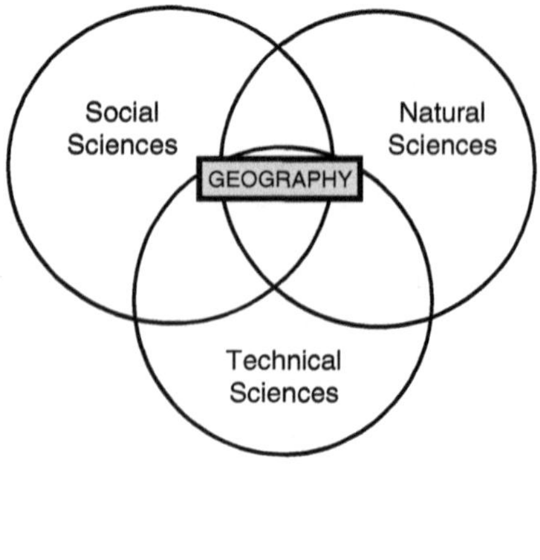

Michaeli and Madziková 2013) was used in Slovak didactics of geography and by, for example, Čipková and Karolčík (2015), and in Polish geography this term is used, for example, by Osuch (2014).

5.4 Inquiry-Based Education of Physical Geography

IBE enables learners to work in situ and in time using scientific processes, simulating actual scientific work, and if the subject matter is chosen correctly, the learning process becomes more effective, and it is more likely that natural sciences become more attractive for learners (Abell 2000; Janoušková et al. 2008; Linn et al. 2004; Papáček 2010b; Petr 2010; Prenzel et al. 2009; Stuchlíková 2010). Inquiry-based education is, in its many forms as introduced by teachers (Eastwell 2009 in Stuchlíková 2010, s. 135), connected to the use of modern scientific procedures, independent identification of environmental problems, working with data and literature, development of communication skills, etc. These are exactly those competences Czech students are missing, according to PISA research (Czesaná et al. 2009; Bičík 2009).

The IBE concept can be very well applied in teaching geography. Partly, it can eliminate a shortage of practical hands on activities in geography lessons, insufficient development of instrumental skills and insufficient focus on problems which the learners encounter in their everyday lives. IBE represents a modern educational direction which, if applied in geography lessons, increases students' motivation and interests, development of their critical thinking, creativity and ability to solve problems.

Teaching physical geography gives students a basic theoretical idea about the subsystem of the physical and geographical sphere of the planet Earth, focusing on individual processes and phenomena that are happening in the sphere (Strahler 2001). The teaching process should outline the mutual links and interconnections between the processes, clarify the causes and mechanisms of their functioning, use a holistic and partial view and adjust the learning process to students' needs and age in accordance with the Framework Educational Programme for Basic Education (FEP BE, Kol. 2013; Hofmann 2006). It is exactly the above-mentioned holistic approach and understanding of the functioning of mutual relationships and feedback of the physico-geographical sphere that are missing during physical geography lessons at Czech primary and secondary schools and sometimes even at universities. The holistic approach, correlation and understanding of functioning of the physico-geographical sphere as a unit are not easy for a majority of students; in other words, students do not know how to create the approach and how to understand and interconnect the given issues. This situation is caused by bad habits which the students acquired during physical geography lessons at all levels of Czech primary education. Primary education is based on memorising of facts and on frontal teaching and not on understanding the context. Physical geography is still taught in a fragmented manner, where students are given the same, nearly identical information about individual components of the physico-geographical sphere (lithosphere, atmosphere, hydrosphere, pedosphere, biosphere) at all levels of primary and secondary education, i.e. about the structure of the Earth, atmosphere, volcanoes, soil composition, etc., without any need to understand the mutual links. This approach is called, for example, by Perkins (2008), "elementitis" when learning is completely separated from skills and is fragmented into individual pieces of information. Learners are inundated with terms and descriptions of various geomorphological shapes and reliefs, geomorphological divisions of the territory, which do not mean anything to them. There is no further explanation or demonstration, and as a result they are not able to understand the connections and continuity of a functioning physico-geographical system. This results in disengagement of students and learners from physical geography. Generally, students consider physical geography to be a demanding subject, where they need to memorise many terms without any connections or links between the acquired information and practice, which, quite logically, makes physical geography uninteresting and demotivating for students. It could be said that students will not learn to understand the landscape (physico-geographical sphere) as one unit; they will not be able to understand it, to have a relationship with it, which might result in their negative attitude to physical geography, but also to their negative attitude towards the environment and their surroundings as a whole. By including the methodology procedures of inquiry-based education (IBE) into physical geography lessons, students will become more motivated, the form and method of learning will improve and lessons will become more attractive, which might motivate students to learn and appreciate the subject more.

Using IBE in physical geography lessons brings many undisputable advantages, but on the other hand, it is more time demanding, teachers will need more time to prepare for the lessons, and they will also need to arrange for additional equipment which will be needed for the practical lessons. Teachers must know how to apply IBE principles in geography lessons and to have the practical skills and knowledge to organise and lead the lessons (either in the field or at school or a classroom). The potential advantages and disadvantages, or risks (Table 5.1), which are connected to inquiry-based education were discussed by, for example, Rychnovský (p. 5–6, 2010).

Despite the fact that Table 5.1 looks at IBE issues from the students' perspective, the outlined aspects are closely connected to the teacher's work. The most prominent *positives* of using IBE methodology procedures for teaching physical geography are, as mentioned above, understanding the basis of the presented issues and supporting efficient active learning. Students gain information through their actual work (in the field or in the classroom) testing the acquired knowledge, gaining personal experience with the respective issues and gaining skills which help to cement the acquired knowledge. Physical geography is now taught not through frontal, static teaching methods but through simple, practical and visual presentations and demonstrations that are sometimes even managed by the students themselves, which makes the actual learning process even more efficient and attractive. Using the IBE methodology processes in practice reflects the quotation of Chinese philosopher Confucius: "I hear and I forget. I see and I remember. I do and I understand".

Table 5.1 Advantages and disadvantages of inquiry-based education of natural sciences from the student's point of view

Subject	Advantages	Disadvantages/risks
Field of didactics	Independent work	Highly time demanding
	Searching for and acquiring information	More extensive preparation necessary
	Higher motivation	Teaching pace is slower
	Ability to sort out information	Risk of unequal involvement
	Understanding the context of the curriculum	
	More activity	
	Different teacher-student communication	
General	Deeper memory footprint	Possible one-sidedness
	Development of creativity	Unsophisticated learning
	Respecting other people's opinion	Nonsystematic learning
	Deepening interdisciplinarity	Suitable for only a small number of learners
	Improving the ability to form and defend one's point of view	
	Suitable for other activities	

Source: Rychnovský (2010)

Similarly, using IBE methodology procedures in teaching physical geography brings many obstacles, big or small, and also possible *negatives*. The most significant one is the fact that teachers must be thoroughly prepared for the lesson. As stated by, for example, Perkins (2008), when implementing IBE it is essential to carefully plan the lesson and the actual learning structure. When using the IBE methodology procedure, it is necessary to have the essential materials and tools at hand and to select a suitable environment for the actual research activity. Inquiry-based education of physical geography requires a sufficient number of devices of good quality that will enable the acquired theoretical knowledge to be employed in practice. This can be done either on an individual basis or in teams, "in situ" and "in time" in the field or in a classroom. It is important to understand, as stated by Stephenson (p. 6, 2007), for example, that IBE cannot be confused for a technique, teaching practice or teaching method that will save the teacher. As stated by Perkins (2008), IBE should be seen as something which enables "playing the whole game". Students are given an opportunity to experience "the entire game" of resolving a scientific problem; they can break away from learning isolated pieces of knowledge (e.g. how to throw a ball) or learning about the game (e.g. studying statistics or history of the game). Also Douglas and Brown (2011) compare IBE to a game in the sense that there is a creative tension between rules and freedom, i.e. between the known and unknown. Just like every game needs its rules, research needs structure and limits that will make it more effective. Research, compared to traditional teaching models, focusing on theoretical knowledge and skills, is open to the unknown to the very end. It is vital to let students "play" with the subject matter and to emphasise what is known (basic concepts and keywords), but also leave space for the unknown where students can construct, interpret and actively participate.

For well-managed lessons using IBE methodology procedures, it is vital that questions are posed which can be formulated by the students themselves who can express what they are interested in and what is critical for the given subject matter. As stated by Fischer (1997), students are often inundated by questions which do not have any learning function. Questions in IBE should be open-ended to make learners think. Fischer also points to the necessity of planning. Learners should be able to define a problem, collate information, create their own strategy which is applicable and to monitor results. Errors are used during inquiry-based education, as students can learn from their false hypothesis and presumptions, just like in real scientific work. Students find out that it is not important in practice to know the exact answer straight away, but to know how to arrive at the correct answer, how to verify the answer and how to analyse it. This develops their critical thinking and enables discussion, planning, analysis and synthesis of the information acquired.

We propose teaching activities based on the methodology of inquiry-based education according to Perkins (2008). He points out the necessity to distinguish the term "inquiry" from "discovery learning" when students are left to explore and develop on their own. Just like Perkins (2008), we also consider inquiry more efficient, in particular with regard to younger learners at primary schools, providing inquiry is presented as a combination of structured learning and intentional opportunities for learners. Learners can consequently build up, present and develop new opportunities, abilities and skills. The current teaching practice in Czechia sees "inquiry" and its incorporation into the learning process as a synonym for "making students find their way back to subjects they have lost an interest in" as stated by, for example, Alvardo and Herr (2003). However, it is not enough to simply incorporate inquiry-based education into lessons, despite the fact that IBE is often seen as a process initiated by students, or the teacher, when students ask questions and the teacher guides students through the learning process. Due to the prevailing standard teaching patterns, learners at Czech primary schools are not able to integrate and independently grasp the principles and basis of inquiry education. Czech learners are not accustomed to this; they do not know how to learn in this way and how to think critically. And yet, new teaching trends, and inquiry-based education methodology, are gradually assimilated into the current learning practice at Czech schools. Czech teachers and their students are only gradually learning new attitudes during the education process, slowly deviating from the traditional ways of teaching and learning.

5.5 Suggestion from Czech Practice on How to Use Inquiry-Based Education for Grades 5–9 at Primary Schools in Physical Geography: Example

In order to illustrate the basic theoretical ideas, we enclose our ideas on how to use inquiry-based education in physical geography. We will try to outline and characterise the teaching of physical geography by a selected topic (paedology) using IBE methodology procedures. Despite the fact that it is impossible to communicate the atmosphere of physical geography lessons taking place in the field in this chapter, as the experience of students and teachers is not transferrable, we will attempt to provide the most faithful portrayal of the learning practice and its evaluation.

The suggestion for teaching selected geographical topic demonstrates the method of leading students towards inquiry. Currently, the time allocation is two hours of geography per week for grades 5–9 at primary schools. This shows that geography is not valued as much as other subjects and is currently in the background. By applying suitable methodology procedures of inquiry-based education, geography lessons can become more attractive, and natural sciences can be made more relevant for students.

Inquiry-based education makes students more interested not only in geography, natural history and other natural sciences but also in their surroundings, countryside and landscape they live in. During inquiry-based education, students learn many attitudes, develop practical knowledge and skills and use theoretical knowledge in practice. Also during inquiry-based education, the thematic unit of geography fieldwork, practice and application is implemented, which forms a major part of geography in FEP BE (e.g. Hofmann 2003b). In real practice, however, it is being constantly ignored and pushed to the background. And yet, due to its multidisciplinarity resulting from narrow links with all cross-sectional FEP BE topics, geography is more than suitable for the application of methodology principles of inquiry-based education. There is a scope to increase the number of geography lessons to anchor and modernise the geography curriculum at Czech primary schools by integrating the subject with cross-sectional topics using IBE methodology procedures.

The use of IBE methodology procedures for teaching selected thematic topics moves the teaching process into a completely new dimension (Hofmann 2003a; Kučerová 2005; Karvánková and Popjaková 2015). The teaching activities that we propose take place directly in the field. But it is not necessary to travel with students far; any outside school premises will do if we want to have a geographical laboratory (Wilczyńská-Wołoszyn 2003), i.e. a geolaboratory (Kučerová 2005; Řezníčková et al. 2008). Sometimes only a change of the environment (from their desks to the field) helps to motivate students if the teacher plans the lesson correctly (Hofmann 2003b).

5.5.1 *Teaching Activity: Soil, What Is It That We Are Actually Walking On?*

Paedology falls under the educational topic Humans and Nature within the Framework Educational Programme for primary education in Czech. However, it is taught at primary schools in two subjects: natural history and geography. The subject matter is partly taught in grade 6 (in geography lessons) when the time allocation is quite small, and then it is taught again in grades 8 and 9 at some more depth, within the scope defined by the educational programmes of individual schools (SEP BE). As shown in Table 5.2, the curriculum is divided between different subjects by

Table 5.2 Incorporation of the monitored curriculum into Framework Educational Programme for Basic Education (FEP BE)

Educational areas	Thematic area	Expected outcomes	Subject matter
Natural sciences	Inanimate nature	Compare the significance of soil-forming factors for soil formation, and distinguish between the main soil types and classes found in the nature of the Czech Republic	Soils: composition, properties and importance of soil for plant nutrition, its economic importance for society, the dangers and examples of its devastation, the possibilities and examples of recultivation
Geography	A natural image of the Earth	Distinguish between and compare the components and elements of the natural (physiographic) sphere, their interconnectedness and conditionality; recognise, name and classify formations on the Earth's surface	Landscape area: natural (physiographic) sphere, social and economic spheres, components and elements of the natural sphere
	Environment	Compare various landscapes as a component of the continental part of the landscape area, and distinguish between specific features and functions of landscapes on concrete examples	Landscape: natural and social environment, types of landscape
		Provide specific examples of natural and cultural landscape components and elements as well as the spatial distribution of the main ecosystems (biomes)	The relationship between nature and society: global, ecological and environmental problems of mankind
			The principles and fundamentals of natural and environmental protection

	Czech Republic	Assess at an adequate level the natural, economic and cultural conditions in the local region and the possibilities for further development, and analyse the connections of the local region to higher territorial units adequately	Local region: geographical location, criteria used for defining the local region, relations with the surrounding regions, basic natural and socio-economic characteristics focusing on the region's specifics important for its further development (potential vs. barriers)
		Assess and compare at an appropriate level the position, natural conditions and resources and the human and economic potential of the Czech Republic in the European and global contexts	Czech Republic: geographical location, area, terrain segmentation, natural conditions and resources
			Population: basic geographic, demographic and economic characteristics, settlement patterns; distribution of economic activities, sector and branch structure of the economy; social, political and economic transformational processes and their territorial manifestations and impacts; economic and political position of the Czech Republic in Europe and the world, its participation in international division of labour and trade
	Field geographic education, practice and application	Apply practical methods in the field when observing, depicting and assessing the landscape	Field exercises in and observations of the local landscape, geographical excursions: orientation points, phenomena, aids and devices; standpoint, determining cardinal and intercardinal points, navigation using a map and an azimuth, estimating the distances and heights of objects in the field; simple panoramic sketches of the landscape, layout plans, schematic sketches of a route axis, evaluation of natural phenomena and indicators

Source: Kol (2013)

topics. The basics, covering key terms and definitions, describing the properties and origin of the soil, etc., are covered in natural history classes according to FEP BE (Kol. 2013). Students should utilise what they have learnt about the soil in natural history in geography lessons at a practical level, in context, for example, in connection to regional geography, talking about the specific properties and natural conditions of the selected region and understanding the links and continuity of the landscape sphere. Nevertheless, considering the standard teaching patterns where memorisation and a lack of practical lessons prevail, the learners are not able to link the information acquired in natural history classes with the topics covered in other lessons, such as geography, and they are not able to use the information.

In view of the current, descriptive method of teaching paedology, soil is often considered by students as something intangible and not really useful. It is a topic which requires them to memorise (often without understanding) answers to the following questions, which are most often posed incorrectly and in a simplified manner: *What is the pedosphere? What is the composition of soil? What is a soil type and a kind of soil? What is the composition of humus? Which soil types do you know? Which kinds of soil do you know?*. The learning process does not emphasise understanding of the topic and interlinking the curriculum with practice, despite the fact that according to FEP BE (Table 5.1) this is one of the main teaching objectives at primary schools. Consequently, students, to a large extent, often repeat the same information which they were taught during natural history classes. They do not think about the context and the connection with real life, and they do not understand the importance of soil.

Exploring the soil and its properties provides an enormous number of inquiry topics that can be explored either in the field or in a laboratory in many classes, ranging from examination of the soil origin and its composition to the selection and localisation of paedology sites and determining basic soil characteristics and features. In terms of methodology, many suggestions were made by, for example, Bergstedt et al. (2005), the Tereza Association (Sdružení Tereza 2012a, b) or CENIA (2013) on how to use methodology procedures of inquiry-based education for soil/paedology. The inquiry activities that we present focus on the integration of the knowledge acquired, in particular, in natural history and partly in geography, and the ability to apply the knowledge during (45 min) practical geography fieldwork at primary schools. The key part of fieldwork is focusing on practical work with soil probes at a selected site and determining the soil type in the monitored location. The main teaching goal is to understand the differences between individual soil types and their geographical locations on the Earth. Individual teaching goals, forming an integral part of the main teaching goal, are to learn skills and gain experience with using soil probes and sledgehammers, to reinforce cartography skills and to gain experience with working with the soil map of the monitored area, i.e. Czechia.

The methodology list for teachers (Table 5.3) presents basic ideas about the concept and anchoring of the monitored thematic unit of paedology as part of FEP BE and represents an integral part of the teacher training programme for inquiry-based education.

Table 5.3 Methodology list for geography teachers teaching paedology in line with IBE

Topics: Soil – What is it we are actually walking on?
Motivational introduction to the lesson:
Soil is one of the main natural resources. It provides people with livelihoods and space to construct their houses; it enables plants to grow and is crucial for animals, providing them with a living environment, food sources and a territory to live in and find shelter. People have a significant impact on soil resources; they transform and change the soil, mostly negatively. Irresponsible human interaction can destroy something that has been created over thousands of years, and sometimes the destruction process can occur very quickly.
During our activity, we will find out the following: How is soil created and how is it degraded? Which factors influence the origin of soil? Why do we need to know the soil composition and its chemical and physical properties? Why do we distinguish soil types and kinds? What is the practical significance of soil classification? Is there life in soil?
Target group: primary school students (grades 8–9)
Students' age: 14–15 years
Keywords: paedology, soil, soil type and kind, basic soil-forming factors, use and protection of soil, soil degradation
Spatial requirements: school surroundings, school premises, geolab
Main objectives: students understand the operation of the pedosphere; they try to characterise and analyse soil; they understand the connection between using and protecting the soil
Students' competencies: students build up a relationship with soil, they understand its importance for humans, they learn how to work with soil probes, they learn to work with the soil map of the world, of Czechia and of a respective area
Tools: soil drill + sledgehammer, soil map of the area, catalogue of soil types, geological map of the area
Budget for devices and equipment: approx. CZK 8000 (approx. EUR 300)
Allocated time: 45 min
Form of teaching: teamwork (five to six students in a team)
Cross-curricula relationships (link to FEP BE): geography
Cross-sectional topics: environmental education – ecosystems, basic conditions for life, man's relation to the environment

Source: Authors

The actual lessons are very different from standard lessons from the very beginning. The teaching activity starts with a *phase of thinking* and *posing questions* in reverse order compared to a standard lesson, when the lesson often starts with questions like: what is the pedosphere, paedology and soil? When students arrive at the field site, they are always intrigued by the soil auger (probe) and the sledgehammer. The first question is always about what the tools are for and what they can be used for in the field. At this phase of thinking, students must work together and agree on the use of the tools, their importance and function. Students are in the phase of *asking questions and providing answers*, which they either confirm or dispute in their discussion. To determine the purpose of the tool and the functioning of the soil auger and the sledgehammer, the students often move from the phase of thinking and asking questions to wondering: what am I going to find out when I hammer the soil auger into the ground? What are we going to examine? Students think and ponder and the teacher does not interfere in this phase. His/her role is to encourage students in the thinking process and their train of thoughts.

Students work out that after taking out the probe, they get a soil sample, i.e. the lesson is going to pose questions about the soil profile, its importance and characteristics, i.e. simple questions, why is it important in practice to know the soil profile and acquire a soil sample? It is fundamental not to forget about the issue of soil origin, its composition and justification of the importance of the process in terms of the characteristics of the respective soil. During the thinking stage, it is advisable to guide students to imagine how big the world's arable land is and in fact how small is the part on which most of the planet's agricultural production depends. The Table 5.4 might provide inspiration for this short motivational phase.

After a short experiment (10 min), attention is paid to issues concerning the importance of the soil, its use and differences, asking questions about the difference between the soil texture and soil structure in terms of their importance. Students must first think about which soil properties are important for humans and which properties we need to know to be able to use the soil for farming or other purposes. This brings students to understand the principles and importance of defining soil types, in particular in the agricultural context. At this part of the activity, a map of soil types present in the location is introduced, together with a map of soil types present in Czechia. Students work with the map and the map key and start thinking about the different soil types in Czechia (Table 5.5).

The role of the teacher is to be in the background at this stage; the teacher guides the students' trains of thought by asking leading questions. It is advisable to make students think about which factors can influence the origin of a particular soil type, i.e. the soil type in the given location. A *story* is introduced into the teaching process when students are asked to imagine that they themselves are the soil forming anywhere on Earth, and they are asked to think about the factors that can influence the process of soil formation, i.e. of their formation. They can use the soil map of the world and try to localise different soil types, e.g. by occurrence, properties, etc. At this stage, students must work together and agree on five major factors (main soil-forming factors) that can influence the origin of the respective soil type. At the end of the story, students *must defend their ideas*, i.e. factors, and explain why they think these factors are important. During this activity students must come up with and justify soil-forming factors: relief, climate (individual factors such as water, rain, wind, frost, etc. are mentioned), subbase (parent material) and organisms (often caused by humans); however, a problem is to determine the last factor important for soil formation, i.e. time.

After this activity, students work with maps for the last time. They use the maps and the map keys to select important soil types in Czechia and in the region, and they try to characterise the soil types. An emphasis is put on the importance of soil types and their occurrence in the respective area (additional questions are asked, making students think about this phase of the activity as stated in Table 5.6). All these questions can be set up by the teacher, but it is often students themselves who ask the questions, based on their work with the map, putting things into context, arguing and discussing. The teaching process has moved away from memorising

Table 5.4 Example of a motivational activity

Materials	*An apple, a knife, a paper towel*	
Procedure	*Slice the apple according to the instructions, narrating as you go*	
Apple	Planet Earth	Narrative
Whole apple	Planet Earth	1. Hold the apple
		"This apple represents our planet"
3/4	Water	2. Cut the apple into quarters. Hold out ¾ in one hand
		"What do these ¾ represent?" (water)
1/4	Land	3. Set the three "water" sections aside and hold out the remaining quarter
		"What fraction of the apple remains? (1/4)
		This ¼ represents the total land surface"
1/8	Uninhabitable and nonarable land	4. Slice the land (the remaining ¼) in half, lengthwise. Hold out one of the pieces
		"This 1/8 represents the half of the Earth's surface that is inhospitable to people and to crops: the polar regions, deserts, swamps and high or rocky mountains"
1/8	Habitable land	5. Set that 1/8 aside and hold out the other
		"This 1/8 represents the other half of the Earth's surface. These are the areas on which people can live, but cannot necessarily grow food"
3/32	Habitable land, but nonarable land	6. Slice this 1/8 crosswise into four equal pieces. You have 4/32. Hold out only 3/32 in one hand
		"This 3/32 represents land on which people can live, but cannot grow food. Some of it was never arable because it's too rocky, wet, cold and steep or has soil too poor to produce food. Some of it used to be arable but is not any longer because it's been developed – turned into cities, highways, etc."
1/32	Arable land	7. Set 3/32 aside and hold out 1/32
		"So, only 1/32 of the Earth's surface has the potential to grow the food needed to feed all of the people on Earth"
1/32 peel	Topsoil	8. Carefully peel the 1/32 slice of Earth
		9. Hold up the peel
		"This tiny bit of peel represents the topsoil, the dark, nutrient-rich soil that holds moisture and feeds us by feeding our crops"

Source: Population Connection (2004); abridged by authors

facts and frontal teaching to actual practical work using a map in the field, where it is the students who manage the learning process and the pace of learning.

Working with the maps, it is time to start posing hypothesis for the examined areas. Students become paedologists and examine the soil around the school and work with the soil probe. Before starting to work with the soil probes, students are asked to work in teams and use the information acquired, their knowledge of the area and the maps, and to create hypotheses about which types of soil will be present in the monitored area or on the site (Table 5.7).

Table 5.5 Basic questions for students to think about when studying pedology in geography lessons in grades 5–9

Worksheet: Our soil – What is it we are actually walking on?
Describe how to work with a soil probe and a sledgehammer. What can be found out about the soil in situ using the probe? Why do we examine the soil? Why is soil important for humans?
Is clay a soil material?
Can you imagine what it looks like in the soil? What is its composition? How do you envisage a soil profile? What does it consist of *(layers – soil horizon)*? Can you compare it to something from your practical life *(hint: imagine a slice of cake or a broken biscuit with filling!)*? What do we use soil profiles for in practice?
Are there any differences in the soil properties around the world? Explain using the world map
In your opinion, why is it important to define soil types in agriculture?
Using the soil map of Czechia, try to define the most common soil types in our territory. Decide which factors have the biggest impact on the respective soil type

Source: Archive Karvánková

Table 5.6 Examples of key additional questions for students when thinking about the soil map of Czechia

Key questions that might help you think
Which soil types can you find in the fertile areas of Czechia and what are the characteristics of these soil types? Why are these soil types linked exactly to the fertile areas of Czechia and the other way round?
Where, in Czechia, can you find fertile areas? Can you name and define some of them? What do you think is a "fertile area" in the context of Czechia and which physico-geographical and socio-economic conditions must be met in such an area?
Which soil types can you find around rivers and what are their specific properties? How fertile are these soil types? How much do we use them for agriculture and why? How does the river flow influence the origin of these soil types?
Which soil types are the most common ones in Czechia and in which areas can you find them? Are we able to use these types for agriculture purposes? Why can we state that this soil type is the most common one in Czechia?
Which soil types are typical for elevated mountainous areas and what are their properties? What is the typical vegetation cover of these soil types?
Do you know any areas in Czechia where you can find marshes and wetlands? Can you determine the soil type in these areas? What are the properties of such soil?
Can you identify certain areas in Czechia which were significantly affected by people and in which the soil cover has negatively changed? Try to locate these selected areas on the map of Czechia and to characterise them briefly from a geographical point of view. Why are or were these areas used by people to such an extent?
Do we have steppes in Czechia? In which areas of Central Europe can you find steppes? Are they cultivated by people or not? Why?

Source: Archive Karvánková

Each of the hypotheses must be first supported by explanation and geographical justification of occurrence of the particular soil type before the students can start working with the soil probe. If working with older students, they can also create a profile of the paedology site in this part of the lesson. When students work with GPS, geological maps of the monitored areas, inclinometer, etc., they localise and

Table 5.7 Example of a part of the worksheet helping students to determine their hypothesis about the soil type in the monitored area

You have become experienced paedologists and now you are facing your hardest task ☺. You must establish the soil type of this site and why is it here. Based on the information and knowledge you have acquired before the start of the actual scientific work, create your own hypothesis about the potential soil types on the site and justify this hypothesis:	
Hypothesis 1:	Justification:
Hypothesis 2:	Justification:
Hypothesis 3:	Justification:

Source: Archive Karvánková

characterise their paedology site. This is followed by the *actual planning of a procedure on how to verify the hypothesis*; students must agree who, in which way and how to work with the soil probe. The teacher's role in this phase focuses on informing the students about health and safety when working with the soil probe and the sledgehammer and how to use the soil probe correctly. This is followed by *the actual experiment* (using the soil probe, creating soil profiles) in which data is recorded and photographed. The next phase is the *evaluation of results, stating of findings and verification of hypotheses.* Students work with the Taxonomic Classification System of Soil in the Czech Republic (Němeček et al. 2001), the Soil Atlas of the Czech Republic (Tomášek 1995) or the soil map of the selected area (ČGS 2012), where they search for soil profiles of the selected soil types which they have defined as part of their hypothesis and which they have compared to the soil profiles actually collected in the field. The territory of Czechia is more than suitable for inquiry-based activities focusing on the soil, since there is a relatively large diversity of site conditions. Thanks to this, there are a large number of soil types in Czechia and many transitions. Students must use the soil probe and define and draw individual parts of the acquired soil profile using professional literature. At the end students evaluate and justify their confirmation or disproof of their own hypothesis concerning the respective soil types in the area. The whole fieldwork activity is concluded by questions, summarising and repeating the subject matter, that are posed by the students themselves.

Instead of using the soil auger and the sledgehammer (auger method) for the respective activity, we can also use a spade and a pickaxe (Pit method) to dig a typical soil profile for the respective area (Figs. 5.2 and 5.3). The disadvantage of drilled probes is space limitation and the size of the drill; in other words, not all morphological properties of the soil can be exactly established from the taken samples, although for the proposed form of the inquiry activity, this tool is perfectly sufficient. The advantage is the speed at which a soil profile can be acquired. A dugout soil probe is bigger, easier for monitoring soil properties and more exact, although it is more time demanding.

The following instructions must be followed during the inquiry:

- The paedology site should not differ from the surrounding terrain.
- The burden on the site, caused by anthropogenic activity, must be taken into consideration. The site should be at least 2–3 m away from the roads, buildings and other areas which could influence the natural soil profile.

Fig. 5.2 Working with the soil probe (Source: Karvánková, photo archive 2014)

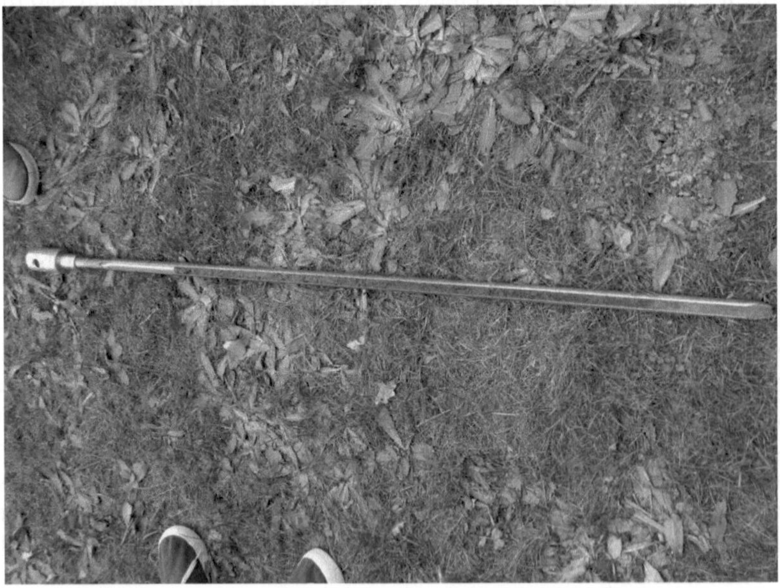

Fig. 5.3 RIDEX soil probe with the soil sample (Source: Karvánková, photo archive 2014)

- It is essential to take into consideration the health and safety of students when working with the soil probe.
- It is important to ensure that there are no utility networks on the respective paedology site to avoid possible damages.

There are many alternative ways of teaching the inquiry activity proposed, for example, taking soil samples and examining the soil properties in a laboratory (pH, volume and density, grain size analysis, soil moisture, etc.) and many other inquiry activities that can be implemented in the field (determining the soil structure, colour, consistency, grain structure, presence of roots, carbonates or skeletons, moisture conditions, etc.). If there is time, students can also measure the temperature of the soil for 1 day at a depth of 5 and 10 cm (at least five measurements per day on site). Each measurement must be made three times at a new, undisturbed place within 10 cm from the previous place of measurement (Kol. 2009). It is advisable to measure both soil temperature and air temperature to compare the results. If long-term measurement can be done with students, it is advisable to focus activities on how the soil temperature can be influenced by the type of vegetation cover. In this case, it is recommended to select more paedology sites with a different vegetation cover or without a cover. Students measure the air and soil temperature every day, inputting their results into a graph. Before the measurements, students should create their hypothesis about the influence of vegetation on the temperature of the soil and the air. The measurements can be taken at selected times (the longer the time sequence of data, the more precise the data and findings are), also as part of extracurricular activities.

5.6 Conclusion

Inquiry-based education makes students more interested not only in geography, natural history and other natural sciences but also in their surroundings, countryside and landscape they live in. During inquiry-based education, students learn many different approaches and develop their practical skills and knowledge by applying theoretical findings in practice and solving issues and problems based on the data acquired through own research (Crawford 2000). Students discover the basis of scientific work through inquiry. For example, they should understand that scientists pose questions, carry out experiments and bring explanation on the basis of their own observation.

Geography proves that it is a discipline which, thanks to its traditional focus on natural sciences, has its irreplaceable place amongst natural sciences and development of IBE at schools. Furthermore, geography as a field, with its depth and focus on studying landscape in a comprehensive manner (nature and culture), has a place in our lives. The ambition is to create a space to develop inquiry-based education and to incorporate social and economic topics into geography lessons in the nearest future. Geography, due to its multidisciplinarity resulting from its narrow links with all cross-sectional FEP BE topics, is more than suitable for the application of methodology principles of inquiry-based education. We can see the possibility of increasing the time allocation for geography lessons if geography gets integrated with other cross-sectional topics at Czech and Slovak primary schools, e.g. incorporation of methodology procedures of inquiry-based education. Also during inquiry-based education, the thematic unit of geography fieldwork, practice and application is applied, which forms a major part of geography in FEP BE, although in real practice it is being constantly forgotten and pushed to the background.

Similarly to other innovative methods, the main vision of inquiry-based education (IBE) is to make the learning process more attractive for students, to activate and motivate students and to arouse their interest and attract their attention. What's more, the ambition and motivation of IBE is to revive students' interest in natural sciences and subjects, such as mathematics, physics, chemistry and biology, but also the interest of their parents and the entire society, by focusing on the landscape and its natural components and geography. The main teaching tool and the main motivational feature of IBE is the simulation of the actual scientific work when students apply adjusted scientific procedures. Primarily, the IBE teaching method leads to more effective understanding of the taught subject, and secondarily IBE has a huge potential to make natural sciences more attractive for students in a purposeful manner (Janoušková et al. 2008; Papáček 2010b).

The suggestions for inquiry-based activities in physical geography, entitled *Soil – do we know what we are actually walking on?*, represent a means how to support the learning and teaching process through inquiry-based education. The activities enable students to simulate scientific work and in some cases also to carry out practical experiments. The prepared inquiry tasks and activities can provide inspiration for teachers who are interested in developing their teaching methods and include the methodology of inquiry-based geography education (amongst others).

References

Abell, S. K. (2000). International perspectives on science teacher education: An introduction. In S. K. Abell (Ed.), *Science teacher education. An international perspective* (pp. 3–6). Dordrecht/Boston/London: Kulwer Academic Publisher. http://bilder.buecher.de/zusatz/22/22438/22438833_lese_1.pdf. Accessed 11 Jan 2015.

Alvarado, A. E., & Herr, P. R. (2003). *Inquiry-based learning using everyday objects*. Thousand Oaks: Cowrwin Press.

Bergstedt, Ch., Ditrich, V., & Liebers, K. (2005). *Člověk a příroda, půda* [People and nature, soil]. Učebnice pro integrovanou výuku. Plzeň: Fraus.

Bičík, I. (2009). Zamyšlení nad proměnami českého vzdělávání [Reflections on changes in Czech education]. *Geografické rozhledy, 19*(2), 24–25.

CENIA. (2013). *Multimediální ročenka Životního prostředí, Vítejte na Zemi* [Multimedia environmental yearbook, welcome to earth]. http://vitejtenazemi.cz/cenia/index.php?p=uvod&site=puda. Accessed 21 Sept 2015.

ČGS Česká geografická společnost. (2012). *Česká geologická služba. Půdní mapa České republiky 1: 50 000* [Czech geological survey. soil map of the Czech Republic 1: 50 000]. http://mapy.geology.cz/pudy/. Accessed 21 Sept 2015.

Čipková, E., & Karolčík, Š. (2015). Bádateľsky orientované vyučovanie s využitím meracích systémov [Inquiry-based teaching with the use of measuring systems]. *Moderní vyučování: časopis pro nové programy v českém základním školství, 21*(5–6), 39–41.

Crawford, B. A. (2000). Embracing the essence of inquiry: New roles for science teachers. *Journal of Research in Science Education, 37*(9), 916–937.

Czesaná, V., Matoušková, Z., Havlíčková, V., Šímová, Z., Kofroňová, O., Lapáček, M., Braňka, J., & Žáčková, H. (2009). *Ročenka konkurenceschopnosti České republiky 2007–2008. Analýza. Část – kvalita lidských zdrojů* [Competitiveness yearbook of the Czech Republic 2007–2008. Analysis. Part – Quality of human resources]. Národní observatoř zaměstnání a vzdělání NVF. Praha: Centrum výzkumu konkurenceschopnosti české ekonomiky.

Douglas, T., & Brown, J. S. (2011). *A new culture of learning: Cultivating the imagination for a world of constant change*. http://sites.duke.edu/arthist110_001_f2011/files/2011/08/Thomas_Brown_A_New_Culture_of_Learning.pdf. Accessed 21 Sept 2015.

Dudášová, E. (2013). *Výbuch sopky v třídě* [The volcanic eruption in class]. Odborná konference 7RP PRIMAS: Objavné vyučovanie matematiky a přírodovědných predmetov. Nitra: Univerzita Konstantina Filozofa v Nitre.

EC European Commission. (2004). *Europe need more scientists*. http://ec.europa.eu/research/conferences/2004/sciprof/pdf/conference_review_en.pdf. Accessed 21 Mar 2010.

EC European Commission. (2007). *Science education NOW: A renewed pedagogy for the future of Europe*. http://ec.europa.eu/research/science-society/document_library/pdf_06/report-rocard-on-science-education_en.pdf. Accessed 21 Mar 2010.

EC European Commission. (2013). *Primas, inquiry-based learning in maths and science Classes*. file:///C:/Users/Popjakova/Downloads/primas_final_publication.pdf. Accessed 10 Mar 2015.

Establish. (2015). *European science and technology in action: Building links with industry, schools and home*. http://www.establish-fp7.eu/. Accessed 12 Oct 2015.

Fischer, R. (1997): *Učíme děti myslet a učit se: praktický průvodce strategiemi vyučování* [We teach children to think and learn: A practical guide to teaching strategies]. Praha, Portál, 172 s.

Fričová, H. (1982). Problémy současné didaktiky geografie [Problems of contemporary didactics of geography]. *Acta Universitatis Carolinae – Geographica, 27*(1), 307–314.

GLOBE. (2015). *Program GLOBE v České republice, Sdružení Tereza* [The globe program in the Czech Republic, the association Tereza]. http://globe-czech.cz/cz. Accessed 20 Sept 2015.

Hájek, J. (2003). *Vybrané kapitoly z didaktiky geografie* [Selected chapters from didactics of geography]. Plzeň: Západočeská univerzita v Plzni.

Healey, M. (2005). Linking research and teaching: Exploring disciplinary spaces and the role of inquiry-based learning. In R. Barnett (Ed.), *Reshaping the university: New relationships between research, scholarship and teaching*. Meidengead: Open University Press.

Healey, M., & Roberts, J. (Eds.). (2004). *Engaging students in active learning: Case studies in geography, environment and related disciplines*. Cheltenham: University of Gloucestershire.

Herink, J. (2009). *Geografie: její postavení a pojetí v národních kurikulech ve světě – shrnutí, závěry a doporučení* [Geography: Its position and approach in national curriculum in the world – Summary, conclusions and recommendations]. http://rvp.cz/detail/id/portal-2917. Accessed 11 Oct 2012.

Hindle, B. P. (1993). The 'project': Putting student-controlled, small-group work and transferable skills at the core of a geography course. *Journal of Geography in Higher Education, 17*(1), 11–20.

Hofmann, E. (2003a). Fieldwork an der Integrierten Facharbeitsstätte der Pädagogischen Fakultät der Masaryk-Univesität Brün. *GW Unterricht, 90*, 71–75.

Hofmann, E. (2003b). *Integrované terénní vyučování* [Integrated fieldwork teaching]. Edice pedagogické literatury multimediální učebnice terénní výuky. Brno: Paido.

Hofmann, E. (2005). *Terénní vyučování* [Fieldwork teaching]. http://clanky.rvp.cz/clanek/o/z/263/. Accessed 11 Oct 2012.

Hofmann, E. (2006). Cíle geografického vzdělávání v ŠVP v kontextu s cíli RVP. In *Problémy kurikula základní školy* [Problems of basic school curriculum]. Sborník z pracovního semináře. Brno: Pedagogická fakulta MU.

Hynek, A. (2002). Výzvy helsinského sympozia IGU pro české geografické vzdělávání [The challenges of IGU Helsinki symposium for the Czech geographical education]. *Geografie, 107*(4), 396–406.

Hynek, A., & Svozil, B. (2011). *Deblínsko: na cestě k trvalé udržitelnosti* [Deblín: Towards sustainable development]. Vlastivědná učebnice. Deblín: ZŠ a MŠ Deblín.

Hynek, A., Svozil, B., Trávníček, J., & Trojan, J. (2011). Best practice example of educational project: Sustainability in Deblín, South Moravia, Czech Republic. In A. Demirci, L. Chalmers, Y. Ari, & J. Lidstone (Eds.), *Building bridges between cultures through geographical education* (Vol. 1, pp. 23–29). Istanbul: IGU Commission on Geographical Education, Fatih University.

Janík, T., & Stuchlíková, I. (2010). Oborové didaktiky na vzestupu: přehled aktuálních vývojových tendencí [Current trends and tendencies in field didactics]. *Scientia in Educatione, 1*(1), 5–32. http://www.scied.cz/index.php/scied/article/viewFile/3/4. Accessed 5 Feb 2015.

Janoušková, S., Novák, J., & Maršák, J. (2008). Trendy ve výuce přírodovědných oborů z evropského pohledu [Trends in the teaching of natural sciences from a European perspective]. *Acta Facultatis Paedagiogicae Universitatis Tyrnaviensis*, ser. D, *2*(12), 129–132.

JU ČB. (2012). Projekt ScienceZOOM [Project ScienceZOOM]. Jihočeská univerzita v Českých Budějovicích. http://www.sciencezoom.cz/. Accessed 20 Sept 2015.

Karvánková, P. (2013). Vývoj didaktiky geografie a nové trendy výuky zeměpisu v Česku [Development didactic of geography and new trends in teaching geography in the Czech Republic]. *Annales Universitatis Paedagogicae Cracoviensis*, Studia Geographica IV, Współczesne obszary badań w dydaktyce geografii, 100–108. http://annalesgeo.up.krakow.pl/article/view/1523/1319. Accessed 11 Sept 2015.

Karvánková, P., & Popjaková, D. (2015). Od tradičního vyučování směrem k badatelsky orientovanému vyučování – příklad výuky zeměpisu [From traditional teaching towards inquiry-based teaching – An example of the teaching of geography]. In P. Hynek (Ed.), *Projektově orientované studium, nové pojetí, nové vize*. Metodika. ESF, MŠMT (in press).

Karvánková, P., Popjaková, D., Krejčí, J., & Soukupová, L. (2013). Integrated thematic teaching of the regional geography in the elementary school. Příspěvek na 21. středoevropské geografické konferenci: *Výzkum a výuka v geografickém vzdělávání*, Jedovice, 11–12 Sep 2013. Brno: PF MU.

Karvánková, P., Popjaková, D., Vančura, M., Blažek, M., & Dvořák, J. (2015). Badatelsky orientované vyučování fyzického zeměpisu [Inquiry-based teaching of physical geography]. In M. Reiterová (Ed.), *Bádateľské aktivity vo vzdelávaní*. Zborník príspevkov z medzinárodnej vedeckej konferencie (pp. 117–132). Bratislava: ŠPÚ.

Kol. (2009). Metodika pedologie [Methodology of pedologie]. Praha: Tereza, Program GLOBE.

Kol. (2013). *Rámcový vzdělávací program pro základní školy RVP ZV* [Framework Education Programme for Basic Education FEP BE]. Praha: Ministerstvo školství, mládeže a tělovýchovy, VÚP. http://www.msmt.cz/vzdelavani/zakladni-vzdelavani/upraveny-ramcovy-vzdelavaci-program-pro-zakladni-vzdelavani; http://www.vuppraha.cz/wp-content/uploads/2009/12/RVP_ZV_EN_final.pdf. Accessed 20 Sept 2015.

Kučerová, M. (2005). *Koncepce Geolaboratoře Albertov/Vyšehrad* [Concept of geolaboratory Albertov/Vyšehrad]. Diplomová práce. Praha: Univerzita Karlova.

Kühnlová, H. (1997). Reflexe světových trendů v pojetí a obsahu perspektivního geografického vzdělávání v ČR [Concepts and contents of geographical education in future: International trends and their reflection in the Czech Republic]. *Geografie-Sborník ČGS, 108*(3), 161–174.

Kuldová, S. (2008). Příklad úkolů z geolaboratoře [Example of tasks from geolaboratory]. http://clanky.rvp.cz/clanek/o/z/2281/. Accessed 11 Sept 2012.

Lacinová Svoradová, M. (2013). *Názornosť – predstava – objavnosť* [Visualisation – Imagination – Heuristic method]. Odborná konference 7RP PRIMAS: Objavné vyučovanie matematiky a přírodovědných predmetov (pp. 11–24). Nitra: Univerzita Konstantina Filozofa.

Linn, M. C., Davis, E. A., & Bell, P. (2004). *Internet environments for science education*. Manwah: Lawrence Erlbaum Associates Inc.

Marada, M. (2008). *Jak na výuku zeměpisu v terénu*? [How on the teaching of geography in the landscape?]. http://clanky.rvp.cz/clanek/o/z/2282/. Accessed 11 Oct 2012.

Matlovič, R. (2006). Geografia – hľadanie tmelu (K otázke autonómie a jednoty geografie, jej externej pozície a inštitucionálneho začlenenia so zreteľom na slovenskú situáciu) [Geography – Search for the bond (Towards the issue of the autonomy and unity of geography, its external position and institutional integration with respect to the Slovak situation)]. In R. Matlovič & V. Ira (Eds.), Vývoj, súčasný stav a perspektívy slovenskej geografie v 21. storočí. Acta Facultatis Studiorum Humanitatiset Naturae Universitatis Prešoviensis, Prírodné vedy, *Folia Geographica, 44*(9), 6–43.

Michaeli, E., & Madziková, A. (2013). Objavné vyučovanie v geografii na príkladu posuvu v polygóne Kapušany [Revelatory learning in geography on the example of a shift in a Polygon Kopušany]. Příspěvek na 21. středoevropské geografické konferenci: *Výzkum a výuka v geografickém vzdělávání*, Jedovice, 11–12 Sept 2013. Brno: PF MU.

Mičian, Ľ. (1988). Problém pozície geografie v systéme vied [Problem position of geography in the system of sciences]. *Sborník Československé geografické společnosti, 93*(4), 292–301.

Němeček, J., et al. (2001). *Taxonomický klasifikační systém půd České republiky* [Taxonomic soil classification system of the Czech Republic]. ČZÚ Praha spolu s VÚMOP Praha. http://klasifikace.pedologie.cz/. Accessed 21 Sept 2015.

Osborne, J., & Dillon, J. (2008). *Science education in Europe: Critical reflections*. http://www.nuffieldfoundation.org/sites/default/files/Sci_Ed_in_Europe_Report_Final.pdf. Accessed 5 Oct 2012.

Osuch, W. (2014). Przykłady innowacji w liceum – wstępna ocena programu nauczania przyrody w ramach projektu "Kształcenie pełne wyobraźni – KPW" [Examples of the innovation in high school – Preliminary assessment of the nature within the framework of the project "Education imaginative – KPW"]. *Annales Universitatis Paedagogicae Cracoviensis*, Studia Geographica VI, Innowacje w koncepcji kształcenia na różnych etapach edukacyjnych (pp. 230–241). Kraków. http://annalesgeo.up.krakow.pl/article/view/2065/1759. Accessed 20 Sept 2015.

Papáček, M. (Ed.). (2010a). *Didaktika biologie v České republice 2010 a badatelsky orientované vyučování* [Didactics of biology in the Czech Republic 2010, and Inquiry-based teaching]. DiBi 2010. Sborník příspěvků semináře, 25. a 26. března 2010. České Budějovice: Pedagogická fakulta Jihočeské univerzity.

Papáček, M. (2010b). Limity a šance zavádění badatelsky orientovaného vyučování přírodopisu a biologie v České republice [Limits and chances of introducing inquiry-based teaching of natural history and biology in the Czech Republic]. In M. Papáček (Ed.), *Didaktika biologie v České republice 2010 a badatelsky orientované vyučování (DiBi 2010)*. Sborník příspěvků semináře, 25. a 26. března 2010 (pp. 145–162). České Budějovice: Pedagogická fakulta Jihočeské univerzity.

Perkins, D. (2008). *Making learning whole: How seven principles of teaching can transform education*. San Francisco: Jossey-Bass.

Petr, J. (2010). Biologická olympiáda – inspirace pro badatelsky orientované vyučování přírodopisu a jeho didaktiku [Biological olympiad – The inspiration for inquiry-based teaching of natural history and its didactics]. In M. Papáček (Ed.), *Didaktika biologie v České republice 2010 a badatelsky orientované vyučování. DiBi 2010*. Sborník příspěvků semináře, 25. a 26. března 2010 (pp. 136–144). České Budějovice: Jihočeská univerzita.

Population Connection. (2004). *Education program, the Earth as an Apple*. http://www.iupui.edu/~ghw/lessons/materials/EarthAppleFarmlandNov02.pdf. Accessed 21 Sept 2015.

Prenzel, M., Stadler, M., Friedrich, A., Knickmeier, M., & Ostermeier, Ch. (2009). Increasing the efficiency of mathematics and science instruction (SINUS) a large scale teacher professional development programme in Germany. Leibniz-Institute for Science Education (IPN) (p. 65). Kiel. https://www.ntnu.no/wiki/download/attachments/8324749/SINUS_en_fin.pdf. Accessed 11 Jan 2015.

PřF UK. (2011). Přírodovědecká fakulta UK v Praze, Přírodovědci.cz. https://www.prirodovedci.cz/. Accessed 20 Sept 2015.

PRIMAS. (2014). *Promoting inquiry in mathematics and science education across Europe*. http://www.primas-project.eu. Accessed 12 Oct 2015.

Řezníčková, D. (2003). Geografické dovednosti, jejich specifikace a kategorizace [Geographical skills, their specifications and categorization]. *Geografie-Sborník ČGS, 108*(2), 146–163.

Řezníčková, D., et al. (2008). *Náměty pro geografické a environmentální vzdělávání. Výuka v krajině* [Suggestions for geographical and environmental education. Teaching in the landscape]. Praha: PřF UK v Praze.

Řezníčková, D., et al. (2013). *Dovednosti žáků ve výuce biologie, geografie a chemie* [Pupils' skills in biology, geography and chemistry teaching]. Praha: Nakladatelství P3K.

Rochard, M., Csermely, P., Jorde, D., Lenzen, D., Walberg-Henrikson, H., & Hermmo, U. (2007). Science education now: A renewed pedagogy for the future of Europe. European Commission, Directorate General for Research, Science, Economy and Society, Information and Communication Unit. European Commission, Brussels. http://www.eesc.europa.eu/resources/docs/rapportrocardfinal.pdf. Accessed 10 May 2015.

Rychnovský, B, (2010). Badatelsky orientované vyučování v biologii a nadání [Inquiry-based teaching in biology and talent]. In M. Janda & J. Šťáva (Eds.), *Nadaní žáci ve škole* (pp. 85–92). Brno: Masarykova univerzita.

Sdružení Tereza. (2010). *Projekt 3V – Vědě a výzkumu vstříc!* [Project 3V – Science and research meet]. http://www.projekt3v.cz/. Accessed 21 Sept 2015.

Sdružení Tereza. (2012a). *Badatalé.cz*. http://badatele.cz/cz. Accessed 20 Sept 2015.

Sdružení Tereza. (2012b). *Program Globe* [Programme globe]. http://globe.terezanet.cz. Accessed 20 Sept 2015.

Stephenson, N. (2007). *Introduction to inquiry based learning*. http://www.teachinquiry.com/index/Introduction.html. Accessed 25 Sept 2015.

Strahler, A. H. (2001). *Introducing physical geography*. New York: Wiley.

Stuchlíková, I. (2010). O badatelsky orientovaném vyučovaní [About inquiry-based teaching]. In M. Papáček (Ed.), *Didaktika biologie v České republice 2010 a badatelsky orientované vyučování. DiBi 2010*. Sborník příspěvků semináře (pp. 129–135), 25. a 26. března 2010. České Budějovice: Pedagogická fakulta Jihočeské univerzity.

Tomášek, M. (1995). Atlas půd České republiky [Soil Atlas of the Czech Republic]. Česká geologická služba, Praha. http://www.geology.cz/1919/historie/publikace/1995-atlas-kompletweb.pdf. Accessed 21 Sept 2015.

Vávra, J. (2008). Czech geographical education: From behaviourist to constructivist learning? In K. Donrt & G. Wall (Eds.), *Future prospects in geography*. HERODOT Conference Proceedings. http://www.academia.edu/2406639. Accessed 11 Oct 2012.

Vávra, J. (2009). Revize amerických Standardů geografického vzdělávání v roce 2009. Může české učitele zeměpisu inspirovat? [Revisions to American standards of geographic education in 2009. Can it inspire Czech geography teachers?]. http://clanky.rvp.cz/clanek/c/Go/16709/. Accessed 11 Oct 2012.

Wahla, A. (1983). *Terminologický a výkladový slovník didaktiky geografie* [Terminology and explanatory dictionary of the methodology of geography]. Ostrava: Pedagogická fakulty.

Wahla, A. (1991). *K otázce hraničních věd didaktiky geografie* [To question border sciences of didactics of geography]. Nitra: PdF-ÚIPŠMT-VÚP, Medacta 91, Zborník z vedeckých seminárov, *4*, 159–163.

Wilczyńska-Wołoszyn, M. M. (2003). The geographical laboratory of half a kilometre space around the school. In A. Kolwalczyk (Ed.), *Theoretical and methodological aspects of geographical space at the turn of century* (pp. 315–320). Warsaw: Warsaw University, Faculty of Geography and Regional Studies.

Chapter 6
Global Development Education: Czech Approach in the Spread of Global Awareness

Petra Karvánková, Dagmar Popjaková, Jana Kubelková, Petra Skalická, Pavel Horejš, Štěpán Klučka, Markéta Martínková, and Martin Kochánek

6.1 Why Do We Support and Need Global Education in the Czech Republic?

The world around us has been changing at record speed. Young people are constantly facing the need to make decisions which can have a major impact not only on them but also on the entire world. Education should enable us to be able to monitor new trends and worldwide changes but also to be able to make sense of it all, to understand and see the connections, to learn new skills and to be able to re-evaluate our attitude and respond to changes. In Czech society, which shows tendencies to xenophobic moods and nationalistic stances, this type of education is crucial. Firstly, we feel that by introducing the attitudes and principles of global education (GE) in Czechia, we enable learners and students, but also other citizens, to learn how to work with new information, to understand the need of cooperation and to change the way we think to have a more open society.

The current way of teaching geography in Czechia is still fragmented, frontal teaching based on memorising basic facts and information. However, the current world is completely different. Due to the impact of modern technology and the Internet, it is not possible to capture all facts and information and to teach students everything there is to know about a given subject. It is crucial to teach students to think, to search for actual facts, and if possible not distorted information, and to verify the information. As mentioned by, amongst others, Miléřová (2015), it is necessary to open our students' minds to the wider world and to prepare them for

P. Karvánková (✉) • D. Popjaková • P. Horejš • Š. Klučka • M. Martínková • M. Kochánek
Faculty of Education, Department of Geography, University of South Bohemia in České Budějovice, Jeronýmova 10, 371 15 České Budějovice, Czech Republic
e-mail: karvanko@pf.jcu.cz

J. Kubelková • P. Skalická
People in need, o.p.s., Šafaříkova 635/24, 120 00, Praha, Czech Republic

P. Karvánková et al. (eds.), *Current Topics in Czech and Central European Geography Education*, DOI 10.1007/978-3-319-43614-2_6

major changes in the future. Therefore, we believe that geography is an ideal subject which complies with the objectives of global education.

The Czech Republic is a Central European country with its own particularities. In connection to global education topics, Czech society represents a relatively young democracy which started to be rebuilt after the Velvet Revolution of November 1989 in the former Czechoslovakia after nearly 50 years of communism that came to power after the Second World War. This is not a very long period of time to create new social values. Despite the fact that, based on a public survey (FoRS 2014), most Czechs (85 %) believe that pupils and students should focus more on global responsibility issues and more than 60 % of respondents realise that our lifestyle has an impact on problems in other parts of the world, there is not enough information about these issues available in our society. In fact, there is hardly any information available at all (Mileřová 2015). Czechs are not actively searching for information, even if they are interested in it. It is astonishing, as stated by Dudková and Tillová (2012), for example, that despite the undisputable solidarity which Czechs show in the case of humanitarian crises or natural disasters, there is still no wider understanding of what it means to live in extreme poverty or of the purpose of long-term development cooperation, and there is a lack of understanding of the wider context of global issues.

The findings of these surveys document relatively well the prevailing Czech character. Most of the negative stances and ideas, or xenophobic views in Czechia, are based, in our opinion, on the fear and concerns of Czech citizens about their own future and their own well-being. These fear and concern arise from recent historical events which most of the current population lived through. They also arise from ignorance, not knowing the causes of events and having distorted ideas formed on the basis of the distorted and fragmented information that Czechs get from the media. Paradoxically, the less information we have about a certain situation and the bigger or nearer the "reality" is to us, the more we are concerned.

There is a whole host of reasons for the slow progress and reinforcement of ideas and issues with a global dimension in Czechia and Slovakia. One of these is the fact that although the Czech economy underwent a transformation, the cultural and value transformation has not yet been completed, and it will not be completed for a considerable time. After the revolution in 1989, Czech society firmly rejected all of the ideas of society-wide equality and collectivisation connected to the communist regime. Individualism and individual success have been the prevailing values in the last 25 years. Success, money, power and an important social position have become one of the key topics in Czech society, often to the detriment of the other main principles which are promoted by global education, such as solidarity, civic participation, individual responsibility, openness and social justice. These principles were also promoted by the former communist regime, though of course under completely different conditions and in compliance with the dictatorship of the proletariat, i.e. a totalitarian social configuration.

It will take a long time until the principles of global education and a new system of social values are anchored in the Czech Republic. These particularities of Czech society are significantly hampered by the understanding of the principles and main ideas of global education in Czech society, i.e. the Czech education system. We can find many social issues, from local to national ones, that are very close to the GE topics in Czechia and Slovakia (e.g. the Roma, social differences and poverty, ecological problems, big regional differences, wasting mineral resources and energy, etc.). However, the Czech education system touches on these subjects only marginally. In society, many of the issues are overlooked on purpose and not discussed. This approach is, perhaps unintentionally, reflected in the learning process and by students themselves. It can be difficult to apply global education and teach learners to think globally in such an environment. We believe that learners should be educated and taught global thinking at Czech schools using "Czech" cases from their surroundings and country. Understanding "local" problems and issues, and making an attempt to resolve them, is on par with an understanding of global issues and understanding the world from a global point of view.

6.2 Why Is the Term "Global Development Education" Used in the Czech Republic?

The Czech and Slovak term *global development education* (GDE) is, in its way, specific and more or less unique within Europe. The global perspective in education was first mentioned in 1989 at an international workshop TOUCH 89 (in Seč, Czechia) where representatives from Great Britain, David Selby and Graham Pike, first introduced the concept of *global education* in the Czech Republic. It is clear that this concept originated in Czechia, thanks to the cooperation with foreign organisations and governments. More specifically, the Czech perspective of global education was formed, mainly under the influence of the British school. Already in 1994 the publication of *Global Education*, written by the aforementioned authors, was translated into Czech (Pike and Selby 1994). The publication shares tested practices and experience of a country, which has a long-term global education tradition, by using examples and showing the way towards efficient implementation of global education. However, not all foreign findings can be easily reflected and applied in the Czech environment. Likewise, foreign recommendations are not always the ideal solution. The diversity of organisations dealing with global education is enormous and causes many ambiguities within the concept of global education.

There is not one universal definition of global education in Czechia. As stated by Nádvorník and Volfová (2004), the "Czech" global development education originated (previously called just development education) as a Czech translation of the English term *development education*. To define areas of interest, i.e. defined disciplines, was far from easy, thanks to an overlap with other areas of education. Global education was often used in Czechia as a synonym for development education in the broadest sense of the word. In order to clarify the relationship between the different types of learning and education, Nádvorník and Volfová (2004) used the global education concept of the North-south Centre Institute of the European Commission, which views global education as an umbrella concept, comprising development education, intercultural education, education for sustainability and education for peace and prevention of conflicts (North-south Centre 2005; Carvalho da Silva 2010). These types of social education and learning have a common set of fundamental values, ideas and knowledge and try to develop key skills and ensure creative opportunities and activities, and they differ by thematic focus and objectives which can complement one another.

The fragmentation of Czech terminology is also due to the fact that different organisations favour one or the other foreign concept. In order to unify the concept of global education in Czechia, the specific "universal" term *global development education* (GDE) started to be used officially, however in a nonbinding manner. It was a compromise between the two most favoured terms (global education and development education), in respect to educational professionals and laymen that have been using the term global development education (GDE) for many years in Czechia and Slovakia. The truly "Czechoslovak" term was first used by Nádvorník and Volfová (2004) in their methodology booklet *Common World*. These authors prepared a *Manual of Global Development Education* for the non-profit organisation People in Need. This term was officially adopted by the Ministry of Foreign Affairs of the Czech Republic in 2011 in its *National Strategy for Global Development Education 2011–2015* (MZV 2011).

Recently, the term "global development education" has often been replaced in Czechia by another term that has a corresponding meaning; however, it is much better known worldwide, i.e. *global dimension* (Belgeonne and Crombie 2013). Global dimension is *including key concepts of global citizenship, resolution of conflicts, diversity, human rights, interdependency, social justice, sustainable development, values and perception. It addresses interconnections between the local and global. It focuses on the development of knowledge, understanding, skills and attitudes...* (DFES 2005). The current understanding of global dimension provides an umbrella structure for global education that was introduced in Czechia by Nádvorník and Volfová (2004). The historical background of global dimension responds to a number of social problems that came to the forefront of public political interest (Belgeonne and Crombie 2013) in the last decades. Different tools were created in response to the problems.

Contrary to the aforementioned positive fact that different types of education and learning were created that were formerly united under one common umbrella concept of global education, we can see certain shortcomings in the current education. The disadvantage of different directions of global education, such as environmental education, development education, multicultural education, education for sustainable development, etc., is, according to Belgeonne and Crombie (2013), that just as easily as they become popular they can fall into disgrace, depending on which topics are in the centre of attention and which interest groups manage to secure space for certain topics in the curriculum. Recently, there has been a tendency to combine the main topics or unify some of the above-mentioned types of education under the label global dimension. This creates certain difficulties in Czechia, since global dimension has not been anchored and permanently included in the curriculum.

6.3 Historical Background of Global Development Education in Czechia

Despite the fact that education, in the sense of compliance with the targets of global development education, has not become an official part of the Czech educational process at all levels, from preschool to tertiary education and adult education, many significant steps have been taken in the last decade, mostly by non-profit organisations (Miléřová 2015).

Since 2002, GDE has been included into the Concept of Foreign Development Cooperation of the Ministry of Foreign Affairs of the Czech Republic for 2002–2007 and later for 2010–2017 (Ministry of Foreign Affairs of the Czech Republic). Since 2003, the main leader of the GDE projects has been the Department for Development Cooperation and Humanitarian Aid of the Ministry of Foreign Affairs of the Czech Republic, followed by the Czech Development Agency. One of the most important steps was the adoption of the main GE concept document for Czechia, i.e. the *National Strategy for Global Development Education for 2011–2015* (MZV 2011), which was officially adopted in connection to furthering development aid of Czechia by the Ministry of Foreign Affairs of the Czech Republic. The above-stated document was created on the basis of the *National Report on Global Education in the Czech Republic* (O'Loughlin and Wegimont 2008) created as part of the European Global Education Peer Review Process that took place in some selected countries (e.g. Czechia, Austria, the Netherlands, Finland, Cyprus) and that was initiated by the Global Education Network Europe (GENE). In 2010, an additional report was prepared by the Czech Forum for Development Cooperation (FoRS), and it was issued in cooperation with the North-south Centre of the European Council and the European Commission. The *National Report on Global Education in the Czech Republic in 2008–2010* (Nádvorník 2010) provides additional information on the GDE development trends in Czechia in 2008–2010.

The national GDE strategy for 2011–2015 summarised primary objectives, topics and principles of GDE in Czechia for the first time. The main attribute of the GDE is *for people to accept their share of responsibility for the situation in the world* (MZV 2011, s. 11). The same idea is incorporated in the main objective of the GE by the North-south Centre definition (O'Loughlin and Wegimont 2008, s. 8) which is to *enable people to participate in the development of their community, nation and the entire world.* In order to comply with these objectives, it is necessary to master *critical awareness of the local, national and international situation based on the understanding of social, economic and political processes* as emphasised by Nádvorník and Volfová (2004, s. 13). The national GDE strategy for 2011–2015 (MVZ 2011) has as one of its main objectives the incorporation of GDE into all levels of formal education. However, as stated by Miléřová (2015), no indicators were developed within the strategy that would enable to evaluate the progress of compliance.

Nevertheless, thanks to many NGOs, some of the GE topics and principles were incorporated into selected areas of education (Man and His World, Man and Nature) and into the cross-sectional topics of the Czech curriculum both for primary and secondary schools (Learning to Think Within the European Context, Environmental Education and Multicultural Education). NGO projects are constantly striving to increase public awareness about GDE topics, and many good quality methodology and didactic materials were written covering GE topics. In addition, some of the main GDE principles have recently become part of the new *Long-Term Education and Development Plan of the Czech Republic* for 2015–2020 (Kol 2015) as stated in the *Strategy for Educational Policy of the Czech Republic Until 2020* (Kol 2014b) or the *Concept for Supporting Youth in 2014–2020* (Kol 2014a).

The development of anchoring GDE principles in Czechia, as outlined above, gives us hope that more GDE topics will be represented in the learning process in Czechia. There is hope that Czech education will comply with GDE objectives which *would enable transition from the basic knowledge of development and globalisation issues as a phenomenon to understanding causes and consequen*ces and being able to form ones attitude, values and actively participate… (Suchožová 2010) and, finally, that students' attitudes, values, skills and knowledge will be developed as discussed by Nádvorník and Volfová (2004).

6.4 What Are the Main GDE Topics in Czechia?

There are a general number of topics which should be addressed by GDE in Czechia, arising from the experience with active global education abroad and tested practices. At the beginning, basic topics were based on the eight development goals for the millennium (UN 2000):

1. Eradicate extreme poverty and hunger.
2. Achieve universal primary education.
3. Promote gender equality and empowerment of women.
4. Reduce child mortality.
5. Improve maternal health.
6. Combat HIV and aids, malaria and other diseases.
7. Ensure environmental sustainability.
8. Develop a global partnership for development.

In addition to these goals, topics focusing on development and global issues were included (Návojský and Zajac 2006). The framework of topics for GDE application into the learning process in Czechia was mainly based on the four main topic areas determined in the national GDE strategy for 2011–2015 (MZV 2011): (1) *globalisation and mutual dependency of individual parts of the globe*, (2) *human rights*, (3) *global issues and* (4) *humanitarian assistance and development cooperation*. This was further specified into 22 topics (Table 6.1).

In 2013, Belgeonne and Crombie (2013) used the British Development Education Project to introduce the idea of education towards global dimension in Czechia. This is currently considered as an adequate motion to inspire the creation of a new strategy for global development education in Czechia. Global dimension brings

Table 6.1 Major topics of global development education according to the National Strategy for Global Development Education from 2011 to 2015 in the Czech Republic

Globalisation and mutual dependency of individual parts of the globe	Human rights	Global issues	Humanitarian assistance and development cooperation
Economic globalisation, including world trade and ethical consumption	Basic human rights	Poverty and inequality	Development, its concepts and cultural aspects
Cultural, social and political globalisation	Children's rights and senior citizens' rights	Poor education	Millennium development goals
Global migration	Discrimination and xenophobia	Health issues	History, motives and principles of humanitarian assistance and development cooperation
International organisations	Democracy and good governance	Malnutrition and famine	Actors in humanitarian aid and development cooperation
	Equality of men and women	Lack of drinking water	Activities pursued by the Czech Republic and the EU in the area of humanitarian assistance and development cooperation
		Environment	
		Population growth	
		Conflicts and violence	

Source: MZV (2011), Karvánková et al. (2015)

along a change of the theme framework, focusing on the eight key concepts of global dimension (Table 6.2) in the Czech education system in general, but also in global development education. All concepts are of the same importance and are interconnected. These concepts might have different positions and be of different importance in different contexts and represent a base for other concepts.

The main idea of education in terms of global dimension is to understand the link between our place in the current world and the role we have in the world (Belgeonne and Crombie 2013), i.e. as stated by Miléřová (2015), the necessity to lead students to think within the context, within the dimension "me in the world". In order to understand the ideas of global dimension and to acquire the skills and experience necessary to be able to think globally, first we must get Czech students interested in trying to understand problems around them and to teach them how to perceive problems from different points of view, which is something they have not been accustomed to. Also, as stated by Skalická and Blažková (2012), it is important that students understand that they can also become involved in and influence what is going on around them, but they must also be aware of the fact that they are responsible for their actions.

Table 6.2 The eight key concepts of the global dimension

Concept	Main goal
1. Global citizenship	Gaining the knowledge, skills and understanding of concepts and institutions necessary to become informed, active, responsible citizens
2. Conflict resolution	Understanding the nature of conflicts, their impact on development and why there is a need for their resolution and the promotion of harmony
3. Diversity	Understanding and respecting differences and relating these to our common humanity
4. Interdependence	Understanding how people, places, economies and environments are all inextricably interrelated and that choices and events have repercussions on a global scale
5. Human rights	Knowing about human rights including the UN Convention on the Rights of the Child
6. Social justice	Understanding the importance of social justice as an element in both sustainable development and the improved welfare of all people
7. Values and perception	Developing a critical evaluation of representations of global issues and an appreciation of the effect these have on people's attitudes and values
8. Sustainable development	Understanding the need to maintain and improve the quality of life now without damaging the planet for future generations

Source: DFES (2005); abridged by authors

6.5 Ability to Anchor GDE in the Czech Education System

The GDE methodology principles and topics can be, in our opinion, applied in all areas of education as part of the Framework Education Programmes (FEP) valid in the Czech Republic (Kol 2013) and at all levels of the education system: preschool, primary, grammar school, vocational/technical and tertiary. The issue of incorporating GDE into the Czech education system at the secondary school level was discussed by Nádvorník and Volfová before the 2004 reform of secondary education. Nádvorník and Volfová (2004) state that two of the most suitable fields of education for the application of GDE methodology principles in secondary education are, in particular, *basic social science courses* (today civics) and *geography*. They also point out the importance of *cross-sectional topics* during the application of GDE in Czech education. Later Dudková and Tillová (2012), amongst others, pointed out the potential of using GDE principles in a majority of subjects, out of which some authors, such as Vaňková Klímová and Gajdíková (2013), prefer *geography* as the most suitable one.

However, even 10 years after Nádvorník and Volfová (2004) pointed out the necessity of introducing GE in Czech education, mainly in official school documents valid in Czechia, we are still only a step away from the very beginning.

Nevertheless, mainly thanks to the systematic work of non-profit organisations, we have managed in the last couple of years to include at least some of the GDE topics and principles into several areas of education (e.g. Man and Society, Man and His World, Man and Nature, etc.) and into cross-sectional topics (mainly Education Towards Thinking in European and Global Contexts – for more details see Table 6.4) of the Czech curriculum, both at primary and secondary schools (Miléřová 2015). At the same time, GDE topics and methodology are slowly, yet systematically, introduced into teacher training programmes at universities (e.g. People in Need introduced the following projects: Varianty, One World at Schools, etc.). Public awareness of general GDE topics has also increased. The latest results from the survey *Attitudes and Needs of Teachers with Regards to Global Development Education* in Czechia (Hrubeš et al. 2014, s. 8) show that teachers spent more time on topics such as globalisation, human rights and global problems in their lessons. However, topics such as humanitarian aid and development cooperation are still getting only a minimum of attention, as pointed out by Vaňková Klímová and Gajdíková (2013) or Pánková (2015).

In order to reinforce the principles of GE in the learning process in Czechia, it is necessary to update the current Framework Education Programmes so that they reflect the main GE principles and values in accordance with current events and educational needs. The first step is the publication of *Recommended Expected outputs – Methodology Support for Teaching Cross-Sectional Topics at Primary Schools* in Czechia (Pastorová et al. 2011). The methodology support was published by the Research Teacher's Institute (today the National Institute for Education in the Czech Republic), introducing ideas how to update cross-sectional topics in the Framework Education Programmes at primary schools in the Czech Republic (FEP BE CR), mainly by incorporating the main GE ideas into the learning process. Overall, this strengthens the importance of cross-sectional topics and their inclusion into learning, as it emphasises that using these topics at schools contributes to the complexity of education and develops attitudes and values important for students and their future lives.

Selected areas of education and cross-sectional topics from FEP BE CR, where some of the GDE principles and topics were partly applied, are stated in Tables 6.3 and 6.4. We have been intentionally focusing on primary education, and the practical examples stated below focus on this age category of pupils at primary schools.

With regard to the current learning process in Czechia, it is vital to realise that it is necessary to anchor GDE topics in other fields of education as part of the Framework Education Programmes in the Czech Republic. As stated by Dudková and Tillová (2012) and Miléřová (2015), it is necessary to work with GE topics when teaching mathematics, languages and other subjects (such as music, arts and physical education, natural sciences, etc.) and to make students think and be interested in global topics. The easiest way of implementing GDE in the Czech learning process is to use cross-sectional FEP Czech Republic topics (Table 6.4).

Table 6.3 Selected areas of education and their FEP BE CR fields, applying the global development education principles in practice

Areas of education *FEP BE CR*	Selected subjects/ curriculum	Possibility to apply GDE principles in the given subject
Man and His World	Place where we live	Enriches students' experience with their closest surroundings by "introducing" the global dimension
For students' ages: 7–11 years	People around us	
	Man and time	
	Diversity of nature	
Man and Society	History	Enables to establish how global topics have changed over a course of time
For students' ages: 12–15 years		Provides options to examine the causes of global poverty and conflicts
	Civics	Develops appreciation of diversity and values which are perceived as global challenge
		Deals with topics such as social justice, human rights, community solidarity and global interdependence
		Discovering identity
Man and Nature	Geography	Discovering global interdependence and global diversity
For students' ages: 12–15 years		Critical approach to global issues from various points of views
		Exploring independence between the world of humans and the countryside
		Able to learn and evaluate situations from a local, regional and global perspective

Source: Belgeonne and Crombie (2013); FEP BE (Kol 2013); abridged by authors

The areas of education and cross-sectional topics stated above help to comply with GDE goals, and, to a different degree, they strive for values education (respecting ourselves and others, social responsibility, solidarity, active learning and teaching of others); development of thoughts and knowledge concerning technology changes, inequalities in the living conditions of the world's population, democracy, governments and citizenship, etc.; and development of key skills and knowledge (communication; social, civic and employability skills; competency to deal with problems; learning skills; and ensuring creative opportunities and activities in the learning process (individual and teamwork, shared responsibility for learning results, participation in the learning process)). The areas differ by thematic topics and objectives, which can complete and support one another. Different competencies and development techniques are based on a different focus of these areas.

Table 6.4 Selected cross-sectional topics FEP BE of the Czech Republic where the principles of global development education are applied

Cross-sectional topics *FEP BE CR*	Suitable thematic topics	Why is it advisable to apply GDE principles in teaching?
Education Towards Thinking in European and Global Contexts	We are interested in Europe and the world	Effectively reflects main GDE goals, topics and principles
	We are discovering Europe and the world	Emphasis on European dimension in learning
	We are Europeans	Supports global thinking and international understanding across the whole primary education
		Puts emphasis on comprehensive interlinking of stories and events from our everyday lives happening locally, nationally, regionally or internationally
Multicultural Education	Cultural differences	Emphasising current changes in society
	Interpersonal relations	Understanding social diversity, development of human relationships and ethnicity
	Ethnic origin	
	Multiculturalism	
	Principles of social conciliation and solidarity	Understanding and supporting multiculturalism, principles of social solidarity
Environmental Education	Human activities and environmental problems	Development of responsible environmental behaviour
	Man's relation to the environment	
Civic Education for Democracy	Civil society and school	Development of civic literacy of students
	The citizen, the civil society and the state	Emphasis is put on the development of students' ability to participate, on social solidarity and accepting responsibility for happenings in society and respecting the rights of others
	Forms of citizen participation in political life	
Moral, Character and Social Education	Character development	The idea is to teach students to act not to harm themselves, or others, or the society or the world we live in. On the other hand, students should learn how to look after their loved ones and people around them and, based on their abilities, to improve themselves and the world around them
	Social development	
	Moral development	

Source: Pastorová et al. (2011), FEP BE (Kol 2013); abridged by authors

6.6 Importance of Geography During the Implementation of Global Education

The close thematic link between the global education (GE) and the area of education *Man and Nature* from FEP BE CR has been mentioned before. The education in the area of Man and Nature is primarily based on the development of an open mindset of students who are open to alternative ideas and critical and logical thinking. The clearest thematic link between GE and subjects, which belong to this area of education (biology, geography, physics and chemistry), is between GE and geography. Thanks to its multidisciplinary concept, *geography* is a subject covering both natural and social sciences, and as such it enables pupils to gain a holistic view of the world and nature around us as one unified and functioning system which is mutually interconnected, whose parts can mutually influence and impact one another and their surroundings to a different degree.

Geography provides a structure of concepts to make global understanding possible; these include place, scale (local, regional, national, international and global), interconnectedness (not only of places, but of the physical, economic, political and social contexts in which we live), universality (global systems) and uniqueness (local outcomes). The educational power of geography lies in understanding the relationship between the uniqueness of place and the universal processes that impact on us all. It helps students see how the economic, social, political and environmental contexts operate simultaneously, and interpenetrate to contribute to their overall understanding or impression of a locality. So through geography with global development education, students come to understand our increasingly complex world and develop a positive and responsible attitude towards it (GA 2014).

The strong position of geography, as a subject that will help to introduce GE into the Czech learning process, is supported by a number of newly created Czech methodology and teaching materials, presenting sample geography lessons using the GE principles and methodology, e.g. Nádvorník and Volfová (2004), Pána and Pospíšilová (2012), Skalická and Blažková (2012), Mimra and Hruška (2012), Vaňková Klímová and Gajdíková (2013), SEVER (2013), Karvánková et al. (2013, 2014, 2015), Pánková (2015), Popjaková et al. (2015), etc., and a whole host of materials created as part of the Varianty projects (Varianty 2015a) by People in Need: Nádvorník and Chára (2006), Kratochvíl (2008), Skalická and Kociánová (2009), Lebeda et al. (2009) and Kol (2010). The first research was done, such as how students perceive global problems (Marada and Řezníčková 2015). Despite all this, the GE methodology and principles are still very new when teaching geography and other subjects in Czechia. Global topics are very hard to grasp for Czech teachers. Many teachers are not sure how to incorporate the topics into the curriculum.

Geography lessons, implementing GE principles, can help students to develop a further understanding of cultural diversity and interdependence in the world by understanding their own position in the country (Czechia) and the world, i.e. by understanding their own identity. Everyday problems at home and the world should be critically analysed in the lessons, i.e. *Geography in the News* (GA 2014), and students should be taught how to view events, global topics and information from different points of view.

The way how geography is taught in Czechia must change; in other words "it must catch a second wind" in order to activate and modernise geography lessons and the way and method of delivering information to the students. The inclusion of GE principles into the curriculum will help geography to become an important part of the Czech education system. Geography could become one of the main subjects enabling global citizenship education/education for global citizenship which has been promoted over the last couple of years. This new concept is based, according to Miléřová (2015), on different educational concepts that have been developed in the last 40–50 years in Western European countries – development education, global learning, human rights education and education for sustainable development (Applis 2014; Hicks 2003; Bourn 2008, 2014; Bullivant 2014; Humer 2014; Hunt 2012). On the one hand, these concepts have largely helped global citizenship education; on the other hand, their contents were largely identical and close to geography.

6.7 Example of GE Activities in Geography Lessons in Czechia

The following practical suggestions of how to include GE activities in geography lessons were prepared during a seminar entitled *Global Development Education of Geography* that has been taught at the Department of Geography at the Faculty of Education at the University of South Bohemia in České Budějovice (PF JU) since 2014/2015. The seminar was created in cooperation with the non-profit organisation People in Need as part of their international project *Teachers: Agents of Chance!* (Varianty 2015b). The aim of the project was to strengthen the position of GE in the Czech education system by introducing innovative courses for future teachers at Czech, Slovak and Polish universities and by sharing experience within the academic sector (in Great Britain, Poland and Czechia) and schools. The teaching activities stated herein were tested by the authors when teaching grammar schools' students at the Grammar School in Česká Street in České Budějovice and Máj Primary School in České Budějovice in Czechia, both during geography lessons and as an extracurricular activity, e.g. during geography courses, summer camps and project days organised by the Department of Geography PF JU. In total, more than 120 students have participated in the lessons presented herein.

6.7.1 Activity Title: Let's Get Exposed!

Expected Benefit To get students interested in the issue of nuclear energy. The activity is based on a critical approach, discussion and raising questions. It is important that students are aware of the risk of nuclear waste and act and behave accordingly.

Importance of the Activity in Czechia

- Controversial views about nuclear energy versus two nuclear power stations operating in Czechia
- The issue of storing nuclear waste in selected locations of Czechia – protests of local inhabitants
- Demonstrations in neighbouring countries against Czech nuclear power on one side, but on the other side, these countries buy cheap energy generated at these nuclear power stations

Importance of the Activity from a Global Point of View

- Worldwide issue of testing of nuclear weapons and the threat of nuclear weapons – the need for peace education (Pasqualetti 2011)

Summary of the Activity Atomic energy! An invention which represented a huge step forward for humankind and which is used as a major source for generating electricity. At the same time, however, this power represents a destructive force which can endanger the lives of millions of people. The testing of the atomic bomb itself had a terrible impact on the environment and the people who came to witness this scientific breakthrough without being aware of the possible consequences. The activity "Let's get exposed!" presents different perspectives of the controversial invention of the atomic bomb, also discussing its basic physical parameters. It also takes a look at the areas where atomic bombs were tested and points out the consequences. Nevertheless, the nuclear reaction is not only a source of destruction but also an important source of energy for humankind. According to the World Nuclear Association (WNA 2015a), as of 1 April 2015, there are 437 nuclear reactors installed and operating in 30 countries around the world with an output of 380,770 MW. Sixty-five reactors are in construction in 14 countries, and there is a plan to construct 165 additional reactors around the world. Nuclear power stations generate approximately 11 % of the world's electricity production. In Czechia, the ratio is more than 1/3, and in neighbouring Slovakia, it is nearly 2/3 of the overall electricity production (ČEZ 2015; IAEA 2015; WNA 2015b). What are the advantages and disadvantages of nuclear power stations, where are they located and are they safe? We will be looking for these answers in our activity "Let's get exposed!"

For Students' Ages 11–15 years (grades 7–9 of primary school)

Allocated Time 90 min

Teaching Goals

- Students will learn about all possible ways how to use nuclear energy and about the threats and consequences of any failure.
- Students will learn about the historical and political background preceding the development of nuclear energy and the atomic bomb (Second World War, Cold War, current nuclear powers).

- Students will learn about the ways how to use nuclear energy for peaceful purposes and about the advantages and disadvantages of present-day nuclear power stations and their use.
- Students will develop critical thinking and will learn to work with facts.

In Connection to FEP BE
Area of education: Man and Nature

Subject: geography

Expected Results

> "Students will locate main energy sources on the map"
> "Students will locate the main current geopolitical changes and political problems in specific regions of the world on a map of the continents"
> "Using selected examples, students will state major consequences and risks of natural and social effects on the environment"

Subject: physics

Expected Results

> "Students will evaluate the advantages and disadvantages of using different energy sources in terms of their impact on the environment"

Area of education: Man and Society

Subject: history

Expected Results

> "Students will explain the causes and consequences of the origin of the bipolar world and state examples of conflicts between the two blocks"
> "Students will demonstrate their basic knowledge of problems in the current world"

Cross-sectional topics:

Education Towards Thinking in European and Global Contexts
Media Education

6.7.1.1 Description of the Activity

Introduction
What has to be done before introducing this activity?

- Two teachers: reporter and atomic tourist (the reporter can have a mushroom cut – a haircut in the shape of the "atomic mushroom" which was very popular in the 1950s in the USA)
- Video creating the background

We would recommend selected passages from the following documentary films:

The World's Biggest Bomb. Great Britain, 2011, 53 min
Atomic Bombs – Life at Ground Zero. USA, 1998, 40 min

A. **Introductory Fictitious News Report with an "Atomic Tourist" from the 1950** [10 min]

The "nuclear atmosphere" is evoked in this part of the activity.

There is a 2–3 min video running in the background (no sound, just pictures) where students can see exploding nuclear bombs, astonished people watching the explosions and the disastrous impact and consequences of the nuclear explosions. While watching the video, the "fictitious interview with an atomic tourist" takes place, and the entire interview is practically a running commentary of the video.

The introductory part talks about the time when an atomic bomb was invented, and it was a sensation for the USA. Only some of the students know that people used to travel to Nevada to see the nuclear tests (e.g. they bought VIP tickets) and that "atomic tourism" was the second-highest source of income for Las Vegas at that time after the entertainment industry (PBS 2005). Students can witness an "authentic" statement of the tourist telling them what amazing power and unbelievable energy is hidden in the bombs which can be witnessed from a close distance. However, at the end the video gets blurry (as if unplanned), and students can see the cities of Hiroshima and Nagasaki immediately after the explosion of the atomic bombs. This part of the video should show the other side of the coin to make students aware of the danger that the people with VIP tickets were unknowingly exposed to in Nevada.

This motivating introduction should get students' attention and give rise to some questions. The idea is to introduce nuclear tourism from the past century. At that time it was an unheard of attraction; however, people had no idea about radiation, the illnesses caused by radiation, etc. After seeing the video, the introduction should smoothly turn into an open discussion dealing with questions raised by the students in response to the introductory story.

The most frequently asked questions by students at this point:

Why don't the people in the video have some special protective equipment when they are so close to a nuclear explosion?
What happened to the people with VIP tickets after they watched the nuclear explosion? Were they aware about the health consequences of gamma radiation?
Is there still some form of "nuclear tourism" today?

Main Part

What has to be done before introducing this activity?

Some photography to evoke the Cold War atmosphere (e.g. with explosion, with mushroom cloud and others)

Some photography with big nuclear weapon

B. **Put Your Atomic Weapon Together**: Nuclear Quiz [30–40 min]

The objective of the activity is to cooperate when answering the questions on the nuclear quiz, getting students involved in order for them to understand the issue of using nuclear bombs and the consequences.

At the beginning of the activity, we try to recreate the atmosphere of the Cold War from the 1980s (show the photography). Divide students randomly by drawing lots into two approximately same-sized groups. One of the groups will represent the *allies* (USA) and the other the *opponents* (Soviet Union). You can divide students into groups by letting them draw, for example, flags of the selected countries. Individual teams of students (allies or opponents) must always agree as a team on the correct answers. For each correct answer, they get one part of the nuclear bomb. The final answer is delivered by the selected "general" of the respective team. A nuclear bomb is a printed picture of a nuclear weapon which is cut into X pieces, corresponding to the number of questions on the nuclear quiz. Questions will be prepared accordingly to suit the time allocation, or the nuclear quiz can be made shorter. Teams take turns in answering to guarantee that the game is fair. The generals of each team will play rock-paper-scissors to decide which team will go first. The aim is to actively participate in the game and to be the first to put the nuclear weapon picture together. If students are competitive, the quiz becomes very entertaining (Fig. 6.1). The quiz questions are both open-ended and closed-ended, testing students' knowledge.

Examples of Possible Questions and Correct Answers that Can Be Used in the Nuclear Quiz

When were the first nuclear weapons developed? During the Second World War in the USA, Germany and Russia

Where were nuclear weapons used for a military purpose for the first time and why? August 1945, Nagasaki, Hiroshima (Japan) – at the end of the Second World War, Japan refused to surrender, so the bombs were used to demonstrate the military superiority of the USA

What is needed to make an atomic bomb? Highly enriched uranium or plutonium

Which two countries competed in an arms race after the Second World War? USSR, USA

When did the USSR manage to make its first atomic bomb? 4 years after the USA

Which city used to profit from atomic tourism? Las Vegas, Nevada

Who tested the biggest atomic bomb in the world and when? The USSR in Novaya Zemlya, the TSAR bomb: the sun was blocked by smoke and dust

Which European countries (apart from Russia) have tested nuclear weapons in the past? Great Britain, France

What was a hot line? A direct phone line between the president of the USA and the president of the USSR

(continued)

How is an atomic explosion created? By generating a supercritical volume of nuclear material (uranium and plutonium)

Which radiation is dangerous during an explosion of an atomic bomb? Gamma radiation

What can gamma radiation, generated by a nuclear bomb explosion, do to a human being? Burns, cancer

In which environment is an atomic explosion most effective when it explodes? In the atmosphere, above land or sea

Would you know what an atomic autumn or winter refers to? Temporary cooling off as a consequence of the explosion

What to do if an atomic bomb explodes in your surroundings? Hide immediately in the cellar or a room which is sealed as much as possible

How many atomic bombs do you think have exploded on the Earth? Over 2000

Nuclear testing was officially terminated by an agreement in which year? 1996

Which was the last country in the world that tested an atomic bomb and when? North Korea (2013)

Which professions are most exposed to radioactive (ionising) radiation? Astronauts, scientists

Source: Klučka et al. (2015), *AFSWP* (1954), *Technet.idnes.cz* (2006)

Fig. 6.1 Children are always excited about this activity (during the nuclear quiz) (Source: Karvánková, photo archive 2015)

During the competition, to make the quiz a bit more interesting, you can use an Internet animation called a *Time-Lapse Map of Every Nuclear Explosion in the World from 1945 to 1998* (AC24 2012), which illustrates the increase in worldwide nuclear bomb testing from the end of the Second World War. The competition continues until one of the teams manages to put together their nuclear weapon. The winning team now has an atomic bomb. The teacher can start asking questions: *You are now holding in your hands the most destructive weapon in the world. What are you going to do with it?*

Usually, at this stage of the activity, there are two scenarios:

- Students say that they are going to destroy the nuclear weapon. However, this raises other questions to be discussed:

What if other countries also have a nuclear weapon and want to use it against you? Can you use your nuclear weapon to defend yourself? What could be the consequences of such a war?

Is it easy to destroy a nuclear weapon?

- The second scenario which was more frequent, and more likely especially for a team of boys, was to drop the nuclear bomb on a selected area (usually an enemy state).

The animation enabling an imaginary nuclear bomb to be dropped on a selected place in the world is freely available on the http://nuclearsecrecy.com/nukemap/ website (Nukemap 2015). The application is clearly structured and offers the possibility to select not only a target location but also how powerful the nuclear bomb should be, the calculation of the radioactive cloud or alternatives of bomb explosions in the air or after falling on the ground, etc. Consequently, after the impact of the selected nuclear bomb on the map, the affected area is shown and a number of people who were potentially killed, injured and otherwise affected by the explosion. Students are always excited about this activity and it is clear they enjoy the multimedia application. They keep selecting new potential places for an imaginary nuclear bomb. The enthusiasm for "bombing" can be explained by the fact that students are fascinated by the new Internet application which can be considered to be an interesting game, a type of "numbers game". They soon find out that the more populated the area is (city, agglomeration), the "numbers" (numbers of affected people, size of the contaminated area, etc.) rise rapidly. This is an ideal moment to grasp the disastrous power of nuclear weapons.

At the end it is suitable to ask the following questions: *What would have happened if you had included your native town, village and a place you live as one of the selected locations?* This is usually the turning point of the lesson. Students understand that the actual dropping of an atomic bomb is not a game. The main objective is to make students think about the unbelievable power of nuclear weapons, and the threat they represent to the world if possessed by some countries, and to understand the immeasurable consequences of their actual use (which is demonstrated by the Nukemap).

C. **History of Atomic Armament and the Political Context** [15 min]

This activity is used to open a discussion about the historical context of armament and the political situation of the past century. Students are divided into the same teams as in the previous activity and during a short discussion, which can be controlled by the teacher, state their views about the correct/incorrect actions that were made and the students learnt about in the previous activities, e.g. *Was it right to drop the little boy and fatman nuclear bombs on Nagasaki and Hiroshima?* Students discuss the position of competing powers within the historical context. At this stage emphasis is put on the development of critical thinking, asking the students to come up with their own ideas and raise questions.

Final Part

What has to be done before introducing this activity?
Two teachers
Video: How does a nuclear power station work (Open Science 2014)
School World Atlas (Kartografie Praha 2009)
A stone looking like spent fuel
Luggage for a shelter (Fig. 6.2)

D. **Peaceful Use of Nuclear Energy (Positives and Negatives)** [25–35 min]

The aim of the activity is to encourage students to work with a map and to think creatively.

This activity focuses on the positive use of nuclear power in nuclear power stations.

Students watch a shot (Open Science 2014) which demonstrates the operation of a nuclear power plant in a simplified way. After studying the shot, go through the main principles and operations of the nuclear power plant. Ask students to find

Fig. 6.2 Luggage for a shelter (Source: Klučka et al. (2015))

Czechia on the map and the neighbouring states in their school atlas of the world *(select the area of interest as required – for the sample activity, we use the territory of Czechia)*, and get a piece of paper or an exercise book ready. Students work with the map of the Czech Republic; the teacher sets tasks and asks questions about nuclear energy and the area of interest.

6.7.1.2 Potential Other Tasks

Try to draw a nuclear power plant. Which natural conditions can influence the construction of a nuclear power plant? Where can you find nuclear power plants in Czechia? Do you know their names? Do neighbouring countries have nuclear power plants? What is the percentage of nuclear energy generated by nuclear power plants in the neighbouring countries? What are the advantages/disadvantages of using nuclear energy?

Students work on their tasks and discuss their answers with the teacher, who provides immediate feedback. While the students are working independently, one of the teachers changes into an "anti-chemical" suit in secret. He/she enters the class unexpectedly (*after the students finish their work, during teacher-student discussion*) and brings a spent piece of uranium stored in a special protective box (any *stone resembling a part of spent fuel*) and pretends that the "stone" is still radioactive which should arouse interest, but also makes the students a bit worried if they remember the information about how dangerous radioactivity is (Fig. 6.3).

Fig. 6.3 Teacher in "anti-chemical" suit brings a spent piece of uranium stored (Source: Karvánková, photo archive 2015)

To make the situation even more credible, the teacher can use a false dosimeter (*application in a mobile phone resembling an actual dosimeter which makes peeping sounds when the phone is close to a person*). The teacher in the suit briefly explains to the students the principle of nuclear fission of uranium and the options of storing spent fuel from nuclear power plants. Then the teacher briefly explains the main rules and instructions on how to look at the "spent fuel" (*the stone must be passed on quickly, you cannot hold it for a long time to avoid residual radiation, etc.*), and the stone is passed around the classroom. Students are usually very interested in this part of the activity as they for the most part believe that it is actually spent fuel. The idea is to arouse their interest and to make the final part of the lesson even more interesting. Finally, the students are told that the stone was only make-believe. Students are instructed how important maximum protection is when handling nuclear material.

This is followed by a situation game **evacuation luggage – pack your bag for a shelter** which helps to reinforce the information learned during the lesson. The game starts with an imaginary phone call to one of the teachers for whom the phone call is "unexpected". While one of the teachers is on the phone, the other teacher hands out sealed envelopes to the students with a picture of an evacuation bag (Fig. 6.2) and cut pieces of paper with a description of things to be put into it. *Children can be divided into the same teams as in the previous activity, or they can work independently.*

Students Can Pack Only Eight Items into Their Evacuation Bag Out of the Following

Mineral water, converse shoes, perfume, guitar, ID documents, cash, gemstone, medicine, canned food, yoghurts, pizza, liquids in containers, hygiene supplies, manicure set, toiletries, PS game GTA V, light source (torch), mobile phone, sleeping bag, spare underpants, map of the world, PlayStation 2, PlayStation 3, nail varnish, ballerina shoes, hair straightener, crayons, stylish hat, winter hat, gloves

The teacher on the phone shares the news that he/she has just received: *A minute ago an atomic bomb exploded in a nearby town* (to make the situation even more realistic a siren can be switched on), *an atomic cloud is spreading at an incredible speed towards us and you have only 2 minutes to pack your evacuation bag using some of the items which are in the envelope in front of you. You must pack very quickly. What will you take with you? You can take only eight items, you cannot carry any more. You will have to run to the shelter and your luggage should not be too heavy.*

From this moment there will be a countdown of 2 minutes on a projector (you can use, e.g. http://timer.onlineclock.net, Timer 2015). After the expiry of the allocated time, students are told this was a situational game. After the stressful situation, the teacher initiates a relaxed discussion. Students are asked to share what is in their evacuation bags. Students discuss if the selected items are needed in the shelter where they are going to stay since the area is contaminated by the atomic explosion.

The final part of the activity is devoted to the **Love Canal** affair, demonstrating to the students the impact of bad storage of nuclear waste on the environment and humans. Some basic detailed information about the Love Canal affair, which is considered one of the ten worst environmental catastrophes in the USA, is provided by, for example, DeCarvalho (2013). The teacher should start a discussion about accidents and the follow-up catastrophes during the industrial use of nuclear energy in power plants (Atomicarchive 2015; Greenpeace 2015) – the biggest and best-known accidents are Chernobyl (Ukraine, 1986) and Fukushima (Japan, 2014).

The activity Let's get exposed! is concluded by a joint discussion about the information obtained and about the students' ideas and feelings, focusing on the importance, level and danger of nuclear weapons and bombs in the world. The activity is based on active student participation. Understanding the issue of the destructiveness of nuclear weapons and the potential threat of the industrial use of nuclear power stations is a gradual process, based on our own experience and often on our incorrect judgement.

6.8 Conclusion

Global education should develop students' basic communication, civic and social skills as well as their critical thinking and ability to cooperate and their active approach to problems. The ambition of global education is not to acquire specific knowledge and information about a specific topic as a unit, but it focuses on students' attitudes and values and the process of forming these. A classroom must be a safe environment for sharing knowledge, experience, ideas, thoughts and students' attitudes towards GE topics.

Global learning explores our connections with the rest of the world, challenging learners to engage with complex global issues. Global learners are successful learners who enjoy learning, make progress and achieve and develop skills of participation and responsible action. Global learners understand others, value diversity, develop shared values, etc. (GA 2014). Teachers encourage students to ask open-ended questions, support creative thinking, give students sufficient space to think, encourage active listening and respecting one another, encourage participation of all students and apply various strategies to involve students with different abilities and learning styles. Teachers must monitor if students feel safe and must create opportunities for students to share their knowledge and experience, but also doubts and fears. It is important that the subject matter is suitable for the students' age so that students can understand and ask questions they are interested in. On the other hand, teachers face a difficult task to challenge the stereotypical thinking of their students and pass on knowledge which will lead to a more complex and unprejudiced view of the world. GE lessons focusing on global issues of humankind must be well prepared, and they must not make students feel hopeless or fearful. On the other hand, success stories of how these problems can be handled must be shared, providing information on the work of international organisations, state powers and individual activities. Although many of the GE topics are controversial and students can become worried, such issues must not be avoided at school. Students come across similar problems (conflicts, migration, diseases, hunger, etc.) every day in the media, in the newspapers and in their everyday lives, and it is essential that they understand the issues better and know that there are ways to mitigate these problems, or to rectify the problems, and that they can also help and do something.

Despite all the current efforts which are mentioned herein and despite the methodology support during the introduction of GE into the process of learning in Czechia, global issues are still difficult to grasp for Czech teachers, and many teachers are not sure how to include these issues into the subject curricula. As stated by Miléřová (2015), for example, it is important to try to include GE into education in order to teach students about active citizenship, sustainable development and democratic values (global citizenship education). This will be achieved through updating the Framework Education Programmes of the Czech Republic, creating a new national GDE strategy and by systematic nationwide GDE training for teachers. In order to anchor GDE at schools permanently, it is essential to create a competency GDE teacher's profile, stating the competencies and evaluation criteria. Correct grasp and implementation of GDE methodology at Czech schools will enable better application of cross-sectional topics that will be consequently seen as a unit with some potential overlaps and shared principles, without being separated and dissected into individual topics.

Due to its multidisciplinary nature, geography is one of the most suitable subjects for the implementation of GE methodology principles into learning. GE represents an opportunity to activate and modernise geography teaching in Czechia, to bring a wider application of the subject out of school and to anchor and strengthen the position of this subject at schools. In connection to geographical thinking, GE can help students to understand the world better by connecting their practical lives and actual world events. Students use GE topics to look for common things in our lives, which helps them come closer to the issues that developing countries face. Compassion and solidarity often create the illusion that it is up to us to save the world, or on the other side of the spectrum, there is the feeling that the little we can do cannot be of any use to anyone. However, each of us can help a little, providing we want to help and we know how to help in order to take the right action, action which is based on a critical evaluation of the various pieces of information we get about the world. It is important to "take the right action at the right time and at the right place". Due to these reasons, GE should not be only a subject, but it should extend right across the entire educational programme and represent a holistic educational approach.

References

AC24.CZ. (2012). *Radioaktivita, jaderné výbuchy a rakovina* [Radioactivity, nuclear explosions and cancer]. Mapa jaderných výbuchů od r. 1945–1998. http://www.ac24.cz/zpravy-ze-sveta/514-radioaktivita-jaderne-vybuchy-rakovina. Accessed 10 Oct 2015.

AFSWP. (1954). *Military effects studies on operation CASTLE*. https://archive.org/details/MilitaryEffectsStudiesonOperationCastle1954. Accessed 11 Aug 2015.

Applis, S. (2014). Global learning in a geography course using the mystery method as an approach to complex issues. *Review of International Geographical Education Online (RIGEO), 4*(1), 58–70.

Atomicarchive. (2015). *Major nuclear power plant accidents*. http://www.atomicarchive.com/Reports/Japan/Accidents.shtml. Accessed 11 Aug 2015.

Belgeonne, C., & Crombie, B. (2013). *Globální dimenze ve výuce* [Global dimension in education]. Příručka pro vzdělávání učitelů. Manchester: Development Education Project.

Bourn, D. (2008). Education for sustainable development and global citizenship. *Theory and Research in Education, 6*(2), 193–206. doi:10.1177/1477878508091112.

Bourn, D. (2014). *The theory and practice of global learning* (DERC research paper (11)). London: Institute of Education.

Bullivant, A. (2014). *Examples of good practice in development education from UK. Teachers agents of change*. Liverpool: Liverpool Hope University.

Carvalho da Silva, M. (2010). *Global education guidelines: Concepts and methodologies on global education for educators and policy makers*. Lisabon: North–south Centre of the Council of Europe.

ČEZ. (2015). *Výroba elektřiny* [Electricity production]. http://www.cez.cz/cs/vyroba-elektriny.html. Accessed 11 Aug 2015.

DeCarvalho, P. J. (2013). *Love canal disaster*. http://www.toxipedia.org/display/toxipedia/Love+Canal+Disaster. Accessed 11 Aug 2015.

DFES. (2005). *Developing the global dimension in the school curriculum. Guidance: Curriculum and standards*. DfES 1409-2005DOC-EN. https://www.education.gov.uk/publications/eOrderingDownload/1409-2005DOC-EN-02.doc. Accessed 25 Sept 2014.

Dudková, L., Tillová, K., et al. (2012). *Na stopě začleňování globálních témat do výuky* [On the track of the integration of global issues into teaching]. Olomouc: Agentura rozvojové a humanitární pomoci Olomouckého kraje, o.p.s.

FoRS. (2014). *Závěrečná zpráva z průzkumu Postoje k rozvojové spolupráci a humanitární pomoci* [The final report of the survey. Attitudes to development cooperation and humanitarian aid]. http://www.fors.cz/pruzkum-verejneho-mineni/#.VgUNTZc8owE. Accessed 11 Aug 2015.

GA. (2014). *Geography: The global dimension*. Geographical Association. http://www.geography.org.uk/projects/globaldimension/. Accessed 26 Oct 2014.

Greenpeace. (2015). *Jaderné havárie* [Nuclear accident]. http://www.greenpeace.org/czech/cz/Kampan/klima_a_energetika/jaderna-energetika/jaderne-havarie/. Accessed 11 Aug 2015.

Hicks, D. (2003). Thirty years of global education: A reminder of key principles and precedents. *Educational Review, 55*(3), 265–275. doi:10.1080/0013191032000118929.

Hrubeš, M., Protivínský, T., & Čáp, P. (2014). *Postoje a potřeby učitelů ve vztahu ke globálnímu rozvojovému vzdělávání* [Attitudes and needs of teachers in relation to global development education]. Brno: Masarykova univerzita.

Humer, R. (2014). Examples of good practice in development education from Austria. Teachers agents of change. http://www.wus-austria.org/files/docs/Agents%20of%20Change%20Austria%20DEF.pdf. Accessed 11 Aug 2015.

Hunt, F. (2012). *Global learning in primary schools in England: Practices and impacts* (DERC research paper (9)). London: Institute of Education.

IAEA. (2015). *Nuclear share of electricity generation in 2014*. https://www.iaea.org/PRIS/WorldStatistics/NuclearShareofElectricityGeneration.aspx. Accessed 11 Aug 2015.

Kartografie Praha. (2009). *Školní atlas světa* [School world Atlas]. Praha: Kartografie Praha.

Karvánková, P., Popjaková, D., Krejčí, J., & Soukupová, L. (2013). Integrated thematic teaching of the regional geography in the elementary school. Příspěvek na 21. středoevropské geografické konferenci: *Výzkum a výuka v geografickém vzdělávání*, Jedovnice, 11–12 Sept 2013.

Karvánková, P., Popjaková, D., & Kovaříková, V. (2014). *Náměty na globální rozvojové vzdělávání dětí mladšího školního věku* [Proposal for implementation of global education for children of primary school age]. Annales Universitatis Paedagogicae Cracoviensis, Studia Geographica VI, Innowacje w koncepcji kształcenia na różnych etapach edukacyjnych, 183–196. http://annalesgeo.up.krakow.pl/article/view/2061/1755. Accessed 11 Sept 2012.

Karvánková, P., Popjaková, D., & Kovaříková, V. (2015). Global education or how to most effectively implement cross-disciplinary themes in the curriculum of primary school – Czechia´s example. *Review of International Geographical Education Online (RIGEO), 5*(1), 8–25.

Klučka, Š., Horejš, P., Martínková, M., & Kochánek, M. (2015). *Let's get exposed: Put your atomic weapon together! Teaching activity*. České Budějovice: Department of Geography, Faculty of Education, University of South Bohemia.

Kol. (2010). *Bohouš a Dáša: šance pro rozvoj* [Students chance for development]. Praha: Člověk v tísni o.p.s.

Kol. (2013). *Rámcový vzdělávací program pro základní školy RVP ZV* [Framework education programme for basic education FEP BE]. Praha: Ministerstvo školství, mládeže a tělovýchovy, VÚP. http://www.msmt.cz/vzdelavani/zakladni-vzdelavani/upraveny-ramcovy-vzdelavaci-program-pro-zakladni-vzdelavani. http://www.vuppraha.cz/wp-content/uploads/2009/12/RVP_ZV_EN_final.pdf. Accessed 20 Sept 2015.

Kol. (2014a). *Koncepce podpory mládeže pro období 2014–2020* [Concept of youth support for 2014–2020]. Praha: Ministerstvo školství, mládeže a tělovýchovy.

Kol. (2014b). *Strategie vzdělávací politiky České republiky do roku 2020* [Strategy for education policy in the Czech Republic until 2020]. Praha: Ministerstvo školství, mládeže a tělovýchovy.

Kol. (2015). *Dlouhodobý záměr vzdělávání a rozvoje vzdělávací soustavy České republiky na období 2015–2020* [The Long-term plan for education and the development of the education system of the Czech Republic (2015–2020)]. Praha: MSMT.

Kratochvíl, J. (2008). *Bohouš a Dáša za lidská práva* [Students for human rights]. Praha: Člověk v tísni o.p.s.

Lebeda, P., Chára, P., Skalická, P., & Novotná, M. (2009). *Bohouš a Dáša na tržišti světa* [Students at world marketplace]. Praha: Člověk v tísni o.p.s.

Marada, M., & Řezníčková, D. (2015). *Globální myšlení Evropanů: v čem se liší čeští žáci?* [Global Thinking of Europeans: How do the Czech Students Differ?]. 23. středoevropská geografická konference „Středoevropský prostor v pohledu současné geografie". 8–9 Oct 2015. Brno: Pedagogická fakulta MU.

Miléřová, J. (2015). *Globální rozvojové vzdělávání* [Global development education]. *Proč je nezbytné pro českou společnost a udržitelný rozvoj*. Praha: FoRS.

Mimra, R., & Hruška, J. (2012). *Jak globálně vzdělávat* [How to educate globally]. *Katalog materiálů a nástrojů psaných v češtině určených pro globální rozvojové vzdělávání a multikulturní výchovu se zvláštním zřetelem k tématu fair trade*. Praha: Ekumenická akademie.

MZV. (2011). *Národní strategie globálního rozvojového vzdělávání pro období 2011–2015*. [National strategy for global development education for the period 2011–2015.]. Praha: Ministerstvo zahraničných vecí České republiky.

Nádvorník, O. (2010). *Zpráva o globálním rozvojovém vzdělávání v České republice v letech 2008–2010* [Report on global development education in the Czech Republic in 2008–2010]. Praha: České forum pro rozvojovou spolupráci – FoRS.

Nádvorník, O., & Chára, P. (2006). *Bohouš a Dáša proti chudobě* [Students against poverty]. Praha: Člověk v tísni o.p.s.

Nádvorník, O., & Volfová, A. (2004). *Společný svět. Příručka globálního rozvojového vzdělávání* [Our common world. handbook for global development education]. Praha: Člověk v tísni společnost při ČT o.p.s.

Návojský, A., & Zajac, L. (2006). *Panáčik proti chudobe. Človek v ohrozeni.* [A little man against poverty. People in need.]. Trenčín: Metodicko-pedagogické centrum v Trenčíně.

North-south Centre. (2005). *Programme for strengthening global development education for the Czech Republic, Hungary, Poland and Slovakia.* North-south Centre of the Council of Europe and Ministry of Foreing Affairs of the Netherlands. http://www.coe.int/t/dg4/nscentre/ge/V4_Final_report_17-10-05.pdf. Accessed 26 Sept 2014.

Nukemap. (2015). *Nukemap.* http://nuclearsecrecy.com/nukemap/. Accessed 11 Aug 2015.

O'Loughlin, E., & Wegimont, L. (2008). *The National report on global education in the Czech Republic. The European global education peer review process.* Amsterodam: Global Education Network Europe - GENE.

Open Science. (2014). *How does a nuclear power station work?* https://www.youtube.com/watch?v=2FGIeUDeZmk. Accessed 11 Aug 2015.

Pána, L., & Pospíšilová, K. (2012). *Globální souvislosti, problémy a výchova* [Global context, issues, and education]. České Budějovice: Vysoká škola evropských a regionálních studií.

Pánková, L. (2015). Rozvojová spolupráce – téma ve výuce geografie [Cooperation in development: Topic in geography teaching]. *Geografické rozhledy, 24*(4), 19–21.

Pasqualetti, M. (2011). The geography of energy and the wealth of the world. *Annals of the Association of American Geographers, 101*(4), 971–980. doi:10.1080/00045608.2011.575323.

Pastorová, M., et al. (2011). *Doporučené očekávané výstupy. Metodická podpora pro výuku průřezových témat na základních školách* [Recommended expected outcomes. Methodological support for teaching cross-curricular subjects at primary schools.]. Praha: Výzkumný ústav pedagogický.

PBS. (2005). *Atomic tourism in Nevada.* http://www.pbs.org/wgbh/amex/lasvegas/peopleevents/e_atomictourism.html. Accessed 11 Aug 2015.

Pike, G., & Selby, D. (1994). *Globální výchova.* [Global education.] Praha: Grada. Preparatory Commission for the Comprehensive Nuclear-Test-Ban Treaty Organization. https://www.ctbto.org. Accessed 11 Aug 2015.

Popjaková, D., Karvánková, D., Vančura, M., & Vlažná, D. (2015). *Linking geography of religion topics with global development education at elementary schools.* 23. středoevropská geografická konference "Středoevropský prostor v pohledu současné geografie". 8–9 Oct 2015 (in press). Brno: Pedagogická fakulta MU.

SEVER. (2013). *Přehoď výhybku, nedívej se na svět černobíle! Globální výchovou za spravedlivější svět!* [Global literacy for a fairer world!] Horní Maršov: Středisko ekologické výchovy.

Skalická, P., & Blažková, B. (2012). *Náš společný svět* [Our common world]. Praha: Člověk v tísni o.p.s.

Skalická, P., & Kociánová, K. (2009). *Bohouš a Dáša: klima v tísni* [Climate in need]. Praha: Člověk v tísni o.p.s.

Suchožová, E. (2010). *Využitie pracovných listov v globálnom rozvojovom vzdelávaní.* [Use of worksheets in the global development education]. Bratislava: Metodicko-pedagogické centrum v Bratislave.

Technet.idnes.cz. (2006). *Co dokáží nejstrašnější atomové bomby na světě: mizely ostrovy, vypařoval se beton* [What can engender the most horrible atomic bombs in the world: Disappeared Islands, evaporate Beton]. http://technet.idnes.cz/co-dokazi-nejstrasnejsi-atomove-bomby-na-svete-mizely-ostrovy-vyparoval-se-beton-gjo-/tec_technika.aspx?c=A060831_152144_tec_technika_NYV. Accessed 11 Aug 2015.

Timer. (2015). *Timeronlineclock.* http://timer.onlineclock.net/. Accessed 11 Aug 2015.

UN. (2000). *United Nations millennium declaration.* http://www.un.org/millennium/declaration/ares552e.htm. Accessed 11 Aug 2015.

Vaňková Klímová, Z., & Gajdíková, I. (2013). *Humanitární pomoc a rozvojová spolupráce v hodinách zeměpisu* [Humanitarian aid and development cooperation in geography lessons]. Praha: Charita Česká republika.

Varianty. (2015a). *Actual projects.* http://www.varianty.cz/projects/. Accessed 28 Sept 2015.

Varianty. (2015b). *Projects. Teachers: Agents of change!* http://www.varianty.cz/projects/65-teachers-agents-of-change. Accessed 28 Sept 2015.

WNA. (2015a). *World Nuclear Association*. http://www.world-nuclear.org/. Accessed 11 Aug 2015.

WNA. (2015b). *World energy needs and nuclear power*. http://world-nuclear.org/info/Current-and-Future-Generation/World-Energy-Needs-and-Nuclear-Power/#.UmqsyHDIY6w. Accessed 11 Sept 2015.

Chapter 7
Case Studies in Geography Education as a Powerful Way of Teaching Geography

Eduard Hofmann and Hana Svobodová

7.1 Introduction

Case studies are a way to mediate solutions for real situations to students from the environment in which they live. Case studies can be generally defined as an intense study of one case (one situation, one person or several people, one problem), thanks to which knowledge which is subsequently applied is acquired. They represent qualitative research methods, primarily because they examine certain phenomena in their depth and in the actual context, which is an advantage particularly in the event that the boundaries between the phenomenon and its context are not very clear. Generally speaking, case studies examine two types of phenomena. It either concerns the most frequently occurring phenomenon or conversely a very specific phenomenon that is different from normal.

There is also one additional bonus of case studies among qualitative research methods. Besides being a tool for expanding the theoretical background of numerous disciplines, case studies also ideally serve as an educational tool for practical skills for students and experienced practitioners in the given field (Olecká and Ivanová 2010: 62–65). The foreign equivalent of this form of teaching is represented by the phrase "powerful knowledge" or "powerful teaching" (Tejeda and Santamaría 2010; Hopkins 2000, etc.), because it is a "strongly anchored instruction", which is intended to ensure that pupils and students are able to cope with the rigours of everyday life through geographic education.

Case studies are often used in the fields of geography, for example, land-use planning. They can also be encountered in psychology, pedagogy, political science, social work, law, medicine and criminology, where they are often referred to as case reports. "The options that a case study provides us are truly abundant. However, in

E. Hofmann (✉) • H. Svobodová
Faculty of Education, Department of Geography, Masaryk University in Brno, Poříčí 7, 603 00 Brno, Czech Republic
e-mail: hofmann@ped.muni.cz

P. Karvánková et al. (eds.), *Current Topics in Czech and Central European Geography Education*, DOI 10.1007/978-3-319-43614-2_7

order for the results of our efforts to be valid, it is necessary to observe a certain procedure, like any other type of research"(Olecká and Ivanová 2010: 62–65). Case studies are based on the assumption that a thorough examination of one case will help us understand and clarify similar cases.

We distinguish several types of case studies:

- Descriptive (narrative report, e.g. a client, pupil)
- Exploratory (e.g. pilot projects)
- Explanatory (explanation – revealing the casual chain)

A combination of different techniques for gathering information is a characteristic for case studies (mainly the analysis of documents in connection with direct observation and acquisition of visual and audio recordings).

7.2 Background and Objectives

The aim of this paper is to present why case studies are used for teacher training at the Department of Geography of the Faculty of Education, Masaryk University (MU KGE). Introducing case studies into the curriculum of elementary and secondary school brings certain risks, some of which can be prevented by educating and training students. Case studies, mainly from the local environment, also require expertise coupled with short-term fieldwork. Another important factor is their implementation into the school curriculum. In the case of the Czech Republic, this concerns their inclusion in educational programmes. This requires knowledge on the part of teachers in terms of planning and preparing lessons – see Annex Lesson planning. Another important aspect for the implementation of case studies is setting specific and measurable learning objectives so that the teacher can justify this form of teaching. In the geography curriculum at primary and secondary schools, this skill is particularly important for its too oversized content. Briefly stated, case studies require that prospective teachers have learned techniques of fieldwork to collect and gather data from different sources and also to be able to realise the additional knowledge, skills and attitudes to solve specific situations towards which they are leading pupils and students.

The paper will further present a case study that focuses on the possibility of restoring a ski slope in Brno. This case study is applicable not only in Brno but also in any other areas (possibly with some modifications for local conditions). In addition to the specific environmental situations, the students primarily learn to ask geographic questions and seek adequate responses, using existing knowledge and skills gained in the study of geography. Pupils/students have the opportunity to apply the knowledge acquired during their studies directly to the specific case in the field, without which the teaching of geography is like the teaching of chemistry without experiments in the laboratory. The case study was tested in 2015 by bachelor's and master's students in the geography teaching programme at MU and by successful geographic investigators who were preparing for the International Geography Olympiad.

7.3 Inclusion of Case Studies in Teaching At the Department of Geography At the Faculty of Education of MU

The following case study is included at the end of the second year of the bachelor's degree programme of pedagogical training in the field of teaching geography. From geography, students have already attended geographical disciplines from physical geography and socio-economic geography, while they have also already completed theoretical lectures and field practice from cartography. The realisation of the case study comes at the end of the year and is ranked as one of the activities of field exercises. After its implementation students should be able to:

1. Ask geographic questions from disciplines that relate to the actual task.
2. Search for answers in various sources of information.
3. Work with various research methods including field research.
4. Process the results and draw conclusions.

Another implementation of a case study falls under the master's degree programme and is related to geography didactics and teaching practice. On the basis of the processed case study, students should be able to implement a case study in a particular school educational programme, which involves:

1. Specifying which year the given study is appropriate.
2. Planning all activities in the field and at school, which are associated with the implementation.
3. Laying down specific teaching goals.

In general, the case study has several phases that can be summarised as follows: defining the objectives of the case study, the initial survey and the project, preparing for data collection and analysis and processing the research report. In other words, this form of education can be divided into three phases: preparatory, implementation and final.

The preparatory phase of teaching takes place before the beginning of the instruction. From the perspective of teachers, the determination of learning objectives (what do we want to achieve through the teaching, what will the outcome be?), location (where will the work take place?), process (what will the students specifically do and how?), what devices will they work with and the timeframe for the chosen form of instruction (how long will the work take place for?) is essential. In our case it concerns the short- to medium-term form of education (according to Hofmann 2015, short-duration fieldwork is 1–3 h; medium-length fieldwork involves education, which usually lasts for one school day).

7.4 Entering a Case Study with Commentary on Its Implementation

The city of Brno is thinking about restoring the ski slope, but it does not have enough data (geographic, economic, technical) to make a decision.

The aim and outcome of your activities will be to process data for the decision-making process at the Brno City Municipality with the help of geographic knowledge and skills.

The municipality has given the following instructions:

- The ski slope should be open year-round. In summer you can ski on artificial turf on brushes. In winter, the same surface will have artificial snow. In the morning in winter, the ski slope is used for school skiing, afternoon ski groups and the public.
- The slope has a width of 50 m; landscaping of the slope from the base is necessary (necessary to ensure safe range) up to the top; for beginners it is necessary to build an alternative route to the ski slope with a gentler incline. A ski lift will be built at the slope.

Form: Group work

The main objective is data processing, which will be used to assess the intention of restoring the slope in Wilson's Wood. The assessment should show the method of how to use geographic knowledge and skills in the decision-making process.

7.4.1 Organisation of Teaching for the Second Year of the Bachelor's Degree Study

7.4.1.1 Motivational Introduction

The 1970s is the time when the idea to build a ski slope in Brno was born. This idea was not long in coming. In 1974, the slope in Wilson's Wood was ceremoniously opened. The ski slope was unique for several reasons. It was placed in the largest forest park in Brno, which was founded in 1882 by the landowner Ludvík Odstrčil under the patronage of Emperor Franz Joseph I. It was also unique from a technical perspective. There was no need for snow or sub-zero temperatures. It was also possible to ski during the summer holidays. The slope had an artificial surface where people skied on special brushes instead of snow. It was a unique project in all of Czechoslovakia, and the slope became a favourite among Brno residents, not only because of its location in the centre of the city but also as it was a sort of island of greenery (Adamová 2014).

Although the Brno slope could not be compared with mountain resorts, plenty of ski groups still trained there, whose members competed among Czechoslovakia's elite. The ski slope has been deteriorating since 1992. The lift and special artificial surface were sold to a private company, and so the only slope with a lift in Brno was closed to the public. In the year 2000 a group of councillors raised the idea of restoring the slope in Wilson's Wood and not only restoring it but tripling the length of the slope and building a new two-seat chairlift (Kárný 2010).

7.4.1.2 Brainstorming

After reading the introduction, students do a brainstorming session in which they imagine that they are professionals tasked with preparing the groundwork for the decision-making process of the Brno municipal representatives. Each student writes their ideas on what should be checked in order for the decision to be correct. After gathering these questions, they put five thematic groups together divided according to the issues that will be subsequently addressed. The result looks as follows:

Group No. 1: Evaluation of site suitability in terms of restoring the slopes

- Find out what the requirements are for ensuring the operation of the ski slope in summer and winter.
- Draw a plan of the slope and the necessary landscaping on the map.
- Select and justify the choice of location for the start and finish of the chairlift route and its management.
- Predict and justify the selection of sites for parking cars in the lower and upper part of the slope; mark the parking lots in the map including the recommended roads.
- *Output: Report for the city council assessing the possibilities for building the slopes, including maps and photographs.*

Commentary

The tasks are based on questions such as:

- *Are there conditions nearby for parking cars?*
- *What is the availability of the slope in Brno?*
- *What are the parameters for the construction of a safe slope?*
- *What documents (e. g. development strategy, spatial plan, specialized maps and data) do we need to prepare for the study of the slope area and the surroundings, etc..?*

Involvement of the knowledge and skills from the following discipline:

- *Cartography*

Group No. 2: Technical and economic renewal requirements of the slope

- Select the best location to place the snow cannons. At the same time, assess the possibility of the supply of potable water from the Svratka River.
- Calculate the area of the slope which must be irrigated and to which it is necessary to place brushes.
- Specify what building or other modifications need to be done to realise the slope from the top to the base.
- Predict and justify the selection of sites for parking cars in the lower and upper part of the slope; mark the parking lots in the map, including the recommended roads.
- *Output: Prepare complete tender documentation for the procurement of small-scale civil works for the building modifications of the slope, including accurate map recording the land modifications.*

Commentary

The tasks are based on questions such as:

- *Where can the snow cannons be placed?*
- *How large will the extent of landscaping be and where to intervene?*
- *Where can the facilities for the slope, parking places, etc. be placed?*

Involvement of the knowledge and skills from the following disciplines:

- *Cartography and mathematics*

Group No. 3: Natural prerequisites for restoring the slope

- Measure the incline of the slope, its orientation and duration of insolation.
- Evaluate climatic conditions over the past 5 years using the available data from the Czech Hydrometeorological Institute. Please justify the selection of the indicators and their development.
- Based on maps or your own fieldwork, draw a precise profile of the slope in the terrain, and propose and justify the deployment of five sensors to monitor the weather on the slopes.
- Prepare the table into which the values of the meteorological parameters will be recorded.
- *Output: Evaluation report of the weather conditions that will be used for the decision of the Department of the Environment of the Brno City Municipality*

Commentary

The tasks are based on questions such as:

- *Which data needs to be determined from the Czech Hydrometeorological Institute?*
- *Which data should be measured on the site and with what?*
- *Is the incline of the slope and its orientation suitable? How do we find out, etc..?*

Involvement of the knowledge and skills from the following disciplines:

– *Cartography, mathematics, climatology and hydrology*

Group No. 4: Impact of restoring the slope on nearby ski slopes and the environment

- Create a methodology for observing visitors to the park in Wilson's Wood. Perform observations, and find out what the average daily attendance of the park is and what activities are run by people in the park.
- On the map mark the place in Wilson's Wood where people are most often present.
- Evaluate whether the restoration of the slope will affect the movement of the people in the park. On the map mark any place of conflict and explain the reasons for the conflict.
- Calculate what area of the forest will become a "victim" of the slope.
- Try to assess what impact restoring the slope will have on the activities of the surrounding slopes (especially housing but also others).
- *Output: Evaluation of the environmental impact of restoring the slopes. The report will serve as an argument for the Department of Environment for/against restoring the slope.*

Commentary

The tasks are based on questions such as:

– *What groups of people visit the selected location?*
– *How will interventions to the slope affect the natural environment?*
– *What other possible negative impacts can building the slope have on the residents who live there?*

Involvement of the knowledge and skills from the following disciplines:

– *Cartography, geography of the population and settlements, sociology and the environment*

Group No. 5: PR group

- During the work of the working groups, carry out photo/video documentation.
- Create a PR article in *Brno Metropolitan* on the possibility of restoring the slope.
- Create a questionnaire that will be distributed after the end of the field survey among (a) the people living in the area, (b) the residents of the city of Brno and (c) the schools that could benefit from the slope.

- Output: *Create a slideshow accompanied by a photo/video report for the Brno Municipality general representatives summarising the pros and cons of building the slope.*

Commentary

The tasks are based on questions such as:

- *How can we inform the public about the upcoming project?*
- *How can we find out the views of different groups of people on this issue?*

Involvement of the knowledge and skills from the following discipline:

- *Sociology*

Division into groups and the following stages of teaching:

1. **Preparatory phase:** team members work mainly in collecting data for the upcoming fieldwork, which includes:
    - **Preparation of maps** printed in A3 format (A4 or glued), which will be used for fieldwork (min. topographic map of the area of Wilson's Wood – three copies, aerial photograph, cadastral map and other maps as desired)
    - **Preparing other tools**: ruler, crayons, solid clipboard, camera, GPS (special or integrated in a mobile phone or tablet), compass, paper for notes, clean paper, pencils, crayons and glue
    - **Findings from the available information** on the ski slope in Wilson's Wood

All of the above documentation and tools must be collected before the actual fieldwork.

2. **The implementation phase includes:**
    - Reconnaissance of the terrain
    - Processing of the assignments in the field
    - The creation of photo documentation
3. **The final phase includes:**
    - Processing of a fair copy output – map + text and their presentation.
    - Data processing for the specified output for their part. Creating a **glossary of geographic terms**, which were worked with.
    - Evaluate climatic conditions over the past 5 years using available data from the Czech Hydrometeorological Institute. Justify the selection of the indicators and their development.
    - On the basis of maps or your own fieldwork, draw a precise profile of the slope in the terrain, and propose and justify the deployment of five sensors to monitor the weather on the slopes.
    - Prepare the table into which the values of the meteorological parameters will be recorded.

- *Output: Evaluation report of the weather conditions, which will be used for the decision of the Department of the Environment of the Brno City Municipality.*

7.5 Commentary on the Realised Teaching in the Bachelor's Degree Study

Each group was assigned a task and worked independently, while the individual tasks were mutually interconnected and complementary. The students first worked in the classroom with different sources of information. They studied written and map data from the Internet (articles, maps, data, i.e. secondary sources) and prepared suitable materials from them for fieldwork, followed by a further fieldwork and data processing.

The basis of the fieldwork is one's own observations (Lenon and Cleves 2015). Important elements are observed and consequently should be marked on the maps – preferably detailed, printed black and white topographic maps – so that the colour substrate does not collide with the marked elements. When surveying, students often make some mistakes – the map section does not have map essentials (when cutting out maps from websites, the scale is often missing; Fig. 7.1).

Fig. 7.1 Sample of an incompletely processed map output from a case study (Source: Department of Geography students of MU)

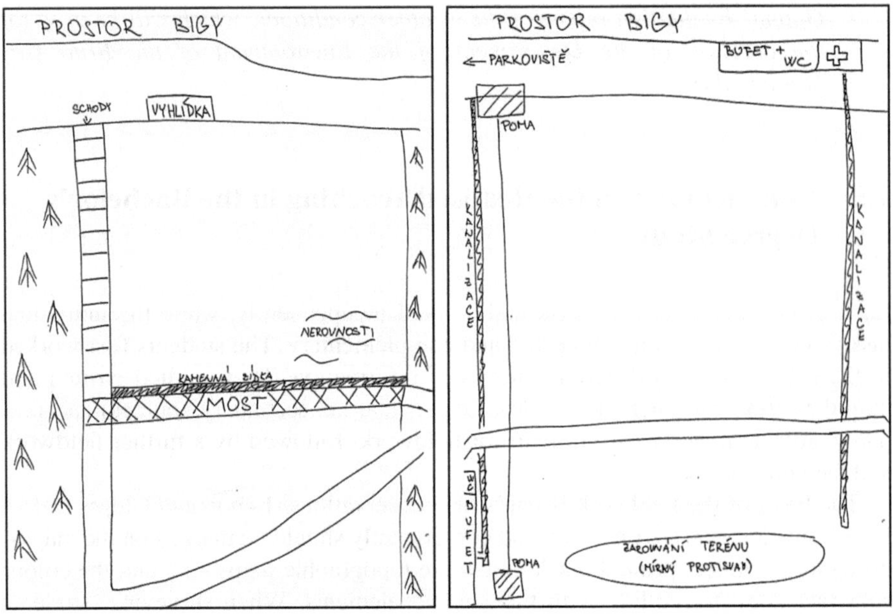

Fig. 7.2 Outline of the current situation of the slope in Wilson's Wood (*left*). Sketch of the modifications to the slope for the slope building variant (*right*) (Source: Department of Geography at the Faculty of Education of MU students)

The actual proposals of the students follow, e.g. on the surface of the slope, its guiding (steeper slope or a milder variant) design of the parking space, the placement of sensors for measuring temperature and so on. It is important that students substantiate their decisions with relevant facts (Fig. 7.2).

An important element which can be created during the processing of case studies is all kinds of sketches (sketches in the teaching of geography were described in more detail in Hofmann and Svobodová 2013 or Bláha and Hátle 2014). Sketches can be classified as non-verbal elements that can be used in teaching. It is expected that non-verbal elements, among which we can also include views of the countryside, speak for themselves. Automatic clarity is expected. Therefore, it is important to use non-verbal elements in the teaching. Pupils should be taught to work with non-verbal elements (according to Janko 2012: 25), not only to read but also to process and work further with the displayed information. As stated by Maňák (1994: 54–55), it is desirable to develop the skills of students so that they are aware of the links between the display and the displayed object and are able to realise retroactive transitions between them, which in the perception of the landscape is extremely important.

The outcome of the case study is a recommendation for the city council. The described form of teaching does not have to end with a letter to the Mayor or PowerPoint presentations. Case studies can be continued, for example, in a role-playing game, which like the case study also represents a strong teaching method.

Role playing, according to Cruess et al. (2008; modified), is also a strong teaching method for the conveyance of knowledge, skills and values.

The report and recommendations on the draft principles of the territorial development of the municipality of Brno

Statutory city of Brno
Brno Municipality
Dominikánské náměstí 1
601 67 Brno

Ing. Petr Vokřál
Mayor of Brno

In Brno on 5 May 2015

Matter: Report and recommendations on the proposed construction of a ski slope in Wilson's Wood in Brno

Dear Mayor,

In connection with the idea of building a ski slope in Wilson's Wood, we have completed an inquiry into the conditions of the proposed construction of a ski slope. The arguments for and against the construction of a ski slope in the centre of Brno are described below.

During the investigation, we concluded that the reconstruction of the ski slope in Wilson's Wood and the extension of the existing slope is not an appropriate solution for the use of the slope. This reconstruction and reopening would require a large initial expense, while construction would be accompanied by numerous problems.
The main problems include building a pipeline from the Svratka River that would bring the water required for the snow cannons. This pipeline would have to be constructed in areas that are densely built-up. Another serious problem is the construction of carparks with sufficient capacity. In the area of the upper part of the slope, a parking lot can only be built on land that belongs to the diocese. Expanding the existing parking area into the space of the park seems problematic due to the terrain. It would also be necessary to modify the terrain of the existing slope. However, the grading of the terrain of Wilson's Wood in the case of the extension of the slope to the base of the hill would be more demanding. The terrain is quite rugged, and its modification would be challenging.
Due to the above problems we do not recommend rebuilding the slope in Wilson's Wood.

(The full text of the students is shown in italics)

7.5.1 Organisation of Teaching for the First Year of the Master's Degree Study, the Subject: Geography Didactics

Students weigh in on the implementation of a case study in the second semester of the follow-up master's degree study of teaching geography, where they should especially learn how to handle preparation with an emphasis on its inclusion in the school programme of a primary school. The emphasis is primarily placed on the formulation of learning objectives so that they can push through this form of teaching, despite its time-consuming nature and content demands. They operate according to a model form of preparation (see Annex).

7.6 Conclusion

There are several benefits to process a case study in teaching geography. If a case study is well prepared, it can conveniently connect individual components of physical and socio-economic geography and cartography. At the same time, a wide variety of methods of geographical research and work with qualitative and quantitative data is practised. This fact is most realised by students of the Faculty of Education during the processing and preparation of this teaching. The model preparation in the Annex documents the breadth of knowledge, skills and attitudes that the case study trains. The third step is the actual implementation of the case study in teaching practice.

The role of case studies, whose preparation takes on the form of fieldwork, also has other positive aspects. Case studies help bring classroom instruction into the real world (Fuller et al. 2006) and help strengthen students' understanding of geographic terminology and processes (Job 1999). The implementation of case studies has a number of other pros, which can be summarised by the following statement:

> *"Well, it's the laboratory, isn't it? You can't prove things without actually doing some concrete data collection and analysis. It brings everything to life" (Clifford School, geography teacher)*

Annex

Lesson planning		
Year of study:	**Thematic unit:**	**Topic:**
Ninth	**Geographic field teaching, practice and application**	**Use of space for the case study solution**
	Geographic information, data sources, cartography and topography	**Time:**
	Natural environment of the Earth	
	The social and economic environment	
	The environment	
	Czech Republic	

Expected outputs:	
Student:	
Master the basics of practical topography and orientation in the field	
Apply practical procedures for observation, imaging and assessment of the landscape in the field	
Apply the principles of safe movement and residence in the countryside in practice	
Organize and adequately assess geographic information and data sources from available cartographic products and guides and from graphs, charts, statistics and other information sources	
Use with understanding of the basic geographic, topographic and cartographic terminology	
Identify and compare natural phenomena, their interrelations and conditionality	
Compare the assumptions and the main factors for the territorial distribution of economic activities	
Present on selected examples of the serious consequences and risks of natural and social impacts on the environment	
Assess on the appropriate level of the natural, economic and cultural relations within the local region, opportunities for further development (*Source: Kol* 2013)	
Skills:	**Suggestions for field teaching:**
Information collection, sorting, processing and evaluation of information	*Sketching, mapping, observations and measurements in the landscape, creating photographic documentation, etc.*
Work with a map, compass, GPS stations	
The creation of map	
The creation of conclusions	
As partial goals, through case studies, the students will learn to:	
Ask geographic questions and search for answers to them	
Prepare maps for fieldwork	
Make a sketch of a defined territory	
Record the upcoming changes in a territory in base maps 1:10 000 or in an aerial image	
Prepare a questionnaire, perform and evaluate it	
Estimate the scope of the work in a given territory	
Introduce and evaluate the pros and cons of the planned grounds	
Compare the traffic and demographic structure of the visitors	
Develop conclusions and send them to the right places	
Syllabus topic/integration into a wider framework:	**Cross-curricular links:**
Connects all the thematic units referred to in the introduction to the table	Mathematics, physics, civics, history
Aids:	**Preparation of the classroom:**
Basic map 1:10 000 aerial photograph, cadastral map, plan for the city of Brno, GPS station or mobile phone, tablet, camera, pens, crayons, tape measure and pedometer	Classical classroom, specialised classroom (computer, access to the Internet) and field
Individual approach:	
It is designed in the framework of group lessons. Groups are differentiated – boys together with girls – in the group of students with excellent academic performance, and poorer are represented	

Lesson structure		
Activities of pupils: preparatory phase	**Time:**	**The activity of the teacher**
This part is completed according to the schedule of the completed study in the bachelor's degree study		
Evaluation of the activities of the pupils in the classroom:		
Primarily, the active approach to tasks and the quality of output materials for the final report are evaluated		
Evaluation of teaching: teacher's self-reflection		
After completing the instruction, think over your role in the management of education. Appraise the strengths and weaknesses that have occurred during teaching		

References

Adamová, K. (2014). *Wilsonův les* [Wilson forest]. Průvodce Brnem. http://www.pruvodcebrnem.cz/wilsonuv-les. Accessed 8 Aug 2015.

Bláha, J. D., & Hátle, J. (2014). Tvorba náčrtů a plánků ve výuce geografie [Creation of sketches and hand-drawn maps in geography teaching]. *Geografické rozhledy, 23*(4), 13–14.

Cruess, S. R., et al. (2008). Role modelling – Making the most of a powerful teaching strategy. *BMJ, 336*(7646), 718–721.

Fuller, I., Edmondson, S., France, D., Higgitt, D., & Ratinen, I. (2006). International perspectives on the effectiveness of geography fieldwork for learning. *Journal of Geography in Higher Education, 30*(1), 89–101.

Hofmann, E. (2015). *Dělení terénní výuky podle různých kritérií* [Dividing of the fieldwork according to various criteria]. Nepublikovaný rukopis/Unpublished Manuscript.

Hofmann, E., & Svobodová, H. (2013). Blending of Old and New Approaches in Geographical Education: A Case Study. *Problems of Education in the 21st Century, 53*(53), 51–60.

Hopkins, D. (2000). Powerful learning, powerful teaching and powerful schools. *Journal of Educational Change, 1*(2), 135–154.

Janko, T. (2012). *Nonverbální prvky v učebnicích zeměpisu jako nástroj didaktické transformace* [Non-verbal elements in textbooks of geography as an instrument of didactic transformation]. (disertační práce/thesis), Brno: Pedagogická fakulta MU.

Job, D. (1999). Geography and environmental education: An exploration perspective and strategies. In A. Kent, D. Lambert, M. Naish, & F. Slater (Eds.), *Geography in education: Viewpoints on teaching and learning* (pp. 22–49). Cambridge: Cambridge University Press.

Kárný, M. (2010). *Sjezdovky v Brně? Zatím zůstává jen u přání* [Ski Slopes in Brno? Yet Remains only in Wish]. Deník 4. 1. 2010. http://brnensky.denik.cz/serialy/sjezdovky-v-brne-zatim-zustava-jen-u-prani20100103.html. Accessed 8 July 2015.

Kol. (2013). *Rámcový vzdělávací program pro základní školy RVP ZV* [Framework education programme for basic education FEP BE]. Praha: Ministerstvo školství, mládeže a tělovýchovy, VÚP. http://www.msmt.cz/vzdelavani/zakladni-vzdelavani/upraveny-ramcovy-vzdelavaci-program-pro-zakladni-vzdelavani. http://www.vuppraha.cz/wp-content/uploads/2009/12/RVP_ZV_EN_final.pdf. Accessed 24 Aug 2015.

Lenon, B. J., & Cleves, P. (2015). *Geography fieldwork & skills: AS/A level geography*. London: Collins.

Maňák, J. (1994). *Nárys didaktiky* [Outline of didactics]. Brno: Masarykova univerzita.

Olecká, I., & Ivanová, K. (2010). Případová studie jako výzkumná metoda ve vědách o člověku [Case study as a research method in human science]. *EMI, 2*(2), 62–65.

Tejeda, R., & Santamaría, I. (2010). Models in teaching: A powerful skill. In *Proceedings of the 7th WSEAS International conference on engineering education* (pp. 77–85). Sofia: World Scientific and Engineering Academy and Society (WSEAS).

Chapter 8
Project-Based Learning in Geography

Petr Hynek

8.1 Introduction

The aim of this chapter is to demonstrate a new approach to project-based learning. This new approach eliminates common errors accompanying the teaching concept as is presented, for example, in publications. It also shows how to use projects thematically very different from geography for teaching. This paper presents the use of a new approach at a small high school. In the project, 30 students from all classes of the high school were involved in order to demonstrate the synergic effects by combining students from different levels in one team. The project covered all subjects. The multidisciplinary aspect was important because of the involvement of teachers and the opportunity they had to try on this new approach in practice. The selected high school promotes and uses project-based learning and an individual approach to students in small classes. Therefore, we assumed that teachers were open to new methods, which could not be confirmed; on the contrary, during the project a common problem in today's education was revealed, which is the contradiction between proclaimed approaches and the reality in the classroom. However, the ability of this method to reveal actual skills and abilities not only in students but also in teachers is one of the strongest benefits at a time when all sides are calling for an enhancement in the quality of education.

Project-based learning is a long-known concept used in education (Bartscher et al. 1995; Bereiter and Scardamalia 1999; Brown and Campione 1996; Diehl et al. 1999). It could be said that this is a concept used since antiquity (Thomas et al. 1999; BIE 2015). From history we can mention the work of J.A. Komensky or even J.J. Rousseau. For this reason alone, we can state that the principle behind project-based learning is familiar to most educators. Finally, most schools also state it in

P. Hynek (✉)
Center for Innovation in Engineering Education, o.p.s,
Na Zlaté stoce 1619, 370 05 České Budějovice, Czech Republic
e-mail: petr.hynek@ciiv.org

P. Karvánková et al. (eds.), *Current Topics in Czech and Central European Geography Education*, DOI 10.1007/978-3-319-43614-2_8

their materials, papers and teaching plans. However, the reputation of the PBL methodology and its broad usage do not mean that educators can cope with the methodology and seamlessly use it. As with many other modern methods and methodology, project-based learning has its supporters and opponents. It is logical and legitimate, but from discussions with teachers, it also emerges that an aversion to or non-usage of this method often stems more from misunderstandings and many ingrained habits and myths. In this chapter, therefore, I want to discuss some of the basic principles of the PBL methodology and the most common misunderstandings and misconceptions. Specific examples will then be used to show how this new approach can be applied to bring joy and benefit to everyone involved.

8.2 A Brief Description of the New Approach

Project-based learning can be defined in a very abbreviated and simplified way as an educational model in which the student works on carefully defined projects as the principal means of acquiring the required professional knowledge and successful adoption programme and subject outcomes prescribed by the learning curriculum or school educational plan. This definition implies that a key means of learning is not lectures, seminars or lessons, in real or electronic form, but the student's own work on pre-prepared projects. Work on the project is also one of the main reasons for the popularity of this educational model among students. It uses the widely known fact that students prefer to try everything themselves than listen to theoretical propositions about how these projects are properly solved; they would rather work on projects than see how the teacher or someone else work on projects. From this perspective, project-based learning greatly improves students' motivation to study. One might therefore expect that this educational model would be widely recognised and used in most schools. And indeed, project-based learning is mentioned and used virtually everywhere.

Many studies have been done about the effectiveness and popularity of this educational model but their conclusions are not as clear as the model seems at first glance. Upon closer exploration, we find that there are a relatively large percentage of teachers who do not like this method and do not use it despite its undeniable advantages. Likewise, there are many teachers who use or have used it but did not achieve better results, or their learning results deteriorated, so they returned to their traditional proven methods of teaching. If you talk to an arbitrary number of teachers in a seminar or lecture about project-based learning, they will always be divided into passionate advocates and passionate opponents, and each of them has countless arguments for and against supporting their view. And if you decide to be an arbitrator and systematically disassemble all the arguments in detail, you will find that most of the arguments are correct and justified, or properly speaking, most of the arguments in this context are correct and justified. The context is important and we have little to suggest what the fundamental problem is. Logically then, we must ask ourselves how it is that this method of teaching, this educational model, has the same problems and the same uneven results as other educational models, how is it that the most beautiful and logical theorems and theories do not work as expected and how is it that even at schools where project-based learning is used extensively, often there is a deterioration in the quality of graduates or students? Exactly these and many other questions at the beginning of our research led to innovation and a new approach to project-based learning.

Consistent application of an individual approach to the student, working with internal motivation and a new understanding and application of the concept of competence are the three principal moments of this new approach.

8.2.1 Individual Approach to Students

The new approach to project-based learning could very simply and provocatively be characterised by the words "everything is vice versa". This apparent paradox is based on a simple fact, which is the consistent application of an individual approach to students. Let's start with another controversial statement. In the traditional educational model, an individual approach to students is not possible. To understand this claim, I provide a short explanation. Individual approach to students means that we take into account the student's individuality and his/her personal psychological and cognitive profile and personal typology and offer him/her such aids and learning methods that are most comfortable and efficient for him/her only. But this would mean in a traditional model that within a single class, one group of students must go to lectures; another group of individuals would do more practical exercises, while others would be holed up with books and self-study with occasional advice given. One teacher is of course unable to cope with that, and whether they want to or not, they must always choose the most convenient teaching method in the classroom. But this means that the indicated teaching methods are not suitable all the time for a large part of the students. And finding a teaching method that would suit all psychological types is impossible. It is obvious that an individual approach in the traditional model is not viable in its full form and is always replaced by a kind of time egalitarian approach where the teacher tries to choose a teaching method satisfactory for the greatest number of students, and those remaining will be helped more in an evenly balanced way if possible. In our new project-based approach to study, the situation is reversed. Thanks to the system of work, it is possible for each student to choose any teaching method that has been provided by the teacher or any other person. Personality typology is a subject for a whole book. We should now use the very frequently used Myers-Briggs Type Indicator (MBTI) psychological types and its 16 basic types. Each type has different ways of perception and different responses to the same stimuli and requires a different way of communication and other teaching methods adequately to it. As everything today relies on communication, it is preferable for students to use the typology division by fear (Riemann 1999). The teacher will communicate completely differently with a depressive personality than a hysterical personality.

The magic is hidden in understanding the needs of individual personality types and finding a combination of all the resources available for the teacher in the classroom. By resources, I do not mean just teaching methods and teaching aids but literally all resources, both material and human. For example, one of the most efficient resources is the students themselves. If we divide theoretical knowledge by difficulty, then we come to the finding that some lighter knowledge can be recited and learnt better by students. The effect is twofold. Firstly, students will adapt themselves better and they will communicate in their own ways, and secondly, this is the best way to strengthen their knowledge by mutual learning. It comes from another well-known pyramid – the pyramid of learning (Sitná 2009).

The teachers must forget today's false dogma of equal access, where they provide all the students with the same resources and provide assistance and resources individually. While the teacher offers one student intensive assistance and support in a specific situation, he/she will recommend to another student in the same situation to learn it himself/herself and find everything himself/herself, because this is more motivating and efficient for his/her personality type (Hynek et al. 2014).

8.2.2 Working with the Inner Motivation of Students

Another frequently mentioned term in education today is student's motivation. Most teachers would agree that it is now very difficult to motivate students to study. Personal interests of students and the content of curricula usually radically differ. In such a situation, it is difficult to find any project which will interest and motivate students towards greater personal activity and a desire to learn. A large number of students see education only as a sort of necessary evil leading to a degree that facilitates obtaining highly rated positions. Education therefore very often turns into a formal affair that both parties, both teachers and students, try to somehow passively endure. And yet motivation itself is the biggest advantage of project-based learning. Ask yourself why this weapon for many teachers fails and project-based learning does not lead to better results. To find the answers, let us analyse some basic concepts and principles in PBL.

The first question is what is motivation. Wikipedia defines motivation as an internal or external factor or set of factors leading to energising of the body. A motive is what prompts the person to act in a certain way or at least develop an inclination to specific behaviour. Another definition highlights motivation as an internal impulse that stimulates human behaviour and can be activated by various stimuli. Motivation is most closely associated with the definitions of the concept of need. Most teachers around the world will certainly have heard of the concept of Maslow's pyramid of needs (Maslow 2014). The need is a state of scarcity or abundance of something that leads us to activities which satisfy this anywhere you like. Needs can be divided into biological or social ones. Similarly, we can divide motivation into internal and external motivation. We can use any definition to any depth. It is important when we talk about education to try to assess all the theories and definitions from the perspective of cognitive psychology. A big surprise occurs at this moment. In none of those theories is curiosity mentioned as one of the basic human instincts, although the hypothesis of curiosity as a natural instinct is generally accepted by psychologists.

From the psychologists' perspective the natural instinct of curiosity is not directed or targeted at any specific or tangible outcome, but from an early age encourages spontaneous exploration and discovery. The response to this natural instinct is very important from the perspective of cognitive psychology. If the reaction around is negative, then the instinct is suppressed in children and students and vice versa.

Close consistency between the instinct of curiosity and intrinsic motivation is evident and can be directly identified as one. It is also one of the points where Maslow was wrong; his understanding of this need places it up at the top of his pyramid, and he claims it to be satisfied only after basic needs are met. The problem with the current education system is that we do not work with intrinsic motivation, and despite the fact that intrinsic motivation is far stronger than extrinsic motivation, we systematically suppress it. In this context, we often say that the need of curiosity is most developed in the schooling period, which is also the most suitable time for its development. This is again a big mistake because the instinct of curiosity develops from birth, and we can also affect it most in the earliest years, in a positive or negative sense. Unfortunately, as parents we often influence it in the negative sense. Simply by making a small observation of ourselves or any parents with children, we very quickly discover that in our communication with children, phrases like "be careful, you cannot do it, do not do that, that's how you hurt yourself", etc. prevail. So we repress the instinct of curiosity from early childhood. And as teachers we often continue this. The need for knowledge and its links with internal interests and needs, which are always highly individual, logically leads to the fact that our ideas about filling this need often sharply disagree with the ideas and thinking of the students. And that is the essence that students often lose interest or get tired of teaching, even if the teacher is passionate about the topic and his/her lectures are perfect from the methodological and content point of view.

Another feature of intrinsic motivation in education is that very often, the situation occurs when internal interest and motivation greatly diverge from the curriculum. This is the moment when teachers say it is difficult to motivate children and they move towards using external motivation. The essence of this problem lies in the methodological guidance of teachers and persistent practice that the teacher selects the task, examples and projects for education. The teacher will not even completely know the self-interest and intrinsic motivation of the student even with the best will in the world. The teacher will not completely know the self-interest and intrinsic motivation of the student even with the best will in the world. In other words, even if the teacher knows it well, often the student does not know it himself. Students often cannot verbally express and describe it. However, if we give space to students, then they definitely consciously or unconsciously choose such tasks or projects that cover their interests and needs. The opportunity to choose a project by a student's interests and needs related to individual approach, another important moment of project-based learning, will be analysed in the following paragraphs.

Extrinsic motivation is the most commonly used motivation method in education. The most famous motivational factors are known as praise or punishment. Motivation is very often in this case simplified to evaluation, praise, rewards and other instruments. In the best case, the teacher tries to increase motivation using modern activation teaching methods. Unfortunately, it does not solve the root of the problem and repeatedly leads to situations where intrinsic motivation still misses each other with learning outcomes. That is why many teachers have a negative attitude towards motivating teaching methods, although theoretically they recognise and respect them.

Techniques for increasing motivation are often built on an appeal to the "consciousness of something". Students should realise their goals and make it clear in their heads what they want; they should do a thousand things and then somehow "pick up" their motivation. These are unrealistic techniques that do not take into account the actual student. Appealing to consciousness and morality is as effective as appealing to gamblers to realise how harmful and devastating gambling is.

Another big contradiction is the phrase "we often have motivation but not the will". This is mixing up concepts with impressions. If we have a real intrinsic motivation, then almost nobody and nothing will stop us achieving our goal, and the question of will is quite useless. Every teacher as well as parent knows the situations where it is almost impossible to pull a child away from the object of his/her interest. And this is the moment which proves the absurdity of the phrase "we often have motivation but not the will". Will is closely linked with external motivation, and only in connection with it does it make sense. Intrinsic motivation has nothing to do with will and in the case of strong internal motivation is a practically useless concept.

Most materials, textbooks, study materials, electronic aids and presentations created for learning clearly show how the topic and teaching look from the perspective of the teacher, not from the student. Materials may be more or less of good quality but always impact the individual personality traits of students and therefore will always be unexciting for most students, demotivating and boring. None of these materials contain the basic question "what do students like and what are they interested in?", and there is no mention at all of students' intrinsic motivation or at least an attempt to find a relationship to it.

8.2.3 A New Understanding and Application of the Concept of Competence

A third key element in the new approach is a new understanding and especially use of the term competence. Each subject is literally dissected into hundreds of unitary competences, which we cannot divide up more and whose assessment can only be "knows / does not know". Any problem in learning is assessed through these competencies, and thanks to this "evaluation", the teacher is able to diagnose what problem the student has and adequately respond to it. In the traditional model, the student is often classified for something other than is actually required from him/her even if it does not seem so to many teachers. So, for example, he/she gets a tenner in mathematics, although he/she knows it well and has problems with completely different competencies that are in this context only remotely linked to math solving. The new understanding and use of competencies can prevent these misunderstandings, which have a negative influence on studies overall. On the contrary, they help to prevent those misunderstandings and solve them to the benefit of the students.

Competence in our new concept is not a vaguely and broadly defined skillset, but a closely and as accurately as possible defined knowledge or skill whose mastery can only be evaluated positively or negatively (i.e. learned/did not learn, knows/does not know). It is at the same time a key tool for assessing whether the competencies are defined correctly. If the question of whether the student mastered the competence cannot be answered unequivocally, then in most cases this means that the competence can be further divided into sub-competencies. This approach is then applied until no further division is possible. Such an understanding and use of competence concept leads to a markedly improved and "cleaned-up" view of teachers' own field of specialty.

8.3 Practical Example of Implementation

To illustrate, we will demonstrate the difference between the standard approach and the new approach used in CIIV (Center for Innovation in Engineering Education, Hynek et al. 2014) using a specific project which has been implemented at a private grammar school. The project, reflecting the new concept of project-based learning, was interdisciplinary and covered most of the subjects. Due to the number of students at the grammar school, most of the students of all years were involved in the project. We have decided to involve students from different years and take advantage of the interdisciplinarity aspect in order to demonstrate the biggest advantages and the biggest synergy effect of this type of learning.

First, the procedure used in a standard educational model will be described, and then the negative sides of this approach will be discussed, followed by the description of the new approach and its advantages. However, if the standard approach is better and more effective and, above all, works in practice, then we would clearly recommend using it.

Right at the beginning of the project, at the stage of defining the project, we came across the biggest difference between the standard and new approach. When following the standard approach, teachers are given a topic stipulated by the respective curriculum or the school's educational plan, for which they find materials and prepare a project. Teachers prepare a certain number of projects, but every year it becomes more and more difficult to find new and interesting projects for the same topic, and eventually teachers start repeating the same projects and follow a routine. At the same time, students start sharing solutions to these projects and most of the students will try to copy the solutions as much as possible. Gradual disinterest and difficulties with finding new, interesting projects are also something which is troubling the standard approach. Requests for new, innovative and interesting projects which a teacher should generate every year cannot be "physically" met.

Issues of the standard educational model:

- Gradual disinterest – it is not easy to find new projects.
- Routine solutions to projects without any emotional involvement.
- Projects are repeated and their results are copied by students.
- Both the teacher and the student lose their motivation.
- Restricted number of projects for a large number of students.
- Time demanding.
- No reflection of students' personalities.
- Students do not become intrinsically motivated.

It takes a lot of time and effort to arrive at solutions to these problems, and sometimes the effort takes over the actual process of learning, which surely is not the objective. To find or think of a project and to prepare its implementation, both theoretically and practically, are very time demanding, and therefore it is no wonder that the teacher is not able to manage more projects. The number of projects offered to students is always smaller than the number of students. Most projects will not suit students' interests, and students' motivation to actively search for a solution will be decreased.

On the other hand, the new approach of project-based learning leaves it completely up to the student to select a project. Ideally, students define their projects themselves or choose a project from a wide spectrum of projects collated by the teacher. If teachers prepare ideas for projects for students, they never prepare the projects as would be the case in a standard educational model. It is the student who prepares the project. The student will master many facts and skills when preparing the project. We keep reminding and managing students accordingly so that they select projects that will correspond to their hobbies and interests, or things they are currently interested in, regardless of specific subjects.

Each actual project is interdisciplinary, consisting of many activities which we are often not aware of and which require knowledge and skills from other areas than the actual project topic. How to ensure that we learn and master what is prescribed? The answer lies in what has already been said. First, each project is inherently interdisciplinary. Second, each project can be extended in the required direction. All teachers should be able to find a scope for their own subject and specialisation within the project and to extend the project in this direction.

Figure 8.1 graphically shows the old and new approach to defining projects. Under the old approach, a project or a real problem is selected, the part corresponding with the respective subject is identified and the project is then defined based on it. The actual project is reduced to a school or academic problem. In the new approach, the actual project is extended, and the required topics are added, sometimes from several subjects. This means that the project is extended and remains connected to reality. Moreover, thanks to the extension, it gains other perspectives too.

In their first lesson, students were presented with the new concept of project-based learning and its main principles. It is necessary to explain that students can come up with a project from any area of their interests, and the project does not

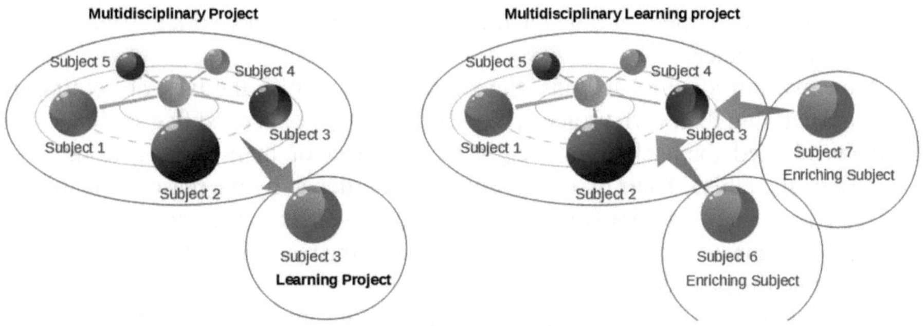

Fig. 8.1 Old and new approach to creating projects

necessarily have to be connected to the respective subject. First, students try to come up with projects which are connected to the respective subject. Therefore, it is very important when defining the first projects to discuss with the students if the project they have defined is really based on their hobbies and interests. Some students are so amazed that they can freely choose and define their own projects that they need intense support. The well-known Stockholm syndrome comes into play here. Students, who have been critical of the school system's rigidity, academic stiffness and detachment from reality, become frightened and want the old system reinstated.

The key question is why. The first and the most important reason for such behaviour is the fact that they lose certainty. Students grew accustomed to the old system, which is familiar and predicable. The transition to project-based learning is a step into the unknown. For the first time ever, students can act independently and must take decisions for themselves and be responsible for their decisions. Students do not find solace in the fact that the teacher is there, ready to provide help and support. This is a negative reason, and if students' behaviour and requests to return to the old methodology are based on this reason, the teacher should not oblige. At this stage teachers should help students to make their first steps into the unknown and to point out that the thing the students are worried about is good-quality studying.

The second reason why students behave like this when project-based learning is introduced to them is positive, and it is their justified request to an individual approach, study aids and teaching methods. If this is the reason, then the teacher should provide students with the requested tools and teaching methods.

It is highly difficult to decide which reason it is and it greatly depends on the teachers' experience. These two main reasons lead to two responses which are in direct contrast. It is more than certain that teachers will encounter both responses in one group of students so they must be prepared to oblige some students and not others.

During this project it was the first time when students encountered this type of learning, and thus the first two hours were devoted to a thorough explanation of the methodology and mostly the requirements placed on the students and the teacher.

In particular, the moment of transferring the responsibility and work to students was of prime importance.

After two introductory lessons, the students were given the task of coming up with and defining their own projects, working in groups of at most five students. The students divided themselves into eight teams; most of them were four-member teams (see the table). Their first proposals were burdened by the reasons mentioned above. The reasons were mostly justified by fear. In order to prevent this risk when defining projects, we strictly followed the rule that the teacher could inspire but was not to define the project.

In addition to problems connected to defining projects in the first lessons, we could also see that students lacked competencies and skills which, in standard learning, are often required only formally, but in project-based learning, it is essential that students truly master these competencies and skills. Specifically, students lacked the following competencies:

- Communication
- Critical thinking
- Project management
- Teamwork

Students had problems communicating with one another. We therefore tried to address communication problems during the first lessons. It was necessary to take students into situations where it would be a problem for them if they were not able to communicate and agree on something. Again we should point out and emphasise the individual approach to students, seeing them as individuals.

Competencies in terms of communication are quite naturally connected to team cooperation. When communication started to work, teamwork also improved. Communication and teamwork were soon functioning in all teams without the need for the teacher to be present; however, most teams experienced a problem of insufficient competencies and experience with sorting out their ideas and thoughts.

8.4 Projects

After the first four hours of directly working with the students at the school, we got into a phase when students started to define their projects within their teams. Each project underwent many changes, and it was widely discussed both as part of the team and with their schoolmates in the class and the school. Consequently, a project plan was prepared. It was recommended to students to prepare a project plan with clearly defined tasks and milestones, dividing their roles and tasks, including all things necessary for a project plan. Some teams managed to prepare their project plans absolutely independently and some teams needed a little bit of advice and help. At this stage the role and work of the teacher changed. Instead of presenting or teaching, the teacher was now more in the background, focusing on monitoring students' work and procedures they followed. The teacher turned into a coach.

Compared to the standard teaching model, when the teacher presents the respective topic in detail or in general, in project-based learning, the teacher tries to lead students to the topic and to help them understand or learn about the topic. However, the actual learning is up to the student and the teacher only helps if necessary and requested by the student. In this way several things are achieved:

- We do not take away from the students the moment of discovery, but we empower them to discover.
- We help students to be active.
- We lead students to be independent.
- We lead students to be responsible for their education.
- We do not force students to be interested in something that they are not interested in and we work with their intrinsic interests/motivation.

It is extremely important how projects and project plans are presented. The ability to present and defend one's findings, even partial, is crucial. Also, for project-based learning, it is essential to be able to define objectives in public. Once the student presents their project intention and project plan in public, including milestones, it becomes more important to meet these milestones. The third reason is evaluation; students determine clear evaluation criteria. Evaluation in project-based learning is not linked to written or oral exams and tests, but to fulfilling the project, i.e. its respective parts.

This places one major requirement on the teacher, namely, the ability to manage students to be able to define their projects and to incorporate the educational objectives determined in the school's educational plan or curriculum. This brings us to the central thesis of this chapter and teachers' most common fears or misunderstanding of the application of project-based learning. This worry could be summed up in the following question: "Is it possible to teach students a specific subject as part of a project that is not related to the subject?"

At first sight the answer seems easy, but therein lies its difficulty. At the beginning we mentioned one of the key moments of the new approach to project-based learning, i.e. the requirement to define projects based on students' interests. It goes without saying that these interests will hardly be identical with a subject matter covered by one particular subject. This is logical and inevitable, just like it is logical and inevitable that all projects are multidisciplinary and contain and require, to a greater or lesser degree, knowledge from different areas, i.e. different subjects. Practically, in any project we can find elements of math, physics, chemistry, geography or linguistics. But when we find the smallest connection between the respective project and a specific subject or competency, there is nothing to prevent us from developing the link and extending it so that it covers the required subject competency. Teachers' skills are not just about thoroughly knowing their subject but also being able to see the subject as part of a greater whole. And this is how a teacher should work – as a specialist in their field.

The following projects were presented, none of which primarily focused on geography.

Team 1	5 students	Film – student horror film
Team 2	4 students	League of legends – programming PC games
Team 3	2 students	Autistic spectrum disorders
Team 4	4 students	Social empathy and interaction
Team 5	3 students	Putting together a desktop PC
Team 6	2 students	Contemporary architecture – designing their own palace
Team 7	2 students	Immune system in the human body
Team 8	2 students	Football – history and present
Team 9	3 students	Watching films
Team 10	3 students	Robot car

We can see that the topics were varied and heterogeneous. We shall not discuss every single project, the respective projects plans and the method of managing students, but we will focus on how the respective project was linked to geography and competencies defined in the school educational plan.

8.4.1 Project 1 – Student Horror Film

The objective was to shoot a student horror film. This is a very inspirational topic, offering many opportunities. During a brainstorming session, we gathered ideas about students' motifs and ideas. The following topics connected to geography arose during the discussion:

- Race and ethnic specifics of students in education
- Different cultural approaches to education
- Learning about the local countryside around the school and in the region
- Comparing different state establishments and cultural differences in Africa, America and Asia in terms of education

The teacher's role is to lead students to respective topics during brainstorming. The teacher should inspire students to incorporate the ideas stated above into their project plans. This is the foundation of the teacher's professionality and specialisation.

8.4.2 Project 2 – League of Legends: Programming PC Games

The aim of the project is to think of and program a PC game. Another objective is to write a script for the game. Students tried to incorporate geography features into their project plan – creation and visualisation of the environment, countryside typology and geography. Students immediately recognised that the scenes would be more realistic if they were close to reality, and the students themselves incorporated into their project plans geology, geography and landscape formation. The discussion about project management focused on how to distinguish, decide and select what is important and essential and to leave behind less important ideas and project moments. It was not difficult to lead students into the areas clearly defined by the school educational plan.

8.4.3 Project 3 – Autistic Spectrum Disorders

This was the hardest project to find a link with geography. The teacher identified an interesting moment during the discussion that many specialists connect the increase in the number of autistic spectrum disorders to harmful substances and environmental damage. After the students identified the issues connected to the environment and landscape, the project was linked to geography. Thanks to the fact that this topic had not been widely researched, the link was sufficiently efficient. Students were fascinated by their adventure when they might potentially make new discoveries and focused their attention on geography.

8.4.4 Project 4 – Social Empathy and Interaction

The title of the project is clearly linked to geography topics. Students set up a project plan without any major help or support from the teacher. The project plan practically covered the social environment topic, with an overlap into the environment and the natural environment.

8.4.5 Project 5 – Putting Together a Desktop PC

This was one of the projects where teachers could not envisage how to link this project to geography. However, it was the students themselves who, after a short discussion about acquiring parts and components, got onto a geography topic. All was needed was a slight suggestion that students should incorporate topics such as geographical division of producers, transportation and logistics issues.

The procedure was identical for other projects. The work consisted of defining a project plan which is a key for several reasons. One of them is that students plan their work themselves. If they create a plan, they feel responsible for fulfilling the plan. The teacher makes sure that the student correctly sets up a project plan and that the project plan is feasible. The project plan should reflect the student's individual abilities. The teacher also ensures that the project plan is interconnected with required competencies and knowledge stipulated by the school educational plan, curriculum plan and study plans. Last but not least, the teacher ensures that the project plan clearly defines milestones (control checks, evaluation).

8.5 Conclusion

In our project, we examined all project teams preparing and defining project plans. At the beginning students cannot envisage putting together a meaningful project plan corresponding to the needs of the project and studies. Most of the teams produced project plans which were poor and unusable from the PBL perspective. The students' inability to prepare a project plan is understandable and legitimate. Teachers must spend time with students and teach them this competency. We have incorporated this fact into our project, and the first two teaching lessons were spent on how to present projects plans and on discussing the plans. Students presented their project plans, and at the end of the presentation, all teams decided to redo them. The second version of project plans was acceptable and usable for the projects.

A lot of time was devoted in PBL to the first phase of the project. The time was spent on major issues, i.e. improving students' independence and increasing their motivation and engagement. It is quite obvious that the time demand placed on the teacher was slowly decreasing and the teams were becoming more independent. The teacher could spend more time addressing minor requests which created and nourished a feeling of certainty in students and increased their motivation. Time saving is one of the biggest positives of project-based learning.

Undoubtedly, there were times while working on the projects when students' motivation and performance were low. However, thanks to having more time, the teacher could focus on these moments and spend more time with the specific team or student. This is one of the strengths of project-based learning. The form and scope of teacher's help always depend on the nature of the problem and the personality of the student or the team, and it should never be the case that the teacher mentors, lectures or criticises, which could be demotivating for the students. This is also one of the reasons why model projects or procedural examples cannot be created in PBL. These are always individual, depending on the situation, type of student, their current situation, etc. The objective of this chapter was to present the obvious benefits of the new approach to project-based learning.

References

Bartscher, K., Gould, B., & Nutter, S. (1995). *Increasing student motivation through project-based learning*. Master's Research Project. Chicago: Saint Xavier and IRI Skylight (ED 392 549).

Bereiter, C., & Scardamalia, M. (1999). *Process and product in PBL research*. Toronto: Ontario Institutes for Studies in Education/University of Toronto.

BIE (2015) Buck Institute for Education. www.bie.org. Accessed 10 Sept 2015.

Brown, A. L., & Campione, J. C. (1996). Psychological theory and the design of innovative learning environments. On procedures, principles, and systems. In L. Schauble & R. Glaser (Eds.), *Innovation in learning: new environments for education* (pp. 289–325). Hillsdale: Lawrence Erlbaum Associates.

Diehl, W., Grobe, T., Lopez, H., & Cabral, C. (1999). *Project-based learning: a strategy for teaching and learning*. Boston: Center for Youth Development and Education, Corporation for Business, Work, and Learning.

Hynek, P. et al. (2014). *Projektově orientované studium – nový pohled na vzdělávání* [Project-based learning – a new approach to education]. České Budějovice: CIIV.

Maslow, A. H. (2014). *O psychologii bytí* [Toward a psychology of being]. Praha: Portál.

Riemann, F. (1999). *Základní formy strachu* [Basic forms of the fear]. Praha: Portál.

Sitná, D. (2009). *Metody aktivního vyučování* [Active teaching methods]. Praha: Portál.

Thomas, J. W., Mergendoller, J. R., & Michaelson, A. (1999). *Project-based learning: a handbook for middle and high school teachers*. Novato: The Buck Institute for Education.

Part III
Information Technology in Teaching Geography

Chapter 9
Use of Information and Communication Technology and Resources of the Internet in Education Natural Sciences

Małgorzata Pietrzak

9.1 Introduction: Education in the Information Society

The media explosion, that is, the creation and rapid popularisation of a full range of completely new tools to store, process and transmit information (Goban-Klas 2005), makes it possible to provide information about events, explain and comment on information, coordinate social activity, create and maintain relations within groups and communities, facilitate learning and development and organise work, entertainment and leisure time. By combining different forms of information transfer, e.g. text, sound, image, graphics, animation and video, the so-called multimedia finds application in many domains, such as advertising, art, education, entertainment, engineering, medicine, administration, business and academic research. Interactive media, e.g. the Internet and mobile communications, apart from providing access to digital resources, makes it possible to exchange information and enables back-and-forth communication.

We are witnessing the emergence of a society that commonly uses information and communications technologies and which draws knowledge from data sources made available through broad and easy access to media. In the modern world, it is media that organises the communications environment of society – it is media that creates hierarchies and prioritises the events presented in news programmes, exalts or disparages politicians, creates our aesthetic tastes and "plants" issues that should be taken care of "to make the world a better place". It is media that creates and glorifies authors and artists, and it is media that tells us what films we should watch, what music we should listen to, which radio programmes we should choose, what newspapers or books we should read and why these authors and not others, etc.

M. Pietrzak (✉)
Institute of Geography and Spatial Management, Jagiellonian University in Kraków, ul. Gronostajowa, 730-387 Kraków, Poland
e-mail: m.pietrzak@geo.uj.edu.pl

P. Karvánková et al. (eds.), *Current Topics in Czech and Central European Geography Education*, DOI 10.1007/978-3-319-43614-2_9

Teachers are the ones who offer the young generation their assistance in assimilating the latest developments advanced by the information civilisation. It is teachers, for the most part, that are responsible not only for the method, but, most of all, for the quality of the education of the future "citizens of the information society". One of the most fundamental dilemmas that teachers are faced with in their work is figuring out the proper method of transition from information and facts through actions and the build-up of knowledge and the contemplation of values – to the world of wisdom.

The inundation of data, sounds, images and information can overwhelm a young person, who is unable to systematically acquire and consolidate knowledge or even to reflect on the newly acquired facts in order to establish cause and effect relationships. The discussions on the challenges faced by modern schools are frequently accompanied by numerous appeals for sensible media education, for building on universal values, for implementing technology as a tool to support teachers and for the transition of schools into institutions that teach students to discover and create knowledge (Goban-Klas 1999; Morbitzer 2010).

The present work consists of three parts. In the first part, the state of the development of the information society in Poland in the context of education is presented, on the basis of the available statistical and project-based research. The second part describes the implementation of information and communications technology (ICT) in light of the theory of teaching. The third part discusses some examples of the use of the education cloud, mobile technologies, social network and a distance learning platform in the education of university and secondary school students.

The described teaching methods and techniques are implemented during preparatory vocational courses for trainee teachers of natural sciences in the Institute of Geography and Spatial Management (IGiGP) at the Jagiellonian University. Moreover, they were used during workshops for teachers of natural sciences teaching in secondary schools (the PiNaP project – Programme of the Innovative Method of Teaching of Science for High School, conducted between 15 January 2013 and 30 September 2015, cofinanced by the European Union, as part of the European Social Fund, Human Resource Development Operational Programme, Pietrzak et al. 2015).

9.2 Basic Concepts

Description of educational activities that use ICT requires that certain terms related to the media used in this article be defined, namely, social media, mobile technology, e-learning and the Internet cloud.

Social media are media that exist under social control and which may be used freely in any scope. Their common features are open access to contents and the possibility of co-creation of the information and references to the transmitted information (Królewski and Sala 2013). Blogs, Wikipedia, YouTube, Flick, SlideShare, Google+, Facebook, Twitter and LinkedIn (Kaplan 2012) constitute a group of applications that allow presentation and exchange of opinions, content sharing, communication and cooperation. Among the listed applications, social networks were mentioned, namely, groups of Internet web pages that function within a certain social group (e.g. professional subject services aimed at specific groups and which allow creation of own profiles). Inherent functionalities of the media allow for the development of educational content and cooperation of the students.

Mobile technology is used for wireless connections between the devices. The twenty-first century saw the evolution of mobile devices that started off as pagers, mobile phones and GPS navigation systems followed by laptops, smartphones with web browsers and finally tablets. Mobile devices enable coordination of the work students perform and communication among the participants of the course as well as provide access to the information content dedicated to one course at any given time and from any place one wishes.

Currently, thanks to a rich offer of open online courses (MOOC – massive open online courses) (List of MOOCs offered by the Best Universities and Entities 2015), **e-learning** that uses computer networks and the Internet has grown in significance. Information contents of the Internet as well as available online educational movies, text, presentations, animations, simulations or educational games create a very useful base for realisation of any project, formulation of problems, asking questions and searching for solutions.

Remote learning platforms, on the other hand, constitute a learning environment with limited access to gathered educational materials. These are virtual learning environments installed on local servers (e.g. university servers) or hosted and available via a web browser. However, each of the e-learning platforms is limited and has a manner of operating that does not always cater for the needs of the taught subject. Despite that fact, such platform may constitute a form of a library or a guide for the course taken up by the students.

Cloud computing serves a similar purpose, however, without the limits on the sizes or formats of uploaded files. To simplify the matters, one might say that the cloud is a service, whereby you may avail of hardware and software on request, from any given devices with access to the Internet, which allows co-sharing, gathering and creation of information/data in the Internet environment. According to GUS (Polish Central Statistical Office) statistics, using cloud computing services is understood as using scalable ICT techniques via the Internet (GUS – Główny Urząd Statystyczny 2015a; IDC Cloud 2015). Service may encompass software access, availing of a given computing power and data storage (GUS, cloud computing electronic documents). Basically, storing data in the cloud or on a virtual disc means the same. The Internet cloud for educational purposes is linked with e-learning and use of applications, web 2.0, and – in the perspective of a couple of years – web 3.0 (Kelly 2012). According to Weaver (2013) and Katz et al. (2009), the main advantages of teaching in the cloud are as follows:

1. Autosave of all created content
2. Storage of several types of data, including films, music, e-books, photographic images and applications, including students' work

3. Accessibility from mobile devices
4. Option to edit the documents by multiple users at the same time
5. Availability of Internet content (without the need to copy and carry media storage devices)
6. Simplicity of organising tasks and routine knowledge tests
7. Use of online resources as means of cost-cutting, especially on equipment and software
8. Time-saving aspects – free management of infrastructure by service provider

9.3 Information Society in Poland: The Availability of Digital Technology, Digital Competences and ICT Use by Students and Teachers

Extensive research is being conducted to better exploit the potential of ICT. On the basis of this research, a brief description of selected aspects of the development of the information society in Poland has been prepared, focusing mainly on education and the use of new technologies by the young generation (*Information society... GUS – Główny Urząd Statystyczny* 2015b, *Social diagnosis...Czapiński and Panek* 2013, *Innovative applications...Jackowska* 2013, *Do teachers use... Mikołajczyk and Pietraszek* 2012, *Modern Technologies...Kwiatkowska and Dąbrowski* 2012).

In the first half of 2013, 70 % of households in Poland were equipped with a computer, while 66.9 % had access to the Internet. Both computers and Internet access were more frequently found in the households where children lived. This percentage was also slightly higher in big cities and in central Poland. Over 87 % of Polish people have mobile phones. The basic reasons why new technologies are absent from Polish households are of a motivational and psychological nature rather than a financial or a technological one.

For many Poles, television is a much more significant medium than the Internet. The percentage of people watching TV for at least 2 h a day is systematically increasing and, between 2007 and 2013, it had increased from 55 to 59 %. Meanwhile, the percentage of people using the Internet for over two hours a day on average is 18.6 %. The Internet is used by most young people (97 % of people aged 16–24) but not by many elderly people (14 % of people aged 65 and more). Almost all schoolchildren and students (99 %) and well-educated people (91 % of people with higher education) use the Internet. Ninety-five percent of children attending primary and middle school have a computer and 90 % have Internet access. Ten percent of 3-year-olds, half of 5-year-olds and as many as 82 % of 7-year-olds use computers.

In 2014, the Internet was regularly used by 95.4 % of people aged 12–15. The most popular reasons for accessing the Internet were using social media, which was the case with 77.2 % of young people, using email (64.5 %), playing computer games, downloading files with games and playing or downloading music and films

(64.0 %), while 59.2 % of people accessed the Internet looking for information to help with their studies. In 2014, 13.0 % of teenagers aged 12–15 used services offered by the cloud, 11.6 % stated that they used them to save files and 4.6 % – to share files with other users (Berezowska et al. 2014).

Unfortunately, computer skills are now worse than they were in 2012 (approx. 20–60 % of the surveyed computer users can, for instance, install hardware, organise folders, create a presentation, use spreadsheets, edit texts, copy files), and for many users, the computer has become merely a tool for accessing the Internet.

Detailed information on how ICT is being implemented in schools is provided in the report from the research conducted in 2013 as part of the project "Digital Teaching Laboratory for Schools in Małopolskie Voivodeship" (pol. *Laboratorium Dydaktyki Cyfrowej dla szkół województwa małopolskiego*), carried out by the Małopolska In-Service Teacher Training Centre (pol. *Małopolskie Centrum Doskonalenia Nauczycieli*) and the "Cities on the Internet" Association (pol. *Stowarzyszenie "Miasta w Internecie"*) (Jackowska 2013).

The report has concluded that only a few schools have clearly defined strategies aimed at acquiring new technologies (e.g. by participating in EU programmes, national and local programmes, competitions). For some, the major priority is an electronic class register, for others – getting a projector and a laptop for every classroom. There are also schools that put the emphasis on interactive whiteboards, considering them the primary and the most important indicator of the digitisation of schools. Ten percent of the examined institutions have modern equipment. In 30 % of the institutions, the equipment is outdated and faulty. The reason why schools are so badly equipped is a combination of two factors: the lack of money to buy the equipment and unwillingness to spend money on IT equipment. In the schools which are the best in terms of the level of digitisation, an electronic class register is the norm, supporting the communication among teachers, students and parents. Polish schools occupy a very low position in the European ranking based on the so-called connectedness (i.e. having a website and a virtual education environment). For comparison, in Europe, 61 % of students on average state that their schools have a virtual learning environment, while in Poland, the percentage is only 29 %. Some schools have Facebook profiles, which are used also by some teachers who log in to directly communicate with students. The most frequently employed solution is a shared email address for a given class, where teachers can send homework and various information. Approximately half of the examined schools provided – to a varying extent – a public Wi-Fi network. The low bandwidth of the networks made it difficult to watch films, view maps or use interactive applications. The materials most frequently used by teachers were the materials supplied by publishers, that is, ready lesson plans and exercises provided together with course books. They constituted 80 % of the materials used for teaching, examined during the research.

Rather popular were the so-called multibooks, multimedia presentations and YouTube videos. One-third of teenagers aged 12–15 have smartphones and use them to listen to music, access social media and play games. Almost all the examined schools impose an official ban on using mobile phones during classes. The research has revealed that most teachers do not realise that a modern mobile phone can function as an access point to Internet sources and that there exist mobile educational applications. The conclusions of the report show that the limited implementation of ICT in schools results from the teachers' concerns related to the fear of losing authority, their lack of computer skills, the time-consuming process of preparing classes, organisational and technical difficulties and the unavailability of ready materials compliant with the core curriculum, as well as the possible negative influence that computers and the Internet can have on students. According to the data obtained in the study conducted by the Extramural Education Development Centre at the Warsaw School of Economics, over half of the teachers (53.9 %) that took part in the study use the Internet to prepare for their classes (Mikołajczyk and Pietraszek 2012).

The European Commission report *ICT in Education* shows that Poland occupies a very low position in the ranking of digitally equipped schools. The ranking covered five areas subject to evaluation:

1. Equipment (desktop and laptop computers, e-book readers, mobile phones, interactive whiteboards, digital cameras and projectors)
2. The percentage of fully operational equipment
3. Broadband speed (above and below 10 Mbps) and type of broadband access (ADSL, cable, etc.)
4. Maintenance and technical support
5. Indicators of connectedness (a school website, email addresses for teachers and students, a local area network, a virtual learning environment)

In the ranking of digitally equipped primary and vocational schools, Poland occupies the last position in Europe and one of the last five in the case of middle and high schools. Regardless of the level of education, Poland is below the European mean, which is caused by a large proportion (30–35 %) of type 3 schools, i.e. poorly equipped schools, and a predominant proportion (50–60 %) of type 2 schools, i.e. partially digitally equipped schools, with lower than type 1 equipment levels, slow (less than 10 Mbps) or no broadband and some connectedness.

Similarly to the entire EU, in Poland, home ICT-based activities related to *learning*, i.e. watching or listening to the news and searching the Internet for information on specific topics of interest (but not any practical ones), are more frequent, compared to the school ICT-based activities.

The European Commission report ICT in Education confirms the observations drawn from the studies conducted in the schools in Małopolskie Voivodeship: schools with poor access to technologies use them as frequently as schools with good access to equipment. There exists evidence that improving the opportunities for professional development of teachers is an efficient way to improve the levels of ICT implementation in teaching and learning. According to the European Commission, the EU countries should consider ICT training as a mandatory element of the teacher training curriculum and strive to improve the quality and cohesion of ICT training throughout the entire education system.

The Regulation of the Minister of Science and Higher Education of 17 January 2012 (ISAP 2012), concerning the educational standards for the vocational training of teachers in Poland, clearly states that a teacher should possess IT knowledge and skills and should be able to incorporate information technologies in their teaching in a variety of ways (Dz. U. 1012 poz. 131).

To sum up, it should be emphasised that the vast majority of children starting school have phones and computers with Internet access at home, and, therefore, the education system should make use of this fact to a greater extent. The use of digital equipment and information technologies in teaching students is dependent not only on the availability of such equipment at school but also on the creativity and skills of teachers.

9.4 Examples of the Use of ICT and the Internet in Education and School

Jagiellonian University has applied information and communication technologies (ICT) in conducting their classes for several years. An example of the use of ICT in education is the virtual university called Academia Electronica that has operated online since 2007 within a 3D graphical environment Second Life (Academia Electronica 2015). The aim of the Academia is to present open to public lectures and discussions as well as conduct courses accepted by academic centres nationwide. Since 2008 prof. dr hab. Michał Ostrowicki, who operates in the virtual world under the alias Sidey Myoo, conducts on said portal an official academic course entitled "The electronic environment as the reality of man" (Ostrowicki 2015).

Another example of the use of ICT is the Jagiellonian Virtual Campus which comprises two e-learning platforms based on Moodle technology (Modular Object-Oriented Dynamic Learning Environment): "Smok" *[Dragon]* and "Pegaz" \ *[Pegasus]*. Pegaz is a platform for conducting remote classes, presentation of study materials (such as web pages, presentations and educational films) as well as enabling communication between teachers and students (through message boards and chat rooms). Each UJ employee can create an e-learning course to support direct teaching. Smok platform is designed to provide open courses that are available free of charge to all Internet users (pol. CZN – Centrum Zdalnego Nauczania Uniwersytetu Jagiellońskiego, Centre for Distance Learning, Jagiellonian University 2015). Modern technologies have been used the longest, for about 10 years, in teaching geography. In 2004, the university launched UNIGIS Geographic Information Systems postgraduate studies in the Department of GIS, Cartography and Remote Sensing in the Institute of Geography and Spatial Management of the Jagiellonian University in collaboration with the Paris Lodron University in Salzburg. Classes are taught in the distance learning mode. During the course the students communicate with academic teachers primarily through the e-learning platform and email (UNIGIS 2015).

One of the original educational methods is a combination of field classes, lab experiments and creative techniques and the implementation of ICT in the production

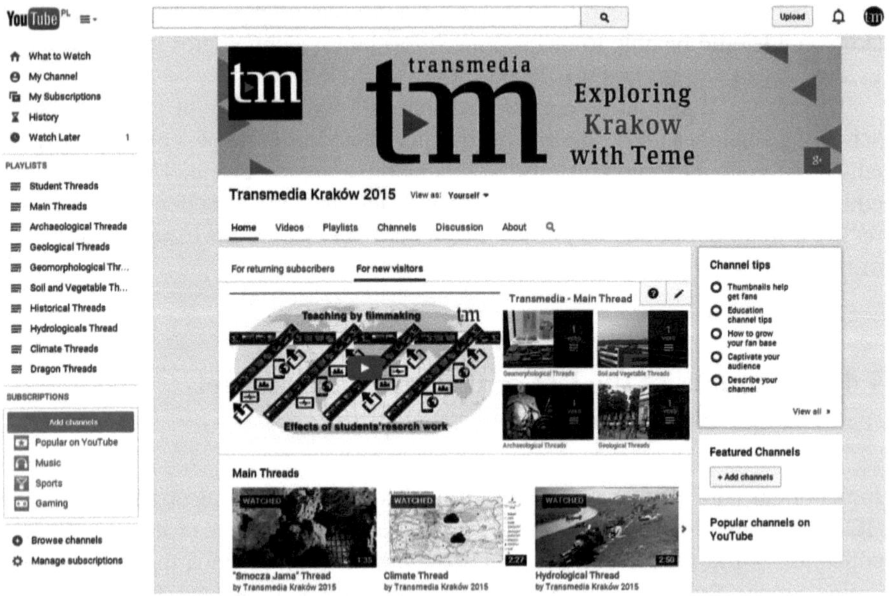

Fig. 9.1 Channel "Transmedia Kraków" created in video-sharing website YouTube Google+ where the clip "Exploring Kraków with Teme" was uploaded. The students watch the main clip with subplots and add further narrative elements created by them

of film narratives during the "Transmedia in Science Education" course. This is what I refer to as *transmedia education*, i.e. gaining knowledge by engaging student *explorers*, through the agency of a teacher, acting as a *guide-narrator*, in the explorations of the scientific, multi-format film storytelling and creating educational content, using ICT. The results and the acquired knowledge are then shared on social media (Pietrzak 2013, 2015).

The film used in the course provides information about different aspects of Kraków, viewed from various perspectives – from the perspective of the geological past of the city area, the state and the changes of the geographic environment, changes related to urban planning as well as changes influenced by economic, political and cultural factors. Students are encouraged by the film to think critically about the new media, and they learn certain user skills related to media, characteristic of both the creator and the viewer. The requirement to produce their own film forces the students to become involved and to share their ideas and gathered materials (Fig. 9.1).

An example of an effective cooperation between academic researchers and secondary school teachers is the PiNaP project. The IGiGP and the Pedagogical University of Kraków have created an innovative programme of teaching natural sciences in secondary schools and have monitored its implementation in schools in Małopolskie and Podkarpackie Voivodeship. In the 2014–2015 school year, 40 teachers from 20 schools conducted lessons using the blended learning method,

Fig. 9.2 Natural and anthropogenic landscape outdoor observations on IGiGP UJ observation deck. The relation of human-environment relationship is supported by digital photographs and maps from the Internet digital libraries

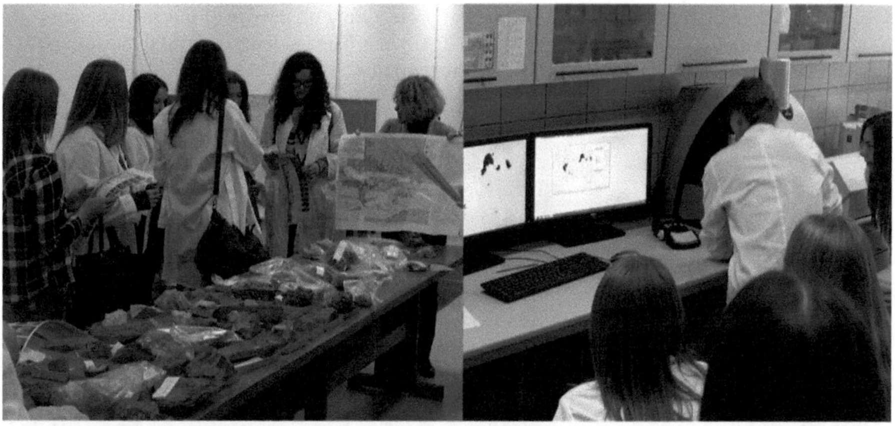

Fig. 9.3 Rock and sediment measurements and observations in IGiGP UJ labs with the use of a scanning electron microscope

with special attention focused on the needs of kinaesthetic learners. An e-learning platform, created specifically for this purpose, was used together with Internet resources available for mobile learning, during the conducting of experiments and doing student projects. Systematic evaluation of classes on the basis of opinion surveys and school visitations has revealed that students have become more interested in natural sciences, and they express enthusiastic opinions about academic classes of knowledge transfer. These were classes held in laboratories and computer rooms at universities, supplemented by direct observations made during fieldwork (Figs. 9.2 and 9.3). The model of these classes, i.e. direct measurements of natural phenomena, combined with the exploration of Internet sources of information, was

Fig. 9.4 Outdoor classes on natural environment protection with the use of information and communication technologies, a secondary school in Zakopane. The students determine the air and soil contamination degree using photographical documentation

Fig. 9.5 A class at a secondary school in Kęty. Using an e-learning platform and selected Internet links to discover light properties. A spectroscope construction and the analysis of a light spectrum produced by various light sources in practice

transferred to schools (Fig. 9.4). Students gladly used the opportunity to check and extend their knowledge through the e-learning platform, and they also used the instructions, films and animations uploaded there for their own projects (Figs. 9.5 and 9.6).

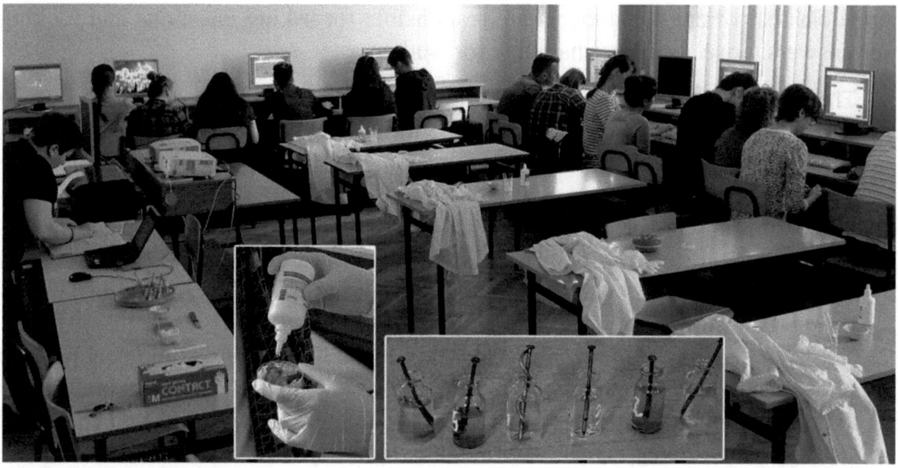

Fig. 9.6 The application of information and communication technologies and experiments in classes on the natural processes' pace at a secondary school in Dąbrowa Tarnowska. Experimental stations are visible in the middle of the room, and the magnification shows the result of weekly observations for various types of corrosion and their effect on metal depending on different environments

9.5 The Use of Information and Communication Technology in Light of the Theory of Teaching

Polysensory education implementing ICT enables various concepts of learning to interweave. By employing various educational strategies, we make it possible for students to take advantage of different learning styles. What follows is a discussion on how to incorporate polysensory teaching into a learning environment: a classroom, laboratory, field or museum but also the world accessible via the Internet. The focuses of attention are the benefits of using the Internet cloud and mobile digital devices. The education cloud is a learning environment where applications and Internet resources have been organised in a manner that makes it possible to achieve the established educational goals. The cloud is the environment, while the Internet provides the resources. Below, I present some concepts of educational strategies, adjusted to the subjects and levels of education mentioned in the introduction, that implement the benefits of working in the Internet cloud.

Interestingly, what is vital here is the way we organise students' work and not just the fact that we use ICT. There is no perfect educational method that will suit every student at every stage of their education. It is evident from experience that the biggest obstacle in organising work in the cloud is not the unavailability of Internet access and mobile devices, but, rather, the change of students' attitudes, that is, their transformation from passive recipients to active explorers of knowledge. Nowadays, when formalised syllabuses and enumerated teaching outcomes are the norm, open classes, where the students are responsible for asking questions and looking for answers, pose a serious challenge for students:

1. The *behaviouristic* concept is implemented by employing curriculum-based teaching organised by the teacher in the form of tasks uploaded to Pegaz, a university interactive platform, and study materials uploaded to the cloud. The *assimilation and reflection* strategy is used to facilitate the perception of information through the performance of given activities. The strategy consists of students preparing essays, databases, multimedia presentations, e-portfolios and mind maps, as well as participating in brainstorming, reading literature and analysing the information made available in the cloud.
2. The vital element of the *cognitive* concept is the internal, psychological process of acquiring and organising information by a student. Knowledge is built up by experience and introspection, that is, by means of the *emotional and empirical* strategy. This strategy consists of conducting analyses of actual cases (available online), recording students' role playing, solving problems and using educational games.
3. The *constructivist* concept consists in students constructing their own system of knowledge through direct observations made during fieldwork and class inspections and through scientific lab experiments. A strategy based on *observation and experiment* is used to identify and define problems and to discover reality through experiments. Crucial elements are fieldwork and work in laboratories, as well as constructing models. The strategy consists of registering presentations and observations made during fieldwork and experiments, which are later subject to critical analysis. The produced films, photos and descriptions are made available in the cloud.
4. The *connectivist* concept is based on the cooperation within Internet communities; for instance, teamwork in the cloud focused on the issues related to the current educational problems, where the crucial element is developing skills necessary to effectively research information on the Internet and share them while working together on papers, podcasts and infographics. The *pragmatic and communicative* strategy employed in this case facilitates the search for ideas for the practical implementation of theories. The strategy is based on discussions held on social media, participation in the information stream and adding comments to the texts available on virtual drive.

Various educational strategies, methods and techniques make it possible to focus on the substantive content of the transferred knowledge, on the student's responsibility for acquiring knowledge and on strengthening their open attitudes in social communication. The idea of education implementing the cloud is to develop competences and not to discard old skills to make room for the new ones. What is important in the education through the cloud is the ability to look for information, ask questions and conduct scientific inquiries, as well as the readiness to share knowledge.

9.6 The Organization of Work in the Cloud: Selected Examples

The examples presented below are taken from the following academic classes held at the Jagiellonian University: "Transmedia in science education", "Higher education didactics in natural sciences", "Natural sciences didactics", "Graduates on the job market", "The rudiments of earth sciences" and "Teaching geography". Since the presented technological solutions require no specialist knowledge and implement free applications and commonly available equipment and digital media (e.g. computers, phones, cameras, the Internet, digital library, websites, computer games), they prove useful for students and unproblematic for all generations of academic teachers (Fig. 9.7).

The Google Drive was used to organise information resources (in the form of arranged folders) and a dedicated learning environment for particular subjects. Those students who registered for a particular course were given access to the folders and resources. Depending on the course, the structure of a folder is more, or less, complex. The fundamental elements of every course are creating a working environment (described in 1), creating a place for storing students' work and their photo documentation (described in 2 and 3), providing access to the literature and conducting evaluations (described in 4 and 5) (Fig. 9.8).

What follows is a description of the resources available in the folders and some examples of the methods employed during classes.

Blended Learning The cloud contains materials used during classes in a computer room and classrooms, during fieldwork and in a laboratory (soil science and geomorphology).

The examples of the files available in this folder are:

- Documents edited together by the students
- Webquests that are solved during classes
- Instructions for lab exercises
- Maps with QR codes for fieldwork

The *documents* available in the cloud are text documents, spreadsheets, graphic mind maps, presentations, infographics and films. What distinguishes working in the cloud from students' individual work using ICT is the fact that groups of several people can work on the same document at the same time. Some of the advantages of working together on the same document are commitment, the necessity to follow the work progress of other people to avoid repetitions, a fast pace of work and creativity inspired (forced) by the other group members. Although students work individually, sitting at their own computers and adding comments, they [are forced to] closely follow the changes and additions to the document made by other people. Such a process induces empathy and sensitivity to the way other people perceive the world, it promotes teamwork and communication, but, most of all, it makes it possible for students to learn from one another (Fig. 9.9).

Fig. 9.7 Educational cloud architecture diagram

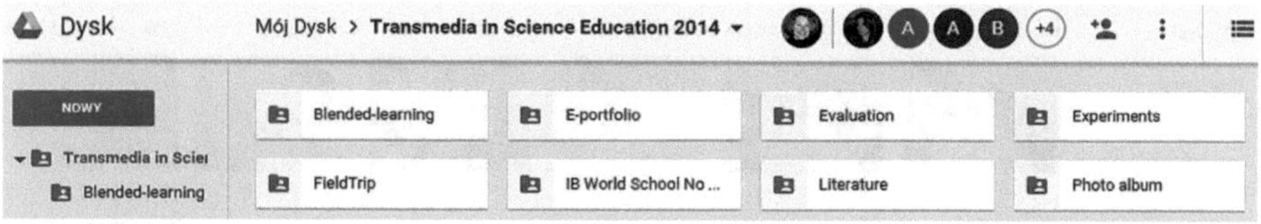

Fig. 9.8 Organization of educational environment in a cloud. The folders are dedicated for work with the blended learning method, collecting an e-portfolio, evaluating and uploading literature and pictures

Fig. 9.9 A concurrent edition of text and graphical documents by students (the arrows indicate locations where we can see how many students are editing the document at a given moment

A *webquest*, in the context of the presented classes, is a document containing a description of a particular phenomenon or a list of questions or issues accompanied by links to valuable Internet resources. They provide access to places where students can look for information or answers, delve further into the issue and ask further questions. A webquest in the cloud leads to the problem analysis, based on the provided Internet sources (e.g. using Borton's reflective model). In the traditional sense, a webquest is a project-based method implementing Internet sources of information. Some of the subjects of the webquests are "alternative evaluation methods", "how to create a media-based learning environment", "does ICT influence behaviour?", "how does the Internet influence the learning process", "back-and-forth communication versus social media" and "how to look for a job on the Internet". The undisputed advantage of working online is the topicality of the discussed issues and the immediate availability of references to actual research, for instance, the Planet Hunters project or the Rosetta mission.

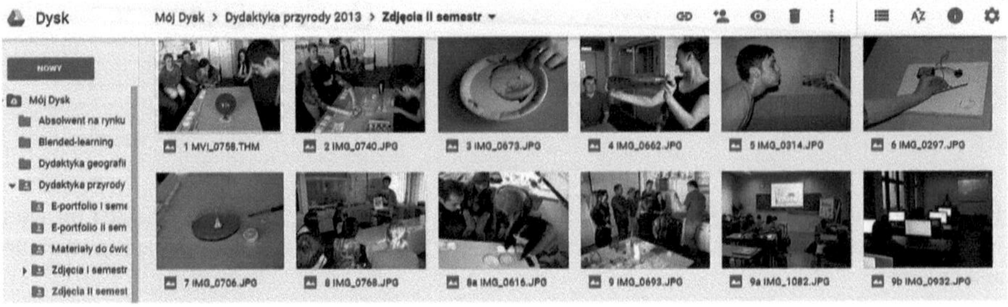

Fig. 9.10 Photographical documentation of natural experiments conducted at classes in the form of pictures, movie clips and animations

Students prepare instructions for exercises and lab experiments before coming to class and upload them to a given folder. After classes, the instructions available in the cloud are supplemented by photos from the presentations and the conducted experiments. Since students can review the conducted experiments, they can prepare new ones for the classes, without having to produce and keep additional records, as everything is available online (Fig. 9.10).

During *fieldwork*, the participants use mobile devices. The materials required for the class are uploaded to the cloud by the teacher, e.g. thematic and historical maps, sketches, section drawings, archival photos, pictures, drawings and diagrams. The teacher creates a *map* with the route of the field trip, where they place *QR codes* to the mentioned resources or websites. For instance, during classes on "Transmedia in Science Education" and the trip around Kraków, while visiting the Wawel Hill, students can browse pictures from the resources of the National Museum and the Historical Museum of the City of Kraków. After visiting Smocza Jama (Polish for "dragon's cave") and observing the tectonic horsts, students verify their field observations against the maps available on the website of the Geological Museum. Links to the Archaeological Museum make it possible for students to discuss the relationships between prehistoric and historic settlements and the geographic environment.

E-portfolio Students upload text, graphic and video works or presentations prepared and presented during classes to the cloud. The subject matter of all the works constitutes a certain whole. A sense of responsibility for original contribution to the classes arises among the students. For example, in 2014, under the subject "Innovations in teaching geography", the doctoral students presented the following works: "podcast as a tool in teaching geography", "augmented reality in teaching geography", "military education", "mentoring in higher education", "use of ICT in cribbing", "holographic learning" and "concept board – shared idea, single idea" (Fig. 9.11).

Photos Video documentation of the classes is managed by an academic teacher so as to analyse the learning process on their basis. Experience shows that students are very emotional about browsing photos and watching videos; they are both interested

Fig. 9.11 An e-portfolio with sample student works on innovation in teaching geography

and excited. Watching photos and videos of the classes allows the teacher to notice and analyse details, to become aware of committed mistakes and to consolidate good habits.

Students also create photo and video documentation, which teaches them thorough observation of phenomena, recording fundamental stages of observed processes and selective processing of the recorded material. The post-production process teaches discipline, concentrates on the facts and thinks about the recipient, and it ultimately verifies the quality of actions undertaken when recording images, as at this stage the student finds certain shots to be missing, pace and sequences to be unsuitable and commentary to be lacking.

Literature is gathered incrementally in the cloud (dynamically/actively). Initially, the resource contains literature recommended by the teacher. The resource grows systematically when it is supplemented by students who are obliged to make a short presentation of the proposed materials during classes. These are, for example, reports from symposia and conferences, publications with abstracts, valuable websites, scientific blogs, online lectures and videos. Students upload short notes on the presented publication including a link to the cloud. By seeking out literature on their own, the students verify their knowledge, asking themselves questions such as what do I know, what do I not know and what information do I need. Such work leads to effective use of information found on the Internet, including the ability to browse through invisible web resources.

Evaluation An important element of education is feedback for the students, showing them the valuable elements of the performed work, their shortcomings and mistakes, but most of all the tips showing how to achieve the determined educational goals. Depending on the level of education, the criteria of evaluation are defined together with the students. Students gain experience in evaluation and self-control. Marks in form of points for particular tasks and comments on works, facilitating correction of committed mistakes and improvement of abilities are available online. Owing to the automatic recording of every activity of course participants, the teacher can follow the progress of each student's work in the document history. This aspect can become a measure motivating to systematic work or reveal its lack.

In the course of classes conducted with students and pupils, it has been determined that an online cloud as an on-demand equipment and software use service available from any device with Internet access enables sharing, gathering and producing information in online environment. Students construct their knowledge system independently owing to mixed teaching supported by e-learning, field observations, museum classes, experiments, visits to education centres and role playing. Owing to the constant availability of educational materials and applications, the students gain the skills of seeking out information online efficiently and learning through cooperation in online communities. The main advantages of education with the use of the online cloud include dynamic learning environment created jointly by the students, reciprocal communication, availability of resources from mobile devices and possibility of creating and storing different kinds of data, including videos, infographics, music, e-books, photos and applications.

The student teaching *experience indicates* that the cloud considered as a service, whereby you may avail of hardware and software on request by way of any given devices with Internet access, allows co-sharing, gathering and creation of information/data in the Internet environment. Students may create their own systems of knowledge using blended learning supported by e-learning and field observations, lessons in the museum, conducted experiments, visitations in educational centres, role play, etc. Thanks to constant availability of educational materials and applications, students acquire the skills of efficient screening for information online and learning via cooperation within social networks.

Dynamic learning environment cocreated by students, reciprocated communication, content accessibility from mobile devices, possibility of creation and storing various types of data like films infographic images, music, e-books, pictures and applications are all among the biggest advantages of educational system supported by the cloud.

9.7 Summary: Axiological Reflection on the Use of ICT and the Internet

The indisputable fact is that modern technologies allow to broaden the horizons of cognition, for example, owing to satellites and telescopes, to probe deeper thanks to electron microscopes and to observe processes thanks to CAT scanners, and in school reality, one can, and should, examine the research results of scientists and follow the progress of science. Only as soon as the educational goals are determined at a particular stage of education for particular students, the teaching strategies should be defined with didactic aids, including the application of ICT. The teacher adjusts the learning environment and tools allowing the observations and studies of the reality to the determined cognitive, educational goals. It is important for the environment to be identical with the environment in which the young person is going to work. A teacher of natural science subjects should therefore utilise the geographical, biological and chemical databases that are publicly available online, including maps, photos, videos, animations, diagrams, statistical data, etc., as in the future, work and life will require young people to navigate publicly available data. Case study methods should be used more universally, as it is more familiar, close and up to date for the students – such examples are easy to find on the Internet. It is good to involve the students in discussions on the current issues, initiate meetings, suggest interviews and projects and arrange plays with role playing; inspirations and topics can be found in current events presented in the media. As far as the attitudes are concerned, this teaches social participation and responsibility. Focusing solely on the technical side of the process of education objectifies students. The answer to the question 'why educate' cannot be found in the equipment but in deep reflection which we wish to encourage in our students. The essential goal of education is to get to know oneself – the development of talents and understanding one's own weaknesses, opening to the needs of the other human being. For the learning process is not only individual but also social in nature.

Visits to the schools where the students of the Jagiellonian University have their teaching practices, to the schools participating in the PiNaP project and multiple reports indicate that it is groundless to expect major teaching efficiency as a result of the use of interactive whiteboards and projectors in classrooms, or from costly implementation of school e-learning platforms. This is because interactive equipment, such as a whiteboard, only stimulates a single person working on it at a particular moment and this usually is the teacher. It is the teacher who goes to no end of trouble in order to flood the students with various animations, websites or multibook exercises. Images and videos shown with the use of a projector bear the same influence on a student in didactic terms as images in a book or a video on a TV screen. As in the case of a whiteboard, all the students remain at their desks and statically watch a visual message. Occasional visits in computer labs where two or three students sit at one monitor deprive them of an opportunity to work individually and become totally involved.

A great alternative for the described educational activities may be the use of students' telephones, smartphones and tablets. These are the devices used universally in many professions for communication, consultations, documentation, work organisation, access to mobile office, making payments and ordering services. E-learning platforms can effectively be replaced by working in an online cloud, which eliminates the costs and equipment maintenance and increases flexibility and accessibility from different devices and any place.

There is a modern myth that the use of modern technologies during a lesson is a sign of innovation in education. Only decision makers and persons uninvolved directly in the work with students can feel good about equipping schools with multimedia projectors and interactive whiteboards, thus contributing to modern teaching. Nothing could be further from the truth. Due to multimedia presentations and showing selected resources from the Internet, we have reached extreme verbalism. Even worse, a lecturer or a teacher does not have to be a good speaker and does not even have to remember the content of the speech, as all it takes is to read or comment on the displayed slides. The skill of creating a drawing, a pattern of thought on the whiteboard or a model from generally available items is vanishing. There is no motivation to organise experiments or go into the field, as it can all be shown on a screen. I claim that there are no modern teaching methods, and the old methods suggested and characterised by Confucius, Socrates and Komeński do very well in the new reality. The reality is of universally available information resources on the Internet and devices allowing access to them. Therefore, apart from the "online" teaching environment, it is also important to use the devices that are popular among the students. However, the point is to use them in direct contact with natural phenomena, not instead of this contact. All it takes is to make a simple assumption that a company or organisation employing people, working with specialist equipment and software, is always going to train its employees. For equipment and software change at a rate faster than the educational cycle of a young person. It is more important to teach them to think and use devices in the real world than to train in using single programmes and operating particular equipment. The greatest obstacle in organising the work with the use of ICT is not access to the Internet and the option of using mobile devices, but a change in the habits of students from passive recipients to their adopting an attitude of responsible knowledge seekers.

Technology can often constitute a barrier to teaching and learning. The multitude of devices, software and applications, their high cost relative to the short life cycle on the educational "market" and frequent software equipment in compatibility may discourage the implementation of particular solutions. Digital technology develops at such a quick rate that there is no future in it for a particular device or application. New solutions keep appearing, and we do not know what digital devices will allow us to achieve in several years' time. In university-level education which trains future teachers, one should teach the skill of learning and cooperation on the Internet, i.e. the global computer network with wide access to resources, with the use of social networks, using open, public online courses and resources created by themselves.

References

Academica Electronica. (2015). *Instytut Filozofii Uniwersytetu Jagiellońskiego* [The Jagellonian University, Institut of Philosophy]. http://www.academia-electronica.net/. Accessed 10 Jan 2015.

Berezowska, J., Huet, M., Kamińska, M., Kwiatkowska, M., Orczykowska, M., Rozkrut, D., & Wegner, M. (2014). *Społeczeństwo Informacyjne w Polsce. Wyniki badań statystycznych z lat 2010–2014* [Information society in Poland. Results of statistical surveys in the Years 2010–2014]. Warszawa: Główny Urząd Statystyczny w Szczecinie. http://biblioteka.mwi.pl/index.php?option=com_k2&view=item&id=347:spo%C5%82ecze%C5%84stwo-informacyjne-w-polsce-wyniki-bada%C5%84-statystycznych-z-lat-2010-2014&Itemid=3. Accessed 10 Oct 2015.

Czapiński, J., & Panek, T. (2013). *Diagnoza Społeczna. Warunki i jakość życia Polaków* [Social diagnosis. Objective and subjective quality of life in Poland]. Warszawa: Rada Monitoringu Społecznego. http://ce.vizja.pl/en/download-pdf/volume/7/issue/3.1/id/295; http://ce.vizja.pl/en/download-pdf/volume/7/issue/3.1/id/346. Accessed 10 Oct 2015.

CZN – Centrum Zdalnego Nauczania Uniwersytetu Jagiellońskiego [Centre for Distance Learning, Jagiellonian University]. (2015). http://www.czn.uj.edu.pl/. Accessed 15 March 2016.

Goban-Klas, T. (1999). *Społeczeństwo informacyjne. Szanse, zagrożenia, wyzwania*. [Information society. Opportunities, threats and challenges]. Warszawa: Wydawnictwo Fundacji Postępu Telekomunikacji.

Goban-Klas, T. (2005). *Cywilizacja medialna* [Mass media civilization]. Warszawa: Wydawnictwa Szkolne i Pedagogiczne.

GUS – Główny Urząd Statystyczny. (2015a). Słownik pojęć, chmura obliczeniowa [Glossary, cloud computing]. http://stat.gov.pl/metainformacje/slownik-pojec/pojecia-stosowane-w-statystyce-publicznej/3086,pojecie.html. Accessed 10 Jan 2015.

GUS – Główny Urząd Statystyczny. (2015b). *Społeczeństwo informacyjne* [Information society]. http://stat.gov.pl/obszary-tematyczne/nauka-i-technika-spoleczenstwo-informacyjne/spoleczenstwo-informacyjne/. Accessed 10 Oct 2015.

IDC Cloud. (2015). International Data Corporation (IDC). http://www.idc.com/prodserv/FourPillars/Cloud/index.jsp;jsessionid=7A0DEB2D99D8C3BF6632162A2E2B4AC2. Accessed 10 Oct 2015.

ISAP – Internetowy System Aktów Prawnych. (2012). Rozporządzenie Ministra Nauki i Szkolnictwa Wyższego z dnia 17 stycznia 2012 r. w sprawie standardów kształcenia przygotowującego do wykonywania zawodu nauczyciela [Regulation of Minister of Science and Higher Education, Act of 17 January 2012, Standards for the teaching profession]. http://isap.sejm.gov.pl/DetailsServlet?id=WDU20120000131. Accessed 10 Oct 2015.

Jackowska, M. (Ed.). (2013). *Innowacyjne zastosowania rozwiązań i narzędzi cyfrowych w kształceniu na poziomie gimnazjalnym i ponadgimnazjalnych w województwie małopolskim, raport z badań 2013* [Innovative application of digital tools in the education at primary and secondary schools in the Malopolska province, Raport 2013]. http://www.mwi.pl/component/content/article/348-badanie-cyfrowej-szkoly.html. http://www.mwi.pl/download/category/3-2007.html?download=190%3Araport-z-bada-nad-zastosowaniem-narzdzi-cyfrowych-w-szkoach. Accessed 10 Oct 2015.

Kaplan, A. M. (2012). If you love something, let it go mobile: Mobile marketing and mobile social media 4x4. *Business Horizon, 55*(2), 129–139.

Katz, R., Goldstein, P., & Yanosky, R. (2009). *Cloud computing in higher education*. http://net.educause.edu/section_params/conf/ccw10/highered.pdf. Accessed 10 Jan 2015.

Kelly, J. M. (2012). Digital humanities: Past, present, future. http://www.jasonmkelly.com/wp-content/uploads/2012/08/Web-1-3.png. Accessed 10 Jan 2015.

Królewski, J., & Sala, P. (2013). *E-Marketing.Współczesne trendy. Pakiet startowy*. [Online Advertising. Contemporary Trends. Starter Pack]. Warszawa: Wydawnictwo Naukowe PWN.

Kwiatkowska, D., & Dąbrowski, M. (2012). Nowoczesne technologie w rozwoju uczniów szkół ponadgimnazjalnych – wyniki badań [Modern Technologies in High School Students Development –Research Results]. *E-mentor, 3*(45). http://www.e-mentor.edu.pl/artykul/index/numer/45/id/930. Accessed 10 Oct 2015.

List of MOOCs offered by the Best Universities and Entities. https://www.mooc-list.com/. Accessed 10 Jan 2015.

Mikołajczyk, K., & Pietraszek, K. (2012). *Czy nauczyciele wykorzystują nowoczesne technologie informacyjno-komunikacyjne w kształceniu?* [Do teachers use information and communication technologies in education? Research report]. Centrum Rozwoju Edukacji Niestacjonarnej SGH, Raport z badań. http://www.ptnei.pl/files/VU2012/prezentacje/2B/2B.2.pdf. Accessed 10 Oct 2015.

Morbitzer, J. (2010). Szkoła w pułapce Internetu [School trapped in the Internet]. In: J Morbitzer & E. Musiał (Eds.), *Człowiek-media-edukacja* [Man-media-education]. Kraków: Katedra Technologii i Mediów Edukacyjnych, Uniwersytet Pedagogiczny w Krakowie. http://www.ktime.up.krakow.pl/ref2010/morbitz.pdf. Accessed 13 Jun 2015.

Ostrowicki, M. (2015). Strona internetowa [Website]. http://www.ostrowicki.art.pl/. Accessed 10 Jan 2015.

Pietrzak, M. (2013). Transmedia i crossmedia w edukacji [Transmedia and crossmedia in education]. In J. Morbitzer, E. Musiał (Eds.), *Człowiek-media-edukacja* [Huma-media-education] (pp. 209–317). Kraków: Katedra Technologii i Mediów Edukacyjnych, Uniwersytet Pedagogiczny w Krakowie. http://www.ktime.up.krakow.pl/symp2013/referaty_2013_10/pietrzak.pdf. Accessed 10 Jan 2015.

Pietrzak, M. (2015). Film transmedialny – doświadczenia realizacyjne z projektu edukacyjnego [Transmedia movie – Experience of educational project realization]. In J. Morbitzer, D. Morańska, E. Musiał (Eds.), Human-media-education (pp. 245–256). Dąbrowa Górnicza: Wydawnictwo Naukowe Wyższa Szkoła Biznesu. https://www.dropbox.com/s/mfqvujmpa7uqfoy/FILM%20TRANSMEDIALNY%20%E2%80%93%20DO%C5%9AWIADCZENIA%20REALIZACYJNE%20Z%20PROJEKTU%20EDUKACYJNEGO.pdf?dl=0. Accessed 10 Oct 2015.

Pietrzak, M., Potyrała, K., Rotter-Jarzębińska, K., Trzepacz, P., & Walosik, A. (2015). *Program Innowacyjnego Nauczania Przyrody dla Szkół Ponadgimnazjalnych z Obudową Dydaktyczna. Przyroda z PiNaP* [Programme of the innovative method of teaching science for High School with Didactic Materials. PiNaP Project]. Kraków: Uniwersytet Jagielloński. Tom 1–11.

UNIGIS. (2015). Studia Podyplomowe Systemy Informacji Geograficznej [Postgraduate Studies Geographic Information Systems]. http://krakow.unigis.net/. Accessed 10 Jan 2015.

Waever, D. (2013). Six advantages of cloud computing in education. http://www.pearsonschoolsystems.com/blog/?p=1507. Accessed 10 Oct 2014.

Chapter 10
Information Technologies in Teaching Geography from the Teacher's Point of View

Martin Blažek, Martin Lána, Vojtěch Blažek, and Jan Dvořák

10.1 Introduction

Nowadays, information technology can be considered one of the main teaching tools which the student and teacher can systematically use to achieve given learning objectives. The use of audiovisual technology and other information technology in teaching has never been so common as today, and the role of technology in teaching has become irreplaceable (e.g. Kerski 2003; Bartha 2009). Mass involvement of information technology, and all other types of media devices, goes hand in hand with modernisation of teaching. Students respond well, they are interested in working with technology that they use in their everyday lives and they are becoming highly skilled (e.g. Baker 2005; Baker et al. 2009).

Geography is a subject interested in changes in the world, in the causes and consequences caused by global, regional and local changes of our natural and cultural

M. Blažek (✉)
Faculty of Education, Department of Geography, University of South Bohemia in České Budějovice, Jeronýmova 10, 371 15 České Budějovice, Czech Republic

UMR PASSAGES, Université Bordeaux-Montaige, 19 esplanade des Antilles, 33607 Pessac, France
e-mail: mblazek@pf.jcu.cz

M. Lána
ZŠ a MŠ Děčín IV, Raisova 688/11, 405 02 Děčín, Czech Republic

V. Blažek
GIS administrator, Regioskop z.s.p.o., Písek; Faculty of Education, Department of Geography, University of South Bohemia in České Budějovice, Jeronýmova 10, 371 15 České Budějovice, Czech Republic

J. Dvořák
Faculty of Education, Department of Geography, University of South Bohemia in České Budějovice, Jeronýmova 10, 371 15 České Budějovice, Czech Republic

Grammar School Jírovcova, České Budějovice, České Budějovice, Czech Republic

P. Karvánková et al. (eds.), *Current Topics in Czech and Central European Geography Education*, DOI 10.1007/978-3-319-43614-2_10

environment, and as such it provides optimal grounds for using modern technology in teaching. This chapter discusses some case studies, capturing the use of information and geoinformation technology in teaching geography in Czechia. The studies are based on pedagogical practices, and each of them captures a specific case of application of the aforementioned technology and also represents certain general principles and tendencies, demonstrated in teaching geography using information and geoinformation means.

The first part of the chapter introduces the option of using new technologies connected to the Internet and sharing information via cloud services (e.g. Google Disc, YouTube, etc.) in combination with social networks. This is presented within the context of Czech education systems. The next part discusses the issues connected to using geographic information systems (GIS) in teaching geography. Attention is paid not to desktop versions, but to versions that are available online, as they provide new opportunities to record and share geographical data. The examples of using telecommunication technology such as GPS in geography lessons that focus on local topics are presented in third part. This part is based on the integration of geocaching principles into geography lessons. Using examples, a whole variety of specific practical applications in teaching is introduced that can be applied by teachers and academic staff who provide further education training for teachers. The diversification and variety of methods, and the attempt of geography teachers and academic staff to respond to the development of modern information technology in European countries, can also be seen in Czechia.

10.2 Using Geospatial Technologies in Teaching Geography

Technological development has accelerated and broadened over the last decades. Technology is now ubiquitous in people's everyday lives, and this trend will surely increase in the future. This expansion, of course, can also be seen in education and teaching in and after schools. For example, the accessibility of navigational technology and its use became quite common and led to huge breakthroughs in this area. As a result, the first map that many children often encounter is a digital one, i.e. the map app running in their parent's car navigation system. A lot of data and information accessible today is constantly transferred into and stored in digital form. Overall we can talk about digitisation. A lot of data and information is created in this form quite commonly, using different types of available digital technologies.

The digitisation of a large amount of data has enabled fast access to a wide range of information to a larger number of users than it could have been before. Because of this and the development of the Internet, the information and data can spread more easily to a wider variety of users. Their recombination in different ways brings us constantly new knowledge and possible applications.

Thanks to the Internet, we not only can search for data and information but also verify, compare and analyse it and then create, sort, save and publish it ourselves. Technology provides us many different tools to do all this. These facts have great

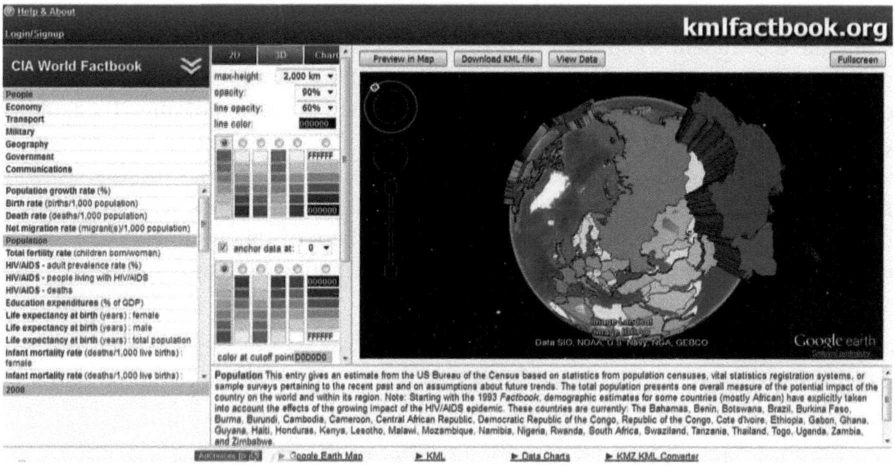

Fig. 10.1 Online application kmlfactbook.org

potential for the teaching of geography, which has traditionally been working with data and information over a long period of time.

Maps are a typical example. Earlier, teachers usually had data in paper form at their disposal, such as atlases and maps. Nowadays, this area has extended, and teachers can use other possibilities thanks to digitisation (e.g. geographic information system (GIS)). In Czechia, digital maps are provided by Google (Google Maps – available on https://www.google.cz/maps/) or the Czech map server https://mapy.cz/, for example.

Students and teachers have not only 2D display available but also 3D display in Google Street View, where they can examine and even walk through places thanks to 360° picture (Google 2015). Or they can use the Google Earth application and take a virtual trip all over the world. These digital tools, which should become common basic aids for current geography teachers, can take students to places that not long ago were nearly unimaginable. This includes not only dry land and closed buildings (e.g. museums), but beneath the seas and oceans or even the planets in our solar system.

Nowadays, there are many other services complementing Google Earth that enable to insert and compare data and relevant information directly on specific surfaces and areas of the Earth. It is possible to create "live" cartograms and thematic maps (see, e.g. Fig. 10.1 representing the online application (http://www.kmlfactbook.org/)).

Geography instruction that integrates digital technologies brings whole new possibilities both for teachers and students. Constructivism, learning by doing, exploration-oriented teaching and many other teaching approaches and strategies aim for active student involvement in the teaching and learning process. The aim is to enable each student to achieve his or her highest individual development and to provide optimal conditions for this. The student becomes the main active element;

the teacher steps aside and provides and creates suitable conditions and situations to stimulate the student's activity in his or her education and the acquisition of necessary knowledge, skills and competence. The teaching should strive particularly to develop geographic knowledge, critical thinking and problem-solving. Technology can be enormously helpful with this approach.

The Internet and its services are constantly and dynamically evolving, placing new demands on continuous teacher education in their field, didactics and of course digital competence. Geography teachers and teachers of other subjects can hardly do without sufficiently developed digital and information literacy.

10.2.1 Cloud

"Cloud computing", or simply the "cloud", is today one of the most frequently used terms in the world of information and communication technologies. It is an Internet-based model of development and using computer technologies. It can be characterised as providing services and programmes that are stored on servers, which users can access via a web browser or certain applications practically from anywhere. All of us are familiar with the concept, since it is the basis of ordinary applications and services, such as the e-mail provided by Gmail.

Thanks to the quick spread of the Internet and its accessibility from different devices (most commonly mobile devices, such as smartphones, tablets and wearables[1]) and at any time and from anywhere, as well as constantly improving web browsers, the possibilities of using web applications have widened mostly in these areas:

- Transfer of traditional desktop (we install them in computer) application into the form of web applications
- Complete office packets, graphic editors, sound and video editors and making notes in presentations
- Communication
- Playing games
- Database systems and cloud operating systems

Web applications and services have started notably competing with "classic" ones. This is due to their rapid development, wide range, easy administration and maintenance (mostly resolved by the provider), easy accessibility and opportunities for sharing and cooperation. The arrival of the "cloud" has created lots of new possibilities and opportunities in the use of digital technologies, including teaching and education.

What does "cloud" bring to school? The school environment has gradually changed from one where there are only a few computers just for a few students and

[1] *Wearables or wearable technology – a term for small electronic devices designed to be worn on the body or clothes (e.g. watch, bracelet, glasses, etc.).*

teachers to one where each teacher is equipped with their own device, and often with several devices, such as laptop, mobile phone, tablet, etc. These days, students also usually have these devices available in and after school. Thanks to the Internet, these devices give their users constant access to various data and information.

What possibilities does "cloud" offer in teaching geography? There are many different possibilities. There were chosen two services from *Google* to demonstrate the general possibilities for using the cloud, specifically *Google Disc* and *YouTube*.

10.2.2 Google Disc

Basically it is a free shared storage space (15 GB of default shared storage space – if the school uses Google Apps for Education, the storage space is nearly unlimited) on the Google server where users can save different kinds of files, such as photographs, pictures, sounds or videos, and at the same time, they have an available online office (Google Documents) where they can create:

- Texts
- Charts
- Presentations
- Forms
- Drawings
- Maps

What can Google Disc offer in teaching geography? There are many possibilities. The students together can create documents in real time (online), whether it is the creation of different documents or charts, presentations, drawings, maps, etc. The students can share these created documents and can choose suitable options, such as edit, comment and view. These materials can be very easily and quickly published.

Here is an example: The teacher presents a new topic into the lesson and he or she would like to know in advance what the students already know about this topic and what they would like to know about it. The teacher creates a document and shares it with the students before the lesson (directly when he/she uses the URL[2] shortener, e.g. *goo.gl*, and he/she tells the students the shortened address). The students can fill in their actual knowledge about this topic (in the form of, e.g. keywords, ten key sentences, etc.) and afterwards can add a question they are interested in. Then the teacher can use this material for his or her further aims in the lesson. Alternatively, the teacher can invite students to answer their classmates' questions (find out and fill in the information, verify what the other classmates have written, etc.).

The teacher can then create a shared presentation in which he or she inserts the questions. The class divides into groups, and they can create for each question, e.g.

[2] *Uniform resource locator.*

three slides into the shared presentation. In these slides they try to find the answers (they can insert texts, pictures, charts, maps, videos…). The following activity can verify the correctness and credibility of these answers. The final presentation serves as common supportive educational material.

Forms in GD have a specific place. With their help we can create, for example, different surveys, questionnaires, researches, etc. The students (teachers) can themselves get various data which they can afterwards easily process (display), for example, in charts, and they can work further with these results. Forms can also be used to create different types of tests complemented with pictures, videos, animations and the like.

The test can be subsequently extended in *Flubaroo*, very easily corrected and evaluated (Flubaroo 2015). The teachers can get immediate feedback on the given topic (lesson) from all students using the digital technologies.

The creation of your own maps – *My Maps Google* – offers the best possible option in teaching geography. The students themselves can create these maps and they can also cooperate. These maps can be supplemented by various objects (e.g. photographs). This area offers a very wide utilisation and the students work themselves in real space. The students can place their own objects and routes in the maps (they can even help improve them, i.e. they become active creators of the map). For instance, they can create memorial trees map in the given area, e.g. in their town (they can mark the position of trees into the map and add photographs students take), black dumps map, etc.

10.2.3 YouTube

It is one of the most common services from Google and enables its users to watch and load (even edit) videos. These days it is one of many useful tools and resources for both teachers and students. We can find many different appropriate videos that can be used in teaching geography. In addition to watching, we can also create videos ourselves and then release them. Many teachers use YouTube to "collect" and create their own *lists* of interesting *videos* that they can share with their colleagues or students. Or they can follow *channels* and users that publish these videos.

Here is an example: An enjoyable activity for many students can be their own presentation of chosen local geomorphological phenomena from their area in the form of a short video report which they create themselves. Google Disc and YouTube have introduced us to the basic possibilities that the "cloud" can offer nowadays: conveniently available data anywhere, any time.

10.2.4 Mobile Technology

Thanks to the expansion of the Internet, mobile devices, such as laptops and especially tablets and smartphones, allow us to access data and information much more easily and continuously. We can search data, create it ourselves and publish it. Before taking a closer look at mobile devices, let's examine searching via the web browser. Google Search can be very useful. If the term *weather Prague is* entered in the search window, we get the current weather as well as a forecast. We can also choose to display *wind*. The following activity in a lesson can be the creation of own chart (wind rose) displaying prevailing wind directions in a reporting period where the students complete the relevant cardinal points. It allows also to search routes from one place to another, not only within Czechia but also between continents. Afterwards we can discuss what transportation we use and why.

Mobile devices are currently a very dynamic technology. These personal devices are finding their way more and more often into the school and after-school environment. Tablets these days offer not only the possibility of installing various apps but are also equipped with a camera, video camera and many other sensors, such as gyroscope, GPS and others, and find broad application in teaching.

Nowadays, there exist many different applications directly aimed at supporting geography instruction, such as various types of atlases, maps and navigations showing the real form of the world (e.g. Google Earth). In addition there are many different games for learning the countries, their symbols, mountains, rivers and towns, "treasure hunts" (in nature and towns) in the form of geocaching (using game elements), GPS drawing or applications showing the sky full of stars or even applications for reading books or magazines. Geocaching is a treasure-hunting game where you use GPS to hide and seek containers, called "caches", together with other participants (Geocaching 2000a). Applications that enable connections to various devices, such as telescopes, are very attractive. The applications on a tablet (phones), such as a camera that the students can use to create 360° spherical and panoramic photographs (where students can name the mountains that they see in the photographs, e.g. or they can name the mountains or points of interest they can see from the window) can be created and afterwards placed into Google Maps.

There are also applications using augmented reality (AR[3]) with which we can draw the contours of an island, for example, and then examine its "real" form using a tablet (application LandscapAR augmented reality, Fig. 10.2). The increasing availability of mobile devices surely provides many possibilities for their use in teaching. Their mobility can move teaching even outside the common classrooms, either to virtual classroom or physically outside the school – teaching in nature (even in towns).

---

[3] *Augmented reality is a term used to describe a real-world environment augmented by computer created objects.*

Fig. 10.2 Application LandscapAR augmented reality

10.2.5 Social Networks and Data Visualisation

Social networks are these days one of the main communication channels for the spread of information. Their users can share discovered information or share their own. They can be widely used even for geography. Many organisations and individuals have created their accounts on social networks and they share data and information with actual use in geography.

The most interesting are the new communities focused on the teaching of geography. The most popular social networks are *Twitter*, *Facebook*, *Google + and Pinterest*. The last of these enables its users to create virtual boards where they can pin interesting photographs, pictures or even videos. We can find many teachers on this social network who create boards focused on teaching geography. We can follow either these boards or these teachers and gain new impulses and ideas that we can apply in our lessons. We can also create such boards (or cooperate with other people) and inspire others. The boards do not have to be created only by teachers – the students can also be involved. For instance, the students can create themselves, or in groups, an animal board, plant board, soil horizon board, etc. near the place they live.

10.2.6 Data Visualisation in Teaching Geography

Thanks to digitisation, we have easier access to various real data from different areas. Geography is a science concerned with gaining, analysing, processing and displaying data in the long term. Modern technologies offer another wide use in all of these areas.

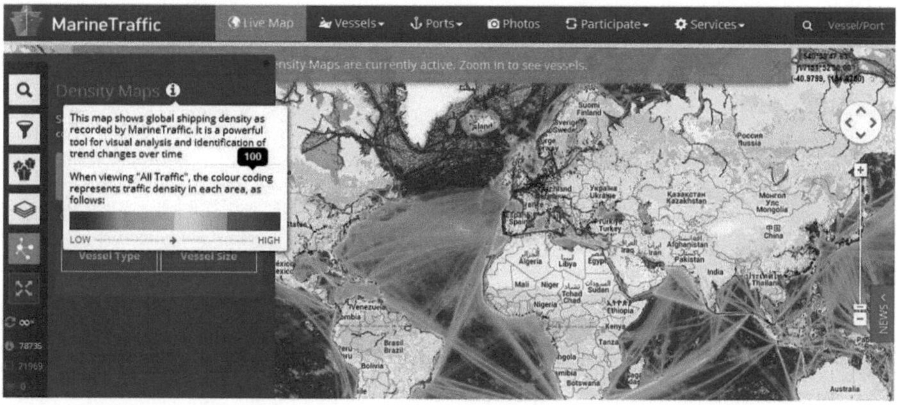

Fig. 10.3 Ship transportation open data visualisation – http://www.marinetraffic.com

These days, we often speak about *open data*[4] (Fig. 10.3) that are subsequently processed using a variety of services and applications, often available on the Internet, into different graphic forms. This includes not only traditional data, such as maps, diagrams and charts, but also the today's more popular infographics. Infographics are a visual form of information publication. They attempt to display the most relevant information in a clearly arranged graphic form.

Geography has a wealth of experience in this area – maps, cartograms and thematic maps. Nowadays technology enables data to be obtained and combined much more easily and, besides statistical outputs, also the use of dynamic interactive outputs. Thanks to online tools, even the students can comfortably work with this data. They can analyse it and combine it; they can even be involved in creating their own infographics.

As indicated above, digital technologies are today strongly integrated into many fields. The school environment should accept this situation and incorporate these technologies into teaching and education. Students come into contact with digital technologies every day and this trend will only continue to increase. Even in geography, we are standing at the very beginning, although it may seem we are ahead, for example, thanks to the GIS. Although we are surely heading in the right direction, it is entirely up to us whether we are able to use the potential and possibilities of these technologies.

[4] *Open data are most often data of public administration, university researches and also individuals. They have to be available online with the least technical obstacles possible (e.g. geographical data, charts, documents), and they have to fulfil the condition of machine readability.*

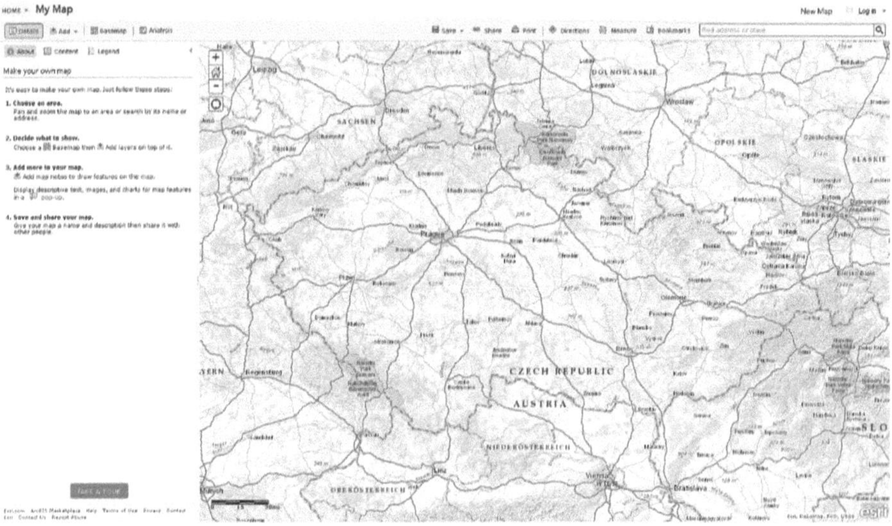

Fig. 10.4 AGO environment with step-by-step guide

10.3 Geography Teaching and GIS in Cloud

ArcGIS Online (AGO) is a cloud geoinformation system (GIS). GIS provides a variety of services on the Internet, whether it is a storage place, the publication of maps and geoprocessing services or even the creation of interactive maps and applications. Emphasis is placed on ease of service and support of effective collaboration of users (Arcdata 2015a). GIS is used constantly in everyday life in most work sectors, such as healthcare, real estate, insurance and government. To date, data have been primarily gathered, and now they will be further processed. We therefore need to involve a greater number of users that are able to access and analyse these geographical data. Currently we are educating only a small number of GIS specialists at universities, so now we are going to focus on primary, secondary and high school education and on the incorporation of GIS into school curricula (Nýdle 2015). Our goal is to gradually make the general public familiar with the GIS tools so people can use them in a simple, clever and effective way. Using AGO in K-12 teaching is therefore a great challenge. This tool, if it is used effectively, can significantly increase students' interest in learning and thus create a positive relationship not only to geography but also to studying in general, because it is an online platform, and nowadays students spend much of their free time on the Internet.

AGO is available via the website www.arcgis.com. The main advantage is the low system requirements. The main requirements are a correct web browser and a stable and good Internet connection. Another big advantage is compatibility with most operating systems. It is suitable for Windows, iOS, Android, etc. In addition, it is suitable practically for any computer, mobile device, smartphone or tablet or can be used for lessons with an interactive whiteboard without the necessity of

	Subscription commercial / non-commercial use	Personal account non-commercial use only
Viewing, sharing, and content management	✓	✓
Insert own data into the maps	✓	✓
Integration of geometric services into maps and applications	✓	✓
Store maps, applications and data in the cloud Esri	✓	Limit 2 GB
Use geocoding services	✓	–
Access to the ArcGIS Portal and Web Mapping APIs	✓	–
Publication of hosted services	✓	–
Esri Maps for Office	✓	–
Manage user roles, approaches and safety	✓	–
Monitoring of services usage	✓	–
Adjusting ArcGIS Online homepage	✓	–
Technical support	✓	–

Fig. 10.5 Comparison of AGO online subscription and personal account (Source: arcdata.cz)

installing any complicated systems. AGO can be used in any language (which has to be set when creating the account). The use is very simple and is accompanied by a step-by-step guide, making it suitable for beginners without prior experience with GIS (Fig. 10.4). Its use for teaching is therefore not difficult at all. The primary technical support in Czechia is provided by the official distributor of Environmental Systems Research Institute (ESRI) products ARCDATA Praha which, among other things, offers free downloads of typical examples and geodatabase ArcČR® 500 on its site. It concerns digital vector geodatabase of Czechia which is created in detail of scale 1:500 000. Its content is well-arranged geographic information about Czechia. For support you can also contact the individual universities or the GeoMentor programme which brings together people who are willing to help with the deployment of GIS (especially AGO) into teaching. The aim of the programme is to create GeoMentor long-term cooperation between geography teachers and geoinformation science experts. More about project is on website www.geomentors.net or http://www.arcdata.cz/oborova-reseni/gis-v-oborech/vzdelavani/Geomentor/.

It is necessary to create an access account. The application allows you to create two types (two levels) of accounts. The first level is an AGO account for the organisation of a full substitute of the desktop version of the ArcGIS programme for desktop (Fig. 10.5). It is offered in an annual scalable subscription adjusted to the size of the organisation and its current needs. The subscription also includes credits, which are a kind of currency of AGO and are used for services such as data storage, creating of map services, spatial analysis and geocoding (Arcdata 2015b). The number of required credits varies depending on the type of analysis. The number of credits needed increases with the extent of the area. In Czechia, annual licensing is necessary at this level. ESRI in the USA for K-12 education releases this licence for free, including education materials (ESRI 2015). The second level is the public (personal) account. This level operates with reduced functionality. It allows you to view data and maps or create your own data and maps, but it cannot be used for analysis. This account can be created free of charge and, as it is less complex, is best for teaching purposes.

AGO is a useful tool not only in geography but also in science, history or biology classes. In fact, almost every field of study is concerned with a subject which is situated on the Earth's surface and which carries some information. Using AGO we can visualise it and show it in a wider geographical context. An example from a history class is the invasion of the Huns in Europe or the political map of Europe in various stages of development, from Samo's Empire to the Cold War. It is a good supplementary tool for interdisciplinary thematic units. Finally, it allows you to instantly react to the latest sociopolitical development in the world so you can provide the students with the latest information in a broader context (Donert 2014). This is especially important since the most recent events will be recorded in maps or textbooks only much later. Recent examples include the crisis in Ukraine, ISIS or disasters such as the spread of Ebola in Africa.

A key element of the platform is the possibility of sharing. In addition to the classic sharing of the map outputs via social networks like Facebook or Twitter, it allows teachers or students to create their own groups. You can, for example, create groups for each grade. You can add maps and map data into these groups from which students can then draw inspiration or they can make their own contribution. The entire content can be used for further study, homework or testing. Groups and applications can be viewed anonymously. A very attractive option for sharing is the web mapping application that can be created with the help of preset templates. These applications can be complemented by a wide range of audiovisual content, such as YouTube, Flickr, Picasa, Twitter, etc. Another possibility is map embedding directly into the web pages. AGO generates a simple HTML code which can be inserted into the page's source code. For example, you can very simply present the work of students on the school's website.

The following example describes how you can teach a regional Czech geography class at the eighth grade of elementary school (age 13–14 years). These are general principles. Particular steps (manuals), how to create map or application or creating and managing groups, are available on various websites. Official websites are http://

video.esri.com/, http://www.arcdata.cz/produkty-a-sluzby/gis-on-line/arcgis-online/ and http://resources.arcgis.com/en/Tutorials/.

The teacher (administrator) creates a group called "population and regions". Then, he/she loads his/her own content into the group, for example, an application about large Czech agglomerations, national parks, mountains, blank maps, etc. Through the external map servers, the teacher can make other interesting material accessible, such as religion and districts, or he/she can create the content from the textbooks. This content can be used in the lessons on the interactive whiteboard (projector), along with traditional maps. For example, he/she can easily show big cities in detail in the form of videos from YouTube. It will significantly increase students' interest, or, at least, it will draw their attention so they do not disturb others. The same method can be applied to national parks or districts. If there are tablets available in school, students can themselves open the map and work together with the teacher. The generated content can be used either on the interactive whiteboard or in print by other teachers in the interdisciplinary topics such as Education to Thinking in the European and Global Context, Environmental Education and Media Education. The most interesting projects can be displayed on the school website, where parents can see them. Students (users) use the contents for studying, for free-time activities or for homework or tests. As a homework they can, for example, create a map diary application called My Holiday Tour of South Moravia. They can share this app via their own profile, social network or their own website, so that classmates or friends can get involved in the topic. Ideally, while spending their leisure time on the Internet, they will themselves improve their maps and applications, thus further educating themselves at home and creating a stronger link to geography.

Several schools in Czechia, especially grammar schools, have already begun using GIS in education. Specific examples are listed in the website of ARCDATA Praha. Large-scale use is mostly hampered by teachers' low qualification and the large upfront costs of the desktop software. AGO could serve as a first step for the further development of this field because of its simplicity and zero cost.

10.4 Geocaching in the Teaching of Geography

Several key topics can be covered using geoinformation technology. In addition to geographic information systems, it is the use of navigation positioning systems which, thanks to processing of signals from Earth satellite orbits, enables the exact geographic location to be obtained of an individual who is equipped with a device that can process the signals. The American Global Positioning System (GPS) is one of the most commonly used systems. This service is available to the public free of charge; all one needs is a receiver with a GPS chip and a suitable application. It is possible to establish your exact location using geographic coordinates, which led to the development of a game called geocaching in 2000. This game has become very popular in Czechia over the last 15 years. Thanks to its concept, the game attracted the attention of some geography teachers, who found it an interesting tool to use when teaching geography.

10.4.1 Geocaching in Czechia

Geocaching is a worldwide game, combining tourism, sport, entertainment, adventure and the innate desire to discover new things. The game is based on boxes (so-called caches), hidden at interesting places around the world for which we know the geographic coordinates. In order to find the boxes, GPS navigation or smartphone applications are used.

The game is dynamically growing. Every day new caches are added and new players get involved. The interaction of players, the willingness to discover new places and the loyalty towards the community are essential. Each player should act inconspicuously, so as not to be seen by nonplayers during the search for a cache or when hiding the cache (Dvořák 2014). The game has exact rules for how to proceed when you find the cache and how to register your find on the joint server. The actual caches can have different appearances and are classified by several different characteristics. The community of players has created their own vocabulary. They have been organising special focus meetings and have created a market for special items and accessories developed just for the game.

The Czech Republic is repeatedly mentioned as a country with a very active base of players (Geocaching 2000b). This is demonstrated by the growing number of caches and players in Czechia as well as by the number of specialised websites. Due to this fact, the game has become a suitable teaching method for introducing students to the principles of GPS navigation and some of the topics taught in geography lessons in Czechia at primary and secondary schools.

10.4.2 Geocaching in the Teaching of Geography

Using geocaching in geography classes is still a relatively new concept in Czechia but it is finding its way into education. There are many questions linked to the use of geocaching for teaching. However, after overcoming the initial stage, teachers find the game an interesting motivational tool (Ančincová 2013). The game can make learning really enticing not only for primary school pupils but also for grammar school and university students. From the teacher's point of view, the game helps teach new facts, gain experience and improve students' skills. Students must search for the required information on the Internet, which teaches them new practical skills for using GPS devices in the field, and they must be able to find their way in the field and have good spatial orientation. Caches are located at various points of interest linked to historical or ethnographical events or at places that are interesting in terms of physics, geography or socio-economics. Students learn new information during the search or from the contents of the boxes. Players/students acquire new facts, skills and knowledge about the local region, and the whole process is strongly connected to very personal and unique experiences when searching for the caches. This helps students remember what they have learned. According to Marada (2006), the experiences that the students gain in the "real" world can help retain knowledge and skills, making the learning process far more efficient.

By using geocaching to teach geography, students can learn about the local region. The learning process combines students' emotions, everyday places, lived experience and location, which are terms used, for example, by Hynek (2009) and Vávra (2009, 2010). The emphasis should not only be on learning and discovering the local region during the learning process but also, as stated by Vávra (2009), on the emotions, attitudes and metacognition, i.e. the "internal" world of the individual/student, which are created during the learning process thanks to geocaching.

Another argument for using geocaching for teaching is the fact that, according to Petty (1993), geocaching uses basic principles to motivate students: *students are immediately rewarded for* learning something by finding a "treasure". The *reward comes immediately after providing the correct response*, i.e. if GPS is used correctly, if students demonstrate good spatial orientation and are observant. Furthermore, *the learning results improve thanks to repeated success.* Students improve their performance in using the GPS navigation, the number of found caches increases and *the past success* (the found "treasure") *motivates students to "learn" more.* The principle of geocaching and the game itself meet the definition of inadvertent learning (Petty 1993), when students are learning without really knowing.

10.4.3 Integration of Geocaching into Educational Process

Geocaching cannot be seen as only hunting for caches. It might be the case that individual caches available on an official server are located far away from one another, or they are not at places which would be suitable for the lesson (e.g. city park, forest). Therefore, it is essential to choose a suitable location where there is no potential danger. The coordinates of such a location must be properly identified, and suitable activities and tasks must be designed for students who will be searching for the caches. Students' age, skills and knowledge must be taken into consideration when planning the work with GPS and when planning the tasks. It is important to determine the objectives of the lesson that can be achieved on the respective route. The authors identified three types of integration of geocaching into the learning process based on how long and how difficult it will be to prepare the game. These can be whole-day trips (lasting 6–8 h), half-day trips (3–4 h) or independent work that can be briefed and evaluated within 1–2 h (Dvořák 2014).

Three geocaching activities were designed as a case study. These activities are listed in the table below, showing the name of the activity, its basic characteristics and duration and the age group for which it was designed (Table 10.1). All activities were situated near the town of Dačice, which is in the southern part of Czechia. It is

Table 10.1 Geocaching activities

Title	Characteristics	Duration	Age group
In the footsteps of robber Grasel *(find out who robbed the castle treasury)*	The activity is located in Česká Kanada (Czech Canada), a natural park, and it focuses on learning about the specific living conditions in this region. The activity covers an 8 km circular route	**6–8 h**	**12–15 years**
Slavonice, a Renaissance pearl	The aim of the activity is to learn about the historical importance of the town of Slavonice. The emphasis is on the historical development and specific living conditions in the countryside near the Czech–Austrian border	**6–8 h**	**14–15 years**
Dačice geocaching	The activity focuses on the town of Dačice, its historical sites, events and personalities. The aim is to learn about the development of the town in this rural area	**3–4 h**	**12–15 years**

a small town of 7500 (ČSÚ 2015) inhabitants with no major regional city nearby. Therefore, from a teaching point of view, this region can be considered a rural area.

A detailed worksheet and methodology sheet, a timetable and educational characteristics have been prepared for each of the activities. The educational characteristics are based on determining the educational attributes of the activity that can have an impact on the progress of the activity, its contents and the target groups. The activity **In the footsteps of robber Grasel *(find out who robbed the castle treasury)*** is specified as follows. Cross-sectional topics, covered by this activity, have been identified. These topics have their special place within the Czech curriculum, as they represent a combination of several traditional subjects and they help to view the world within a much wider and more complex context. The activity comprises the following cross-sectional topics: Personal and Social Education, Education of Democratic Citizenship, Media Studies and Environmental Education. The activity also specifies individual subjects and competencies. In this case it is geography, history and natural science. Students are required to master the following skills: GPS navigation, spatial orientation, teamwork and basic mathematical operations (addition, subtraction, multiplication). Last but not least, specialist educational goals were determined which should be achieved through the activity. These include the development and furthering of cooperation skills, developing students' feelings towards their region, furthering students' knowledge of the local region, incentives for further learning, testing technical skills and working with GPS navigation systems.

10.5 Conclusion

The authors had the opportunity to implement some of the activities at the Božena Němcová Primary School in Dačice in May and June 2014. The whole-day activity (lasting 6–8 h) "In the footsteps of robber Grasel" was organised during a project day to celebrate Earth Day. Two classes ages 14–15, in total 48 students, took part in this activity. The activity was briefly introduced, where the basic principles of geocaching work with GPS and the specifics of the activity were presented. Students were divided into six groups and were issued worksheets and GPS devices. After a brief refresher on how to use the devices, the groups started to search for the caches and perform their tasks.

The results of the study can be summarised as follows. None of the students had problems understanding how to work with the GPS navigation. No group found it difficult to find the locations and solve the tasks on their worksheets. The only technical problem was spatial orientation. The responses of the majority of the students were very positive. It was amazing to see how quickly the students mastered GPS navigation and how interested they were in the game. Their enthusiasm and desire to find the hidden boxes put the organisers into the role of coordinators whose task was to manage the search in the right direction. It was also interesting to see the atmosphere in the class and the student–teacher relationship. Before the start of the activity, the organisers were informed about groups of students who might be problematic as they viewed the activity as boring. Contrary to expectations, however, these students in fact were the most active and were the most eager to share their experiences from the search. This shows that during geocaching, even weaker students (or those who are considered "weaker" students during the regular learning process) had a chance to show their skills.

Based on the case study and practical verification of the selected activity, using geocaching to learn about the local region can prove very successful and is advisable. The main advantages are as follows: easy motivation, interesting and original learning process, application of learning in the field, teamwork, etc.

In conclusion, it must be stated that geoinformation technology serves mostly as a technical and support tool that enables the game. The students do not learn specific technical details about the functioning of the GPS system. To paraphrase, this is a "learning process with geoinformation technology and geocaching" (Sui 1995). In view of the complexity and diversity of topics that can be introduced during the game, it can be discussed whether this tool should be used during geography lessons. The study demonstrated that the biggest strength of geocaching and its principles was the fact that it motivated students and aroused their interest in their immediate surroundings. These surroundings are a reflection of their natural and social environment and their mutual interactions, which are exactly the principles analysed by geography looking at the spatial "imprint". This clearly justifies why this tool should be applied within Czech education in the teaching of geography, as geography is a subject with very strong multidisciplinary features.

References

Ančincová, D. (2013). *Metodické kabinety pro DVPP v Karlovarském kraji* [Methodical Cabinets for DVPP in Karlovy Vary Region]. Geocaching ve výuce. Krajské vzdělávací centrum, http://metodik.kvcso.cz/view.php?cisloclanku=2013020002. Accessed 18 July 2014.

Arcdata PRAHA. (2015a). *Arcgis Online*. http://www.arcdata.cz/produkty-a-sluzby/gis-on-line/arcgis-online/. Accessed 31 July 2015.

Arcdata PRAHA. (2015b). *Licencovani Arcgis Online* [Licensing of Arcgis online]. http://www.arcdata.cz/produkty/arcgis/arcgis-online/licencovani/. Accessed 31 July 2015.

Baker, T. R. (2005). Internet-Based GIS Mapping in Support of K-12 Education. *The Professional Geographer, 57*(1), 44–50.

Baker, T. R., Palmer, A. M., & Kerski, J. J. (2009). A National Survey to Examine Teacher Professional Development and Implementation of Desktop GIS. *Journal of Geography, 108*(4), 174–185.

Bartha, G. (2009). Objectives of GIS Teaching in Higher Education Developing Experts or Training Teachers ? In K. Donert (Ed.), *Using Geoinformation in European Geography Education* (pp. 92–101). Roma: Società Geographica.

ČSÚ – Český statistický úřad. (2015). Počet obyvatel v obcích k 1. 1. 2015 [Population of municipalities to 1. 1. 2015]. https://www.czso.cz/csu/czso/pocet-obyvatel-v-obcich-k-112015. Accessed 19 Aug 2015.

Donert, K. (2014). Building capacity for digital earth education in Europe. In R. de M. González & K. Donert (Eds.), *Innovative learning geography in Europe: New challenges for the 21st century* (pp. 9–20). Newcastle upon Tyne: Cambridge scholars publishing.

Dvořák, J. (2014). *Využitelnost geocachingu ve výuce zeměpisu na 2. stupni ZŠ* [Usability of geocaching game in education of geography for 2nd stage primary school students]. Diploma Thesis. České Budějovice: Departemen of Geography of Faculty of Education of the University of South Bohemia.

ESRI. (2015). *Thinking spatially using GIS*. http://edcommunity.esri.com/Resources/Collections/thinking-spatially---ago. Accessed 8 Aug 2015.

Flubaroo. (2015). *Overview*. http://www.flubaroo.com/. Accessed 9 Oct 2015.

Geocaching. (2000a). *The official global GPS cache hunt site*. http://www.geocaching.com/. Accessed 9 Oct 2015.

Geocaching. (2000b). *The official global GPS cache hunt site*. Geocaching Fact Sheet. http://www.geocaching.com/articles/Brochures/footer/FactSheet_Geocaching.pdf. Accessed 18 July 2015.

Google. (2015). *Street view*. http://www.google.cz/intl/cs/maps/streetview/. Accessed 9 Oct 2015.

Hynek, A. (2009). Prostorovosti: místa, krajiny, regiony [Spatialities: Places, landscapes, regions]. *Acta Geographica Universitatis Comenianae, 52*, 75–86.

Kerski, J. J. (2003). The implementation and effectiveness of geographic information systems technology and methods in secondary education. *Journal of Geography, 102*(3), 128–137.

Marada, M. (2006). Jak na výuku zeměpisu v terénu? [How on the teaching of geography in the landscape?]. *Geografické rozhledy, 15*(3), 2–5.

Nýdle, J. (2015). *Arcgis pro recenze* [Arcgis for reviews]. http://www.gisportal.cz/2015/03/arcgis-pro-recenze/. Accessed 5 Aug 2015.

Petty, G. (1993). *Moderní vyučování* [Modern education]. Praha: Portál, s. r. o.

Sui, D. Z. (1995). A pedagogic framework to link GIS to the intellectual core of geography. *Journal of Geography, 94*(6), 578–591.

Vávra, J. (2009). Místo (place) v regionální geografii a v geografickém vzdělávání [Place (Locality) in the Regional Geography and Geographic Education]. *Acta Geographica Universitatis Comenianae, 52*, 119–127.

Vávra, J. (2010). Jedinec a místo, jedinec v místě, jedinec prostřednictvím místa [An Individual and a Place, an Individual in a Place, an Individual through a Place]. *Geografie, 115*(4), 476–478.

Chapter 11
Reading Satellite Images, Aerial Photos and Maps: Development of Cartographic and Visual Literacy

Hana Svatoňová

11.1 Introduction

The accessibility of servers, websites and the Internet has allowed the public to discover the world of aerial and satellite imagery. The most noticeable source of such information is Google Earth, which, together with Virtual Globe, NASA World Wind and ESRI's ArcGIS Explorer, has become extremely popular for educational and noneducational applications (Schultz et al. 2008). Adults and children are now confronted with new topographical materials, such as aerial and satellite images. While people have been using maps for more than 1000 years, high-resolution digital imagery represents a progressive option and a new experience for viewers. Although both maps and aerial images present a bird's-eye view of the earth, aerial images are not maps. Maps and images differ in many aspects, and therefore the reading of an image differs from the reading of a map. A map is a representation of the earth's surface supplemented with annotations and provides geometrically accurate shapes and distances. Aerial and satellite images contain no explanatory text and have certain distortions which need correction. Although people have just started to discover the advantages of surface imagery of the earth and gain experience in utilising such imagery, this method already has taken its place in daily life. Results of research among children and teenagers indicate an extremely rapid increase in popularity of aerospace imagery supported by development of new technologies, such as smart phones, tablets, laptops and social media. This situation raises questions as to how young people (in this study 11-, 15- and 19-year-old students) can handle various types of images and tasks, if there is some variation in the efficiency of map and image interpretation, and how children judge the images.

H. Svatoňová (✉)
Faculty of Education, Department of Geography, Masaryk University in Brno,
Poříčí 7, 639 00 Brno, Czech Republic
e-mail: svatonova@ped.muni.cz

P. Karvánková et al. (eds.), *Current Topics in Czech and Central European Geography Education*, DOI 10.1007/978-3-319-43614-2_11

Therefore, this research presents the results of a study focused on visual and cartographic skills, as well as the development of imagination at various stages of pupils' growing maturity.

11.2 Image Is Not a Map

Prior to the testing of the students' ability to read and interpret maps and images, it was necessary to clarify the basic differences between these documents. Aerial or satellite images differ from maps of the same area in many aspects, and this fact could affect the interpretation of the content. This issue was described in detail by Kovařík (2012).

An image depicts the real situation of the landscape up to the limit of the camera or sensor's resolution. However, as the images show all the features in detail, it can be difficult to distinguish important objects from those less important, and some features can be completely invisible. An image cannot contain any names, spot heights or grid lines. One single image cannot provide real information about landscape profile. Interpretation aids must be used to identify the features in an image. The fundamental tools for visual image interpretation are as follows: tone, texture, shadow, pattern, association, shape, size and site.

Images show real features, whereas maps depict features using cartographic symbols. The interpretation of images is therefore straightforward, without the necessity to work with a legend as is the case of maps. Images can be classified according to various criteria. Considering the research described in this paper, images can be classified according to viewing angle and colour and by how up to date they are. Based on the viewing angle, images can be classified as oblique or vertical. Based on the colours used, we distinguish between true-colour images, not-true-colour images and black and white images. In terms of date, the images can be recent or historical.

Different types of images place different demands on their readers. Reading orthogonal images is unusual for non-expert readers. The orthogonal view, without information about the object and sidewall height or a landscape profile, is more difficult to interpret compared to the oblique view. The oblique image is more familiar to non-expert readers. It provides a more familiar view of the depicted area. However, a big disadvantage of oblique images is shape deformation and variable scale. The viewing angle of orthogonal images directly affects the use of interpretation aids such as shape, size, site and shadow.

Colour (or a tone) represents another important image interpretation aid. Black and white images generally appear to be more difficult to interpret. In the case of colour images (both true colour and not true colour), the method of interpretation depends on the aim of the particular task and the need to identify the features as having various physical or chemical properties.

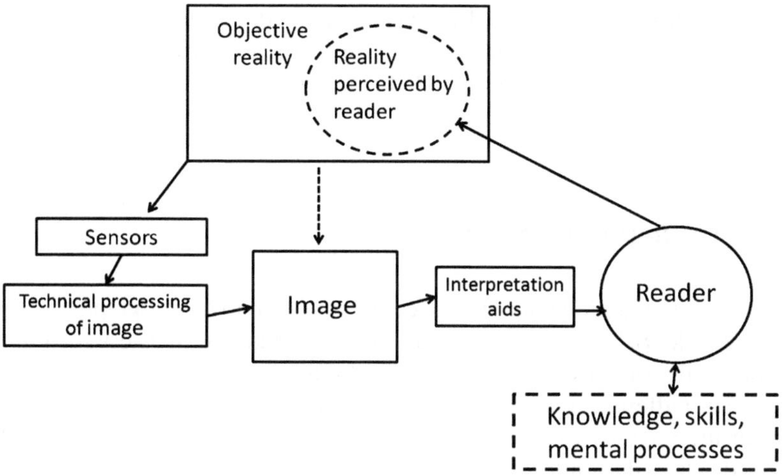

Fig. 11.1 Model of image reading (Source: Author)

11.3 Images as Models of Communication

According to a theory of information technologies, a map and an image can be viewed as a means of communication. The fundamental components of that communication system were presented by Robinson and Petchenik (1977). MacEachren (2004) states that the first models of cartographic communication were introduced by Board in 1967 and by Kolacny in 1969. Particularly, the model of Kolacny inspired us to create the "image-reader" communication model which is shown in Fig. 11.1. Similarly as in the model of Kolacny, the reader creates a mental representation of a reality on the basis of both the received information depicted in the image and his or her previous experience. Such representation is identical with reality to some extent. Unlike map reading, where the communication medium of the process is represented by a map, the image does not offer the subjective view of a cartographer and the use of a coded language. The coding of information in the image can be associated with a colour of the images or with the image classification. It is possible to generate a hybrid product, known as an image map, where the image also contains coded information in the form of cartographic symbols, lettering, etc.

11.4 The Reading of Images vs. the Reading of Maps

The reading of images and maps belongs to cartographic and visual literacy. The International Visual Literacy Association defines the term "visual literacy" as the skill of receiving and evaluating information using a visual medium. According to Pravda (2001), "cartographic literacy" comprises reading, using and creating maps. It can be divided into natural and acquired cartographic literacy. When working with aerial and satellite images, both visual and cartographic literacy are needed. When preparing the experiment, we built on theoretical works and research concerning cartographic skills that included reading maps and, in a broader context, reading images.

11.4.1 Cartographic Skills

The term "cartographic skills" is a specific expression of cartographic and visual literacy. Van der Schee (2000, 218) states that the degree of mastery regarding cartographic skills corresponds to the amount of information that we are able to obtain from a map. Differences in the mastery of cartographic skills in pupils are associated with their age as well as with their own practice, i.e. cartographic skills are developed when the children are working in the classroom with maps and the teacher pays particular attention to these activities (Gerber 1981, 128). According to Catling (1978), children are able to use either an existing map or their own cognitive map at an early age, but their map-reading skills are at a low level. Wiegand (2002, 19) cites three options of using maps, each requiring different skills and leading to the acquisition of different information: (1) searching for particular features and determining their locations in relation to other elements, (2) using the map to find a route and (3) using the map to solve problems with an emphasis on the spatial distribution of phenomena and spatial relationships. Catling (1978) and MacEachren (2004) interpret the concept of cartographic skills from two perspectives: (1) as children's ability to understand the spatial arrangement of their own surroundings and the related solving of problems, such as finding an appropriate route and (2) as children's skill to read and interpret information using thematic maps. Wiegand (2002) divided cartographic skills into three groups: (1) decoding maps (reading, understanding and interpreting), (2) assessing maps and (3) creating their own simple maps. Van der Schee (1987) offers a simple definition of cartographic skills as being map reading, map analysis and map interpretation.

11.4.2 Skills of Working with Aerial or Satellite Images

Analogously, the skills of working with aerial or satellite images comprise reading, analysing and interpreting images. The elements of visual image interpretation – tone, texture, shadow, pattern, association, shape and size – are routinely used when interpreting aerial photographs or satellite imagery (Lillesand et al. 2008).

Children, from 3 years old, begin to understand aerial photographs as a representation of real spaces (Blaut 1997), and their understanding continues to develop into adolescence (Downs and Liben 1997; Liben and Downs 1992). Liben and Downs (1989) and Liben and Yekel (1996) described preschoolers' and older children's understanding of maps and map symbols. Liben (1999) proposed a developmental trajectory for understanding spatial representations. Muir and Blaut (1969) tested 5- and 6-year-old American children with vertical black and white aerial photographs and found that children taught about aerial photographs demonstrated map interpretation abilities above those of untaught peers. Stea and Blaut (1973) asked Puerto Rican 5-year-olds to interpret vertical aerial photographs centred on the children's school. They concluded that the children had no trouble reading the content of the photographs. Blaut et al. (1970) tested 5- to 7-year-old children using black and white vertical aerial photographs at scales of 1:2000 and 1:3000, respectively. The children identified the features and then were asked to trace a map on acetate identifying features from the photograph. They also had to plot a route on their maps. Most children successfully completed the task. These studies concerned children before they had attended school or in the first years of primary school. Kim et al. (2012) investigated whether young children possess the potential to understand map-like representations using aerial photographs, how scale affects children's performance and whether children show interest in and enjoy working with spatial representations. Three remote sensing images of different scales were employed to examine children's ability to interpret spatial representation. The results indicate that young children have the ability to use spatial representation. Most participants (Korean pupils) were able to understand aerial photographs. Adults' visual interpretation was also analysed by Lloyd et al. (2002) and Van Coillie et al. (2014). Lloyd et al. (2002) investigated how people process information from aerial photographs to categorise locations. Three cognitive experiments were conducted with human subjects viewing a series of aerial photographs and categorising the land use for target locations. Van Coillie et al. (2014) analysed the accuracy of image digitisation performed by adults with various degrees of experience regarding processing images and various degrees of motivation. Digitising accuracy varied strongly across monitored participants. Moreover, it was stated that, generally, accuracy was very poor. Svatoňová and Rybanský (2014) evaluated various types of landscape visualisation with respect to reality perception. Testing of research participants proved that the most effective visualisations are simulated flights over the country and 3D visualisations using orthogonal images.

Cartographic skills and the skills of working with aerial or satellite images change over time. They can be developed and are related to the development of geospatial thinking. Schuit (2011) studied the use of exercise cards in a new way of teaching the reading of topographical maps. Lee and Bednarz (2012) demonstrated the efficacy of a set of methods which accurately identify the cognitive mapping strategies of individuals. The relation of geospatial thinking and vocabulary was investigated and experimentally verified by Golledge et al. (2008). Lee and Bednarz (2012) performed an analysis of tests focused on spatial thinking using factor analysis that employs principal components.

The purpose of the study described in this paper is to investigate, analyse and compare the efficiency of reading images and maps, the ability of adolescent children to read images in relation to colours used and the subjective opinions of research participants on the difficulty of reading various types of images and their preference for images or maps as a source for acquiring information.

11.5 Research

The research focused on three main areas:

1. A comparison of visual interpretation efficiency regarding aerial images and topographic maps of the same area
2. A comparison of visual interpretation efficiency for true-colour satellite images and not-true-colour satellite images
3. An evaluation of the subjective perception of reading difficulty for various types of aerial and satellite images, preference for maps or images and respondents' experience with use of images in an education process

Differences in visual efficiency regarding reading aerial and satellite images and the subjective evaluation of maps and images related to students' gender and age were briefly analysed as well.

Research respondents solved selected spatial (and identical) tasks in pairs of documents (imaging and map, true-colour image and not-true-colour image). Part of the task was identical for all respondents: to evaluate landscape development based on a pair of satellite images acquired at different points in time, to identify objects in historical aerial images, to evaluate the materials subjectively and to specify a personal preference for images or maps.

According to Catling's research (1978), images of areas known to the respondents were used for several of the test questions. Specifically, some tasks involved identifying objects in large-scale topographic maps and aerial images which captured the vicinity of the school. Other tasks focused on the identification of types of objects and land use defined in satellite images with a scale 1:100,000 taken of environments neutral (i.e. unknown) to participants.

11.5.1 Research Questions

The main research questions, with sub-questions, were formulated with respect to the following research objectives:

1. Is the **interpretation** efficiency identical **for aerial images and maps** with the same scale and showing the same area?
   (a) Is the identification of objects in topographic maps more successful than the identification of objects in images?
   (b) Is the number of objects identified in maps higher than the number of objects identified in images?
   (c) Is there some variability of the identification score (maps, images) related to the participant's age and gender?
   (d) How successful were participants in solving spatial tasks using historical black and white images showing the same landscape in the 1950s?

2. What is the difference in the **visual interpretation of true-colour satellite images and not-true-colour satellite images**? Is the identification score for types of land use and the purpose of selected objects identical for true-colour satellite image and not-true-colour satellite images?
   (a) Are the respondents more successful in the identification of elements in true-colour images or not-true-colour images?
   (b) What types of land use are distinguished better in true-colour images, and what types are easier to identify in not-true-colour images?
   (c) How well can the students interpret satellite images taken at different times?
3. What is the **subjective evaluation of participants concerning the difficulty of visual interpretation** of documents? How frequently do they work with images? Do they like this activity? Do they use such documents at school?
   (a) How do participants evaluate the difficulty of reading aerial images and maps? Which type of documents do they prefer?
   (b) Is there any difference between girls' and boys' scores? How does their preference change with age (regarding maps, images)?
   (c) What are the participants' subjective evaluations of reading true-colour satellite images and not-true-colour satellite images?
   (d) Do they work with satellite images at school and during leisure time?
   (e) Is this activity interesting and amusing for them?

11.5.2 Test Preparation

Considering the research questions, the test consisted of three parts. The first part of the test contained **tasks focused on the evaluation of efficiency regarding the interpretation of aerial images and maps**. The second part focused on **evaluation regarding the efficiency of interpretation of satellite images**, and the third part of the test contained questions concerning the participants' **subjective evaluations**. The documents used were sorted into two groups, and two equivalent test sets were created: version A and version B.

Seven questions were aimed at **objective differences in the interpretation of images and maps**:

1. Find identified objects in the aerial image (orthogonal, true colour) and in the topographic map of the same area, with the same scale.
2. Find and plot described objects; draw the shortest path from the station to school in the aerial image and in the map.
3. List the objects contained in the map and the aerial image.
4. Plot the school building in the historical aerial image. Write down what was at this school's location 60 years ago.
5. Specify which types of land and objects (water area, river, forest, field, built-up area) can be found in the not-true-colour image and the true-colour image.

6. Match the map cut-out into the appropriate segment in the not-true-colour image and the true-colour image.
7. Analyse two satellite images taken of the same locale 20 years apart and describe how the mining area and its vicinity changed.

The next five questions were aimed toward the **subjective evaluation of participants**:

8. Do you find working with images interesting and amusing?
9. Do you look at images on the internet in your leisure time?
10. Do you work with images at school?
11. Evaluate the difficulty of reading various types of images and maps.
12. Which do you prefer: maps or images?

11.5.3 Test Documents: Maps and Images

Images and maps available in the map server "Národní geoportal INSPIRE" (Geoportal 2015) were used as a main source to ensure the availability of images for participants and researchers. Cut-outs from images taken by the LANDSAT satellite were used to test work with satellite images. The colour combination labelled "742" was selected for not-true colours, which allowed for good recognition of water areas, rivers and forests. The colours of the LANDSAT images combined with the red, green and blue (RGB) "742" option show water areas in blue or black, water streams in blue and forests in green. Fields are displayed by a pink-green mosaic and built-up areas appear in violet-pink. Moreover, a cut-out from the historical (1950s) black and white mosaic of aerial images showing the required area was used.

Test documents:

- Basic topographic maps, scale 1:10 000
- Orthogonal, colour, aerial images, resolution 0.5 m
- True-colour satellite images (LANDSAT 7, RGB "321"), scale 1:100,000, resolution 30 m.
- Not-true-colour satellite images (LANDSAT 7, RGB "742"), scale 1:100,000, resolution 30 m.
- A pair of satellite images from the LANDSAT satellite (RGB "432") dated 1984 and 2005.
- Historical aerial black and white images from the 1950s, cut out from a mosaic.

11.5.4 Participants

Research was performed with students ages 11, 15 and 19. In total, **378 students** participated in the research: 198 boys and 180 girls (Table 11.1). Students ages 11 and 15 were educated at primary school. They were members of eight different

Table 11.1 Research participants

	Age		
	11	15	19
Total number	108	134	136
Girls	59	61	60
Boys	49	73	76
School and class	Elementary school, sixth year	Elementary school, ninth year	Bachelor studies, first year
Number of classes	4	4	2

classes and two different schools. One teacher conducted all the lessons in each school. The 19-year-old participants were first-year university students (teaching of geography, Masaryk University, Brno, Czechia). Ninety percent of the students studied at grammar school in the past. Two subsequent classes whose lessons were conducted by one teacher were tested.

11.5.5 Testing Process

Testing was performed at the schools in the spring and autumn of 2013. Research questionnaires were distributed as individual tests for students. The purpose of the test and organisational instructions was presented to students at the beginning. No time limit was specified; however, students completed the test in an average time of 20 min. Any ambiguities were answered during the test. With respect to the test form and task specification, the score was assigned manually. The answers were evaluated by one person.

11.5.6 Test Evaluation and Scoring

The following subchapters contain results and scores achieved in the test.

11.5.6.1 Evaluation of Results: Comparison of Objective Efficiency Regarding Object Identification in Aerial Images and Topographic Maps

The following conclusions can be made based on the evaluation of tasks focused on the **identification of particular objects in aerial images or in maps**:

- 64–80 % of objects were successfully identified in the aerial images; 53–70 % of objects were successfully identified in the maps (Fig. 11.2).

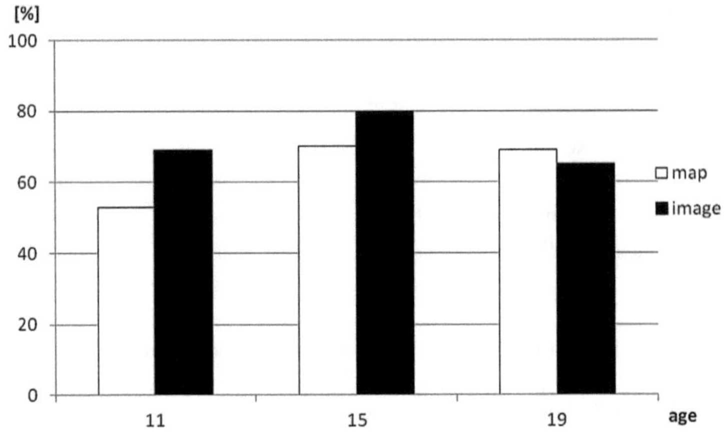

Fig. 11.2 Aerial image and map – reading efficiency in relation to participants' age

- The older the age of the students, the lower the difference between image and map processing efficiency. The highest difference was detected in the group of 11-year-old students. Among the 19-year-olds, the efficiency of map reading was slightly higher (+4 %); see Fig. 11.2 for more details.
- The results revealed interesting information when compared according to gender. Both genders solved tasks with similar efficiency (difference within 5 %). However, a detailed analysis of map-reading efficiency showed significant differences between genders. In the 11-year-old students' group, the girls achieved better results (difference 10 %), but boys were more successful in the 15- and 19-year-old students' groups (+12 %; +16 %). Differences in map reading between genders were also confirmed by Chang and Antes (1987). That article evaluated males as better at map reading.
- Participants were asked to write down objects which they were able to identify in areas displayed in the map or image. The document contained ten objects to identify. Participants identified more objects in the image than in the map (4.9–6.7 objects in map vs. 3.8–5.6 objects in map). This observation corresponds with a better score for object identification achieved in aerial images. The amount of identified objects slightly increased with participants' age. This was in line with our assumption that abstract thinking develops with age.
- The results of plotting specified objects and the shortest path on maps or aerial images correspond with the evaluation of object identification. Younger students can better read aerial images (+15 % and +16 % for images); 19-year-old students are better at reading maps (+21 % for maps). Map-reading efficiency again shows a slightly increasing trend if put in relation to the participants' ages (results for age groups ascendantly 48 %, 70 % and 81 %, respectively).

Besides actual colour aerial images, the **solving of spatial tasks in historical, black and white aerial images showing the area 60 years ago** was included in the test as well. Students were asked to plot the school building and several old and newly established streets in the historical aerial image. The students were asked to find out what was originally located on the school site. Actual aerial images with identical scale and street names were provided as a support document to help solve these tasks.

- Generally, the participants achieved scores of 66–76 % on average. It should be noted that the scores varied with the participants' age. Participants with high scores (min. 85 %) identified the location of their school in the historical image. They also discovered that there were fields in this location in the past. They could plot the neighbouring streets in the vicinity with lower accuracy (ca. 50 %). Participants identified the school building and objects in the vicinity very well (up to 90 %), but more distant objects were more difficult for them (ca. 40 %).
- The achieved scores were very similar across ages and genders (between 66 % and 76 %). Better results were expected from the group of 19-year-old students. Their performance could have been affected by the fact that they solved the task in an area unknown to them, while the 11- and 15-year-olds solved the task in the vicinity of their school. Thus, we believe that knowledge of the area increases the chance to identify the object/element in images showing the same area under significantly different conditions.

Participants were also asked to provide **their personal subjective feelings concerning the difficulty of reading aerial images and maps**. A 1–5 scale was defined for this purpose (1, very easy; 5, very difficult). The results and chart in Fig. 11.3 show that:

- The students consider maps and colour aerial images well readable.
- The perception of difficulty changes with age. The older the participant, the better feedback concerning map readability and the better the assigned score. Eleven-year-old students achieved an average score of 2.28; 19-year-old students achieved an average score of 1.48. The opposite trend was observed for aerial images: 11-year-old students achieved an average score of 1.75 and 19-year-old students achieved an average score of 2.52. Fifteen-year-old students achieved similar scores in maps and aerial images, which means that they consider the difficulty of reading them to be equivalent.
- All the age groups stated that the black and white image was significantly more difficult to read. Scores here varied from 3.48 (min) and 4.05 (max). The highest scores (worst) were achieved by the 19-year-old students. This indicates that the colours in the aerial image are of significant interpretive help and increase the chance of identifying the objects in the image.

Subjective preference (Fig. 11.4) – selection of maps or images for obtaining needed information showed that:

- 19-year-old students prefer maps (67 %), 11-year-old students showed a similar preference for both options (46 % and 49 %) and over 20 % of 15-year-old participants preferred both materials similarly, i.e. one fifth of 15-year-old participants did not indicate a preference for either of these options.

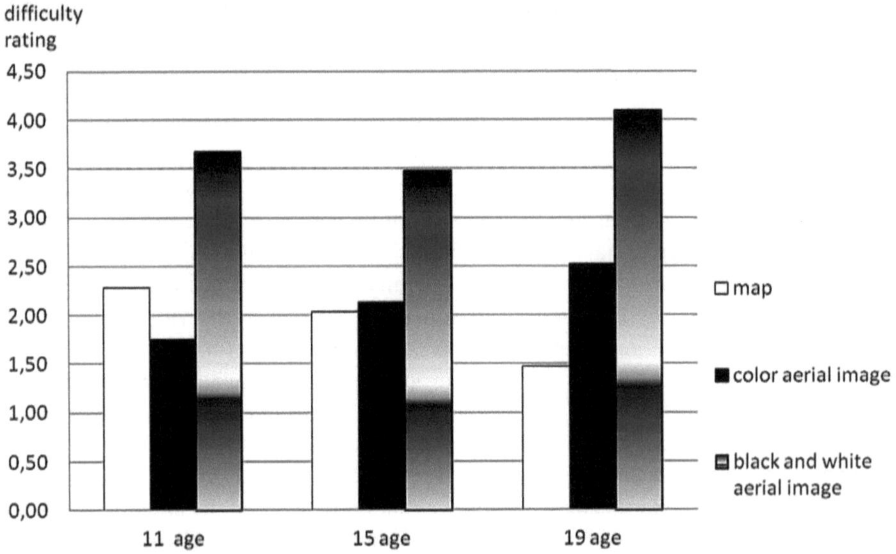

Fig. 11.3 Subjective statements concerning the difficulty of reading aerial images and maps

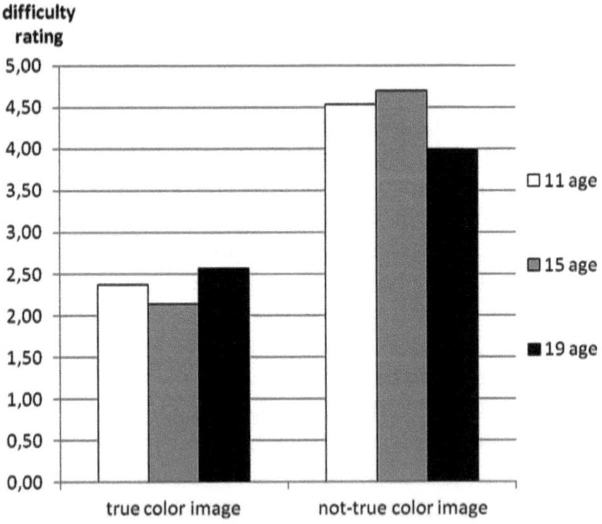

Fig. 11.4 Subjective preference for maps and images

- From the results it is possible to conclude that 19-year-old students realise the information value of maps. Moreover, they achieve objectively better results on maps compared to images.
- Interesting results might be obtained by also comparing the results with orthophotomap preferences, but this question was not included in the test.

11.5.6.2 Evaluation and Results of Comparison of Visual Interpretation Efficiency Regarding True-Colour Satellite Images and not-True-Colour Satellite Images

The second part of the research focused on a comparison of visual interpretation of true-colour images and not-true-colour images by non-experts, specifically adolescents. True-colour images contain clear, needed colour combinations of channels in RGB, evoking an impression of the true colours. Images taken by the LANDSAT 7 satellite were selected for testing. In this image the true colours can be obtained from channels 1, 2 and 3 in green, red and blue. The colour combination "123" is favourable for non-expert interpretation, because the colours of the object match with reality. However, not-true colours are frequently used for practical applications. In this case, combinations of different channels and colours are used and provide better interpretability of objects or their features. The colour combination "742" for LANDSAT satellite images were used for testing. The image colours enabled a very good distinguishing of water areas and flows. These are displayed in blue or black, which is very close to the real colours. Similarly, the forests are displayed in green. Fields are more difficult to find a parallel with reality for non-expert interpreters; depending on the crop or surface, they appear as a pink-green mosaic. Built-up areas are displayed with violet shades.

The participants solved identical tasks using true-colour images or not-true-colour images. With respect to the purpose of not-true colours, we were interested whether untrained interpreters could achieve better results identifying particular objects, even if these colours were not true.

The following conclusions can be made based on an evaluation of tasks focused on the **identification of particular objects in not-true-colour images and colour images**:

- The participants achieved an average score of 66–94 %.
- The average score achieved in images with different colours is not the same. Differences in score were small; differences corresponding to the age groups of participants were at most 5 % (Fig. 11.5). Slightly better scores were achieved with not-true-colour images.
- Image interpretation skills changed with the participants' age. Eleven- and 15-year-old students achieved better results with not-true-colour images. That result is in contrast with their subjective evaluation of image-reading difficulty as related to colours, as the participants considered the not-true-colour images very difficult to read, see Fig. 11.5.
- The evaluation of interpretation efficiency in relation to gender shows very similar values. The only exception was 11-year-old girls, who had significantly better scores for true-colour images (difference 15 %). This supplements and confirms the analysis of the reading of aerial images and maps. In this case, both genders showed similar results with images (there was a difference for maps).
- The efficiency of identification of selected objects and areas varies in relation to colours used in the image provided as a source for identification (Fig. 11.6).

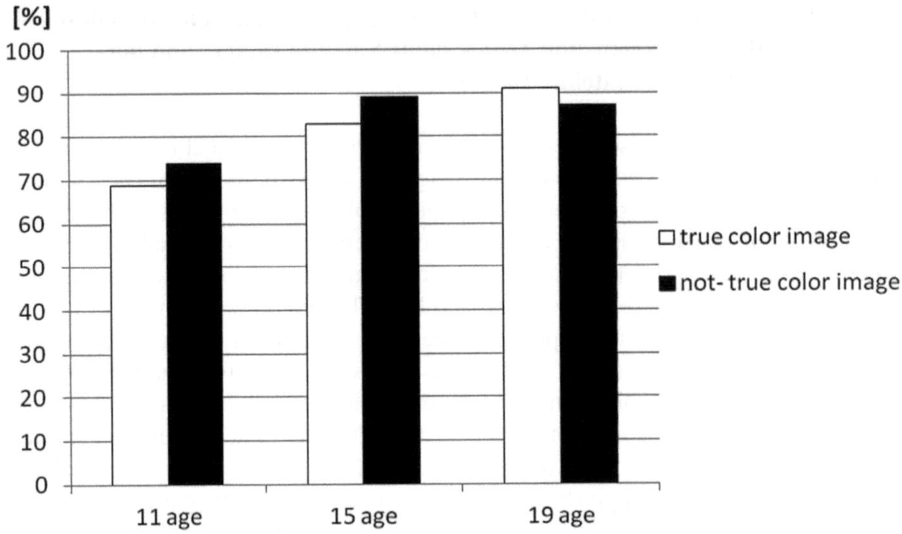

Fig. 11.5 Efficiency of object identification regarding true-colour images and not-true-colour images

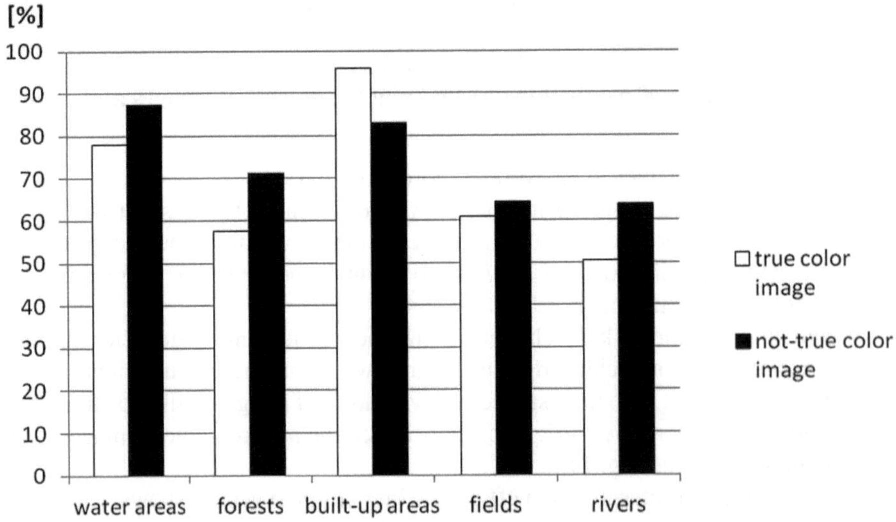

Fig. 11.6 Efficiency of interpretation of types of land in satellite images in relation to image colour type

- Participants achieved very good scores in the identification of water areas and rivers shown in not-true colours (water areas: on average 95 %, 15- and 19-year-old students achieved full 100 %; 83 % for rivers in not-true-colour images vs. 61 % for true-colour images). This confirms the assumption that the selected colour combination, i.e. LANDSAT 7 image with combination "742", enables a very good distinction of water areas and water flows to non-experts. The reason is that these elements appear in expected and predictable colours.
- Participants were slightly more successful in identifying built-up areas, fields and forests in true-colour images. However, the differences were within 5 %. The unusual colours used to display built-up areas in not-true-colour images only confused 11-year-old participants, who achieved a score of 83 % (a very high score all the same) compared to 96 % achieved with true-colour images. Consistent scores exceeding 70 % were achieved for forest identification in true-colour images, and the score increased with age, specifically from 58 to 93 %.
- Generally we can observe that non-experts and children without previous training are able to interpret not-true-colour images very well.

Detailed analysis in relation to participants' age was performed for the task of **identifying the type of land use.** (Note: There were very small differences between genders, so a detailed analysis of results related to gender was not performed.)

- The older participants showed a more efficient identification of land-use type.
- Significant difference was detected between 11-year-old children and the other two age groups. The youngest group did achieve a significantly lower score. The other two age groups were quite equipollent. In other words, there is a big jump in score between 11-year-old students and the older participants.
- Only the score for identification of built-up area types in true-colour image was very similar across all age groups. Eleven-year-old children achieved almost the same score as older students.
- The success ratio remained unchanged for various areas. Eleven-year-old participants showed the same differences in score according to the type of area, but the total results were approximately 20 % lower than the score achieved by 19-year-old participants.

The participants were also asked to **assign a map cut-out to the appropriate location in satellite image (in both true-colour images and not-true-colour images)**. The two cut-outs contained objects with noticeable colours or shapes (e.g. lake and airport runway); the other two cut-outs contained less noticeable objects (suburban settlement ad highway crossing). The evaluation brought a number of results:

- Cut-outs with noticeable objects (e.g. lake, shape and colour; airport, shape) were assigned correctly with a score up to 100 % in several groups of participants, including the youngest participants.
- Colours did not play an important role for map cut-out assignment in aerial images. Figure 11.7 shows that the score is slightly higher for not-true-colour images, but the differences are within 9 %.

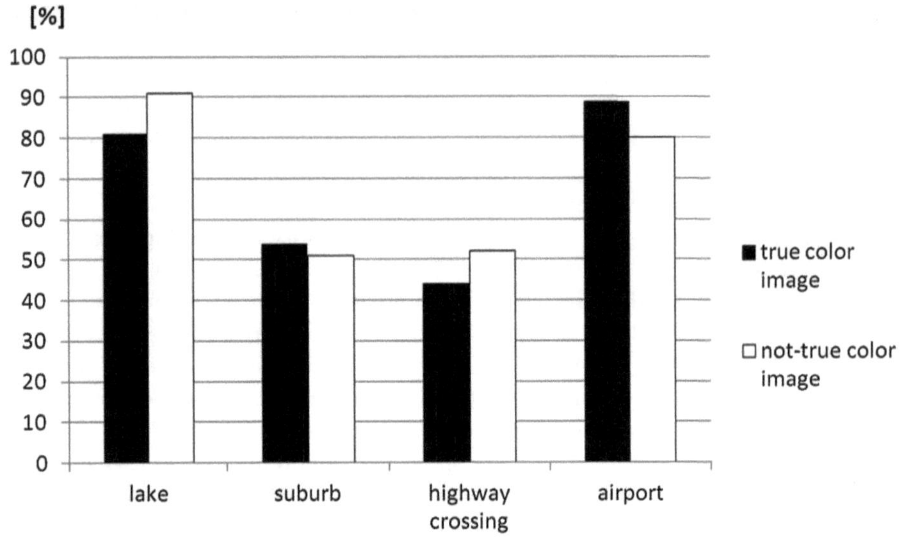

Fig. 11.7 Score for identification of particular objects in true-colour and not-true-colour satellite images

- For the cut-outs containing less noticeable objects, the score was significantly lower – working with suburban settlements: approximately 50 % of participants correctly assigned a round settlement with a forested area, and approximately 44 % of participants correctly assigned a highway crossing (both were presented in true-colour images, see Fig. 11.7).

The identification of landscape changes over time was quite difficult for students. The real situation in LANDSAT 5 and LANDSAT 7 satellite images displayed a landscape dramatically damaged by surface coal extraction, surrounded by damaged forest in 1984. After 20 years, this area is significantly reclaimed, but the extraction still continues in to a limited extent (image dated 2005). However, all the answers containing a formulation mentioning a positive change, such as "forest ratio is increased", "forests are greener", "there are more meadows, fields and ponds", "the damaged areas are smaller", "the landscape looks more healthy, greener", etc., were considered to be correct.

The results show that:

- Participant's age played a significant role in the ability to specify the changes in the area. Nineteen-year-old students achieved significantly higher scores, probably also thanks to a deeper knowledge of environmental issues, as they are confronted with these issues throughout their education.
- Further analysis of results according to gender shows that girls achieved higher average scores than boys. In the 15-year-old students' group, this could be due to girls' faster development to maturity and better formulation and synthetic thinking. Figure 11.8 shows that in the 19-year-old students' group, both genders are much more capable.

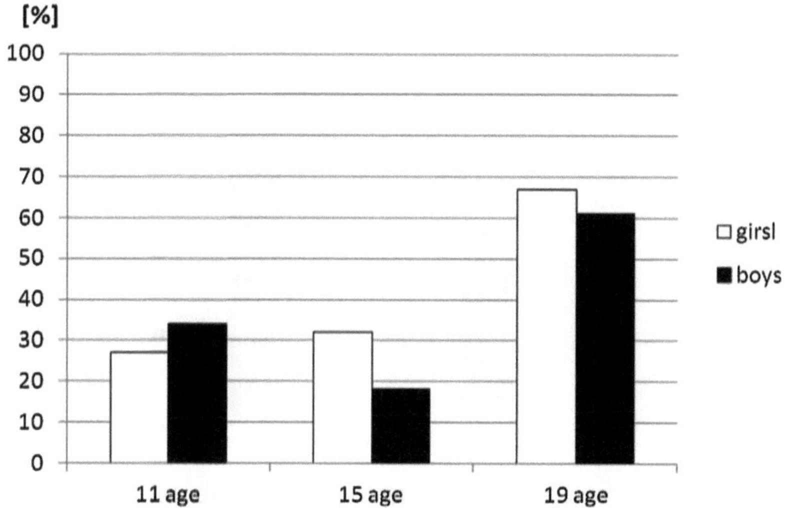

Fig. 11.8 Identification of changes in landscape based on pair of satellite images from different time points, sorted according to gender

Moreover, **subjective classification of the difficulty of object identification in true-colour images and not-true-colour images** was evaluated as well. We wanted to know if the subjective evaluation corresponds with the objective results and how the subjective evaluations change with participants' age.

The following conclusions can be derived from an analysis of the subjective evaluations:

- Although objectively the students achieved better scores with not-true-colour images, eventually the results were very similar (see Figs. 11.5, 11.6 and 11.7). Subjectively they considered reading of not-true-colour images to be very difficult.
- Participants assigned ratings ranging from 4.0 to 4.8 to the not-true-colour (scale 1–5; 5 for maximum difficulty). The true-colour images obtained ratings from 2.1 to 2.6.
- Subjective evaluation of the difficulty of satellite image interpretation in relation to colours used (Fig. 11.9) does not correspond with the objective scores achieved by participants for the particular type of image. Not-true-colour images are considered very difficult to read. Objectively, the efficiency of reading true-colour images and not-true-colour images is very similar. In some cases, the participants achieved higher scores with not-true-colour images.
- From our investigation it is obvious that colours used in the image have a significant impact on the subjective perception of interpretation difficulty.

Just as a curiosity, we mention that during the testing, the students frequently called the not-true-colour image "that map". We asked why they used this expression,

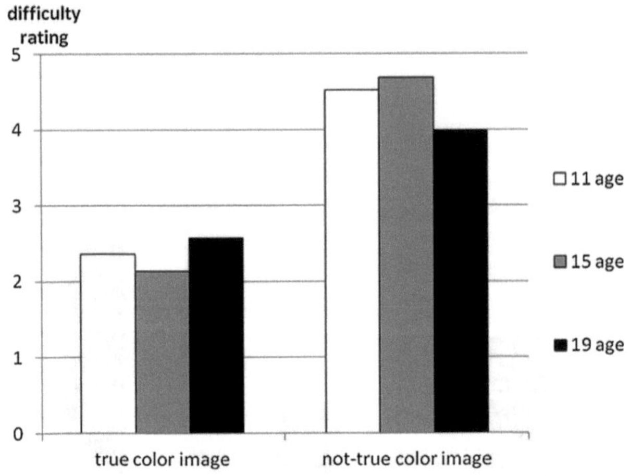

Fig. 11.9 Subjective rating of the difficulty of reading satellite images in relation to colours used

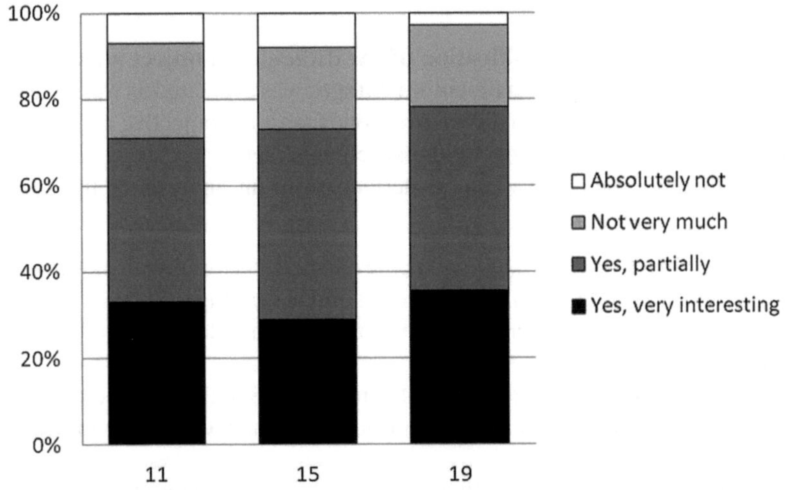

Fig. 11.10 Subjective evaluation of reading images – is it interesting and amusing?

and they specified the colours as the reason. They saw a parallel with the map which also used colours deviating from reality.

- Students find the work with images interesting. More than 70 % of participants selected the answer "yes, very interesting" or "yes, partially" (see Fig. 11.10).
- The answers were very similar across age groups. More than 70 % of students regularly (i.e. at least twice monthly) or irregularly look over images.

One question in the test was focused on **the use of images in lessons from the students' view**, i.e. if they use images and, if so, if the use is detailed or just marginal. **A high degree of evaluation subjectivity** was detected in the answers. Specifically, students from the same class provided different answers, even if they attended the same lessons. Our research shows that the use of images in lessons occurs later and less intensively compared to students' desire to work with images. The use of images at school depends on the teacher's interest. According to students' feedback, the degree of image integration in lessons varies across individual schools. However, it is necessary to mention that the 11- and 15-year-old participants were from two schools only, and thus this result cannot be generalised. More complex information concerning use of images in high schools can be deduced from the answers of 19-year-old participants, who graduated from various grammar schools. Students worked with aerial or satellite images during their studies in high school rather sporadically; 68 % of participants selected the answer "never", 31 % of participants selected the answer "rarely" and only 2 % of participants selected the answer "regularly".

A special questionnaire for geography teachers (45 participants in 2013) showed that older teachers who were not specifically trained on the use of remote exploration in geography integrated a significantly lower number of images into their lessons and were also more sceptical of the students' abilities to interpret images.

11.6 General Conclusion of the Research

Young people explore the world using aerial and satellite images accessible mainly via the Internet. These images might be gradually integrated into lessons as new material for tasks in geography and other subjects.

The purpose of the research performed in 2013 with 378 participants, specifically 11-, 15- and 19-year-olds, was to compare differences in the efficiency of visual interpretation of maps, aerial images and satellite images in various colour presentations. It was found that 11-year-old students achieve better scores with images than with maps and that the scores are very similar for girls and boys. Nevertheless, boys achieved better scores with maps than girls. Differences in map skills related to gender were also confirmed by Chang and Antes (1987). Personal preference for maps or images varied with age. Younger students preferred images, while 19-year-old students preferred maps. These results could be connected with the gradual development of abstract and spatial thinking in adolescence. A comparison of scores for true-colour images and not-true-colour images provided interesting results. Although objectively the students achieved better scores with not-true-colour images, subjectively they considered the reading of not-true-colour images to be very difficult. Participants' age was identified as a significant factor of success for tasks focused on the investigation of landscape changes in satellite images. This specific task was too complicated for 11-year-old students, and they needed the teacher's support. More than 70 % of students view images in their leisure time and find this activity interesting and amusing. Despite that interest, more than half of the participants report that they do not work with images at school. Images produced by high-end technology are still better accessible via the Internet, and young people find them attractive. Their integration in geography lessons could bring new incentives for both students and teachers.

The primary contribution of this research to teaching activities is the finding that students possess high-level skills for work with images, especially younger students, who are better at interpreting images than maps. Children's ability to interpret aerial images was also confirmed in a study performed by Liben and Yekel (1996) and Muir and Blaut (1969). It would be suitable therefore to integrate both document types (aerial images and maps) in education. Students can better realise abstract representation of reality in the map. Working with images helps students develop their skills related to unusual orthogonal views, including the gradual transition from the interpretation of oblique images to the interpretation of orthogonal images and subsequent interpretation of the map, which is in fact also an orthogonal view of the area using symbols.

Work with true-colour and not-true-colour satellite images also helps develop cartographic skills. Using not-true colours also proved to be favourable for primary school students; no increased difficulty of image interpretation related to not-true colours was confirmed. Working with not-true-colour images develops new skills applicable for map interpretation. Students are confronted with documents showing reality in unnatural colours, i.e. the reader must deal with the fact that the picture deviates from reality. This skill can also be used later for map interpretation. Another advantage of not-true-colour expression is the potential for highlighting some objects and features normally hidden from standard sight.

The development of visual and cartographic skills by tracking landscape changes in satellite images taken from different points in time seems very promising. This activity and skill must be trained. It is difficult for younger students, but they achieve very good results with a teacher's support (see also studies performed by Blaut et al. (1970)). Landscape changes and their environmental evaluation represent an important part of an education in landscape and environment protection. The monitoring of landscape changes is also necessary in case of an emergency situation when such changes could indicate a natural or human-induced catastrophe. Working with satellite images, aerial images and maps is an educative activity requiring student involvement. Blaut (1997) confirms that an active approach to working with maps and images supports improved development of new knowledge and cartographic skills. When testing the big group (in total 378 participants), we observed that most students were absorbed by the activity and worked continuously from start to end. This is significant, as students work and learn best when the activity is interesting for them and gives them a feeling of satisfaction and practical usefulness. According to the trend, satellite and aerial images are very useful tools which can be integrated into geography lessons. Our research supports the use of satellite and aerial images in education. The research will continue with other test groups composed of older students as well as senior participants, non-experts and professionals working with images and maps.

References

Blaut, J. M. (1997). The mapping abilities of young children: Children can. *Annals of American Geographers, 87*(1), 152–158.

Blaut, J. M., McCleary, G. S., & Blaut, A. S. (1970). Environmental mapping in young children. *Environment and Behavior, 2*(3), 335–349.

Catling, J. S. (1978). The child's spatial conception and geographic education. *Journal of Geography, 77*(1), 24–28.

Chang, K., & Antes, J. J. R. (1987). Sex and cultural differences in map reading. *The American Cartographer, 14*(1), 29–42.

Downs, R. M., & Liben, L. S. (1997). The final summation: The defense rests. *Annals of American Geographers, 87*(1), 178–180.

Geoportal (2015). *Národní geoportál INSPIRE.* https://geoportal.gov.cz/web/guest/home. Accessed 30 July 2015.

Gerber, R. V. (1981). Young children's understanding of elements on maps. *Teaching Geography, 1*, 128–133.

Golledge, R. G., Marsh, M. J., & Battersby, S. J. (2008). A conceptual framework for facilitating geospatial thinking. *Annals of the Association of American Geographers, 98*(2), 285–308.

Kim, M., Bednarz, R., & Kim, J. (2012). The ability of young Korean children to use spatial representations. *International Research in Geographical and Environmental Education, 21*(3), 261–277.

Kovařík, V. (2012). *Effects and limitations of spatial resolution of imagery for imagery intelligence.* Proceedings of the International Conference on Military Technologies and Special Technologies – ICMT'-2012 (pp. 363–368). Trenčín: Alexander Dubček University of Trenčín.

Lee, J., & Bednarz, R. (2012). Components of spatial thinking: evidence from a spatial thinking ability test. *Journal of Geography, 111*(1), 15–26.

Liben, L. S. (1999). Developing an understanding of external spatial representation. In I. E. Sigel (Ed.), *Development of mental representation.* Mahwah: Lawrence Erlbaum Associates.

Liben, L. S., & Downs, R. (1989). Understanding maps as symbols: Development of map concepts in children. In Reese H. (Eds.), *Advances in child development*, New York: Academic Press.

Liben, L. S., & Downs, R. (1992). Developing an understanding of graphic representations in Children and Adults: The Case of GEO-Graphic. *Cognitive Development, 7*, 331–349.

Liben, L. S., & Yekel, C. A. (1996). Preschoolers' understanding of plan and oblique maps: The role of geometric and representational correspondence. *Child Development, 67*(6), 2780–2796.

Lillesand, T., Kiefer, R. W., & Chipman, J. W. (2008). *Remote sensing and image interpretation.* New York: Wiley.

Lloyd, R., Hodgson, M. E., & Stokes, A. (2002). Visual categorization with aerial photographs. *Annals of the Association of American Geographers, 92*(2), 241–266.

MacEachren, A. M. (2004). *How maps work: Representation, visualization, and design.* New York: The Guilford Press.

Muir, M., & Blaut, J. (1969). *The use of aerial photographs in teaching mapping to children in first grade: An experimental study.* The Minnesota Geographer 22 (pp. 4–19). New York: Wiley.

Pravda, J. (2001). *Výskum niektorých stránok kartografickej gramotnosti.* [Research of certain aspects of cartographic literacy]. Bratislava: Geografický ústav SaV.

Robinson, A. H., & Petchenik, B. B. (1977). The map as a communication system. *Cartographica: The International Journal for Geographic Information and Geovisualization, 1*(14), 92–110.

Schuit, W. (2011). A method for teaching topographic map interpretation. *Journal of Geography, 110*(5), 209–216.

Schultz, R. J., Kerski, J. J., & Patterson, T. C. (2008). The use of virtual globes as a spatial teaching tool with suggestions for metadata standards. *Journal of Geography, 107*(1), 27–34.

Stea, D., & Blaut, J. (1973). Some preliminary observations on spatial learning in school children. In R. M. Downs & D. Stea (Eds.), *Image and environment.* Chicago: Aldine.

Svatoňová, H., & Rybanský, M. (2014). Visualization of landscape changes and threatening environmental processes using digital landscape model. *IOP Conference Series: Earth Environmental Science, 18*, 012018. doi:10.1088/1755-1315/18/1/012018.
Van Coillie, F., Gardin, S., Anseel, F., Duyck, W., Verbeke, L., & De Wulf, R. (2014). Variability of operator performance in remote-sensing image interpretation: The importance of human and external factors. *International Journal of Remote Sensing, 35*(2), 754–778.
Van der Schee, J. A. (1987). *Kijk op Kaarten* [Look at maps] (Dissertation). Amsterdam: Vrije Universiteit.
Van der Schee, J. A. (2000). Helping children to analyse a changing world: Looking for patterns and relationships in space. In M. E. Robertson & R. Gerber (Eds.), *The child's world: Triggers for learning* (pp. 214–231). Camberwell: ACR Press.
Wiegand, P. (2002). *Learning and teaching with maps*. Abingdon: Routledge.

Part IV
Current Topics of Political, Regional and Social Geography and Their Didactic Applications

Chapter 12
The World's Major Regions as Part of Regional Geography Courses?

Jiří Anděl, Ivan Bičík, and Kateřina Zavadská

12.1 Introduction

The division of the world into major geographical regions is a matter of general interest, discussed on a daily basis also in the media. Different concepts towards major geographical regions are suggested by geographers, economists, linguists, ethnologists and experts in political studies. Most of these concepts include their own "unique" and "original" regions; however, explanations how these regions were delimited are often absent. Little attention is given to the methods of forming such regions.

Häufler (1985) was among the first Czech (Czechoslovak) geographers who studied this concept. He argued that "social, economic, and political geographers should create clearly defined, reasonable major geographical regions that would serve as a basis for studying global processes. The division of the world into six (seven) large landmasses and four oceans, traditionally used by 'old' geographers, is nowadays just of historical interest and does not meet current needs".

Though scientific works that deal with major geographical regions do exist (Hampl 2010; Polonský 2012), there is lack of sources that could be applied in the curriculum. The authors of this text have recently prepared a textbook focused on regional geography and major geographical regions (Bičík et al. 2010) intended for secondary schools, but which is also used in a number of university courses. The delimitation of major geographical regions was an important part of the work.

J. Anděl • K. Zavadská
Faculty of Science, Department of Geography, University of J. E. Purkyně in Ústí nad Labem, České mládeže 8, 400 96 Ústí nad Labem-město, Czech Republic

I. Bičík (✉)
Faculty of Science, Department of Social Geography and Regional Development, Charles University in Prague, Albertov 6, 128 43 Prague, Czech Republic
e-mail: bicik@natur.cuni.cz

P. Karvánková et al. (eds.), *Current Topics in Czech and Central European Geography Education*, DOI 10.1007/978-3-319-43614-2_12

The first two sections in the chapter discuss the broader framework of the subject. Geographical paradigms and globalisation processes are discussed as well as possible difficulties of implementation into educational programmes. The third section of the chapter provides the delimitation of the world used by the UN and examples of erroneous categorisation of countries into these units. The fourth and fifth sections evaluate various methodologies of delimitation of major geographical regions prepared by various authors. The sixth section presents the results of surveys of mental maps of major geographical regions of students from three universities: Charles University in Prague (Czechia), Lomonosov University in Moscow (Russia) and the University of Maribor (Slovenia). Conclusions showing the reasons for the macro-regional concept in teaching geography at secondary schools and universities follow.

12.2 Geographical Paradigms: Changing Role of Geography Over Time

Geography has one of the longest traditions among all scientific fields. The beginnings of geography are closely connected with the necessity to secure basic orientation in the landscape. As a result, the very first geographical paradigm can be identified as the art of spatial orientation (including tools needed for mapping). This is why, especially in popular culture, many identify geography with mapmaking. The first paradigm prevailed for a long time – until the end of the nineteenth century and in some cases even until the twentieth century. At present, the general perception that providing the spatial orientation on the Earth is the major task of geography as a science still prevails – and many think that it is the only task. Sadly, geography is often described and taught in this way at primary and secondary schools, especially by teachers without formal geography education.

A new paradigm appeared in the nineteenth and early twentieth centuries. At that time, during the colonial expansion of European powers, geography played an important role, securing information that was necessary for conquering distant parts of the world. This information included maps, knowledge of natural conditions, available raw materials and the character of potential markets. This new paradigm, however, did not fully replace the original one. It was long part of the curriculum, especially at universities focused on economics and related subjects.

Looking at further developments in geographical paradigms, different branches of physical geography gradually became independent as research methods, and techniques were partially adopted from other sciences (physics, biology, geology, etc.). In this way, the basic distributions of climatic, soil and vegetation zones were formulated. This occurred roughly during the first two decades of the twentieth century, when conditions were ready to study the effects of natural conditions on society. The changing character of traditional rural communities was analysed at that time, too. However, geography still could be seen as a unitary natural science.

Rapid advances in statistics, economics and sociology had a direct as well as indirect influence on the development of anthropogeography, which retained its descriptive character. Different branches of social geography began to use a number of theories and methods of the above-mentioned sciences (agricultural geography, geography of industry, transportation and especially population studies). This advance of social geography resulted in the division of geography into two poorly connected academic disciplines.

Society became the major driving force of changing the landscapes, in many cases at least equally important as natural processes themselves. Such an imbalance between natural forces and increasing human pressure over the last 25 years resulted in the formation of a new paradigm in geography as well as in other sciences, one focusing on the interaction between nature and society, on analysing the driving forces behind visible changes and on the search for ways to influence interactions. In our opinion, this is among the great challenges for current research in many sciences, including geography.

12.3 Globalisation, Formation of Major World Regions and the Effects on Geography Education

Globalisation is one of the most important trends over the past 25 years. In the Czech context, a sort of "dual transition" can also be identified. The first effect of this transition was closely linked to the fall of the Iron Curtain, which enabled the free movement of citizens and brought profound economic and social reforms. The second, even more important effect is related to globalisation and its influence on the character of the economic and social transition in the Czech Republic.

How did educational programmes cope with such deep changes and how was geography education changed? Both types of transition were fundamental for economic and social changes and also influenced the interaction between nature and society. These events and topics are at the core of geographical research and education; as such, the changing character of the reality should be reflected in educational programmes at secondary schools and universities, including educational programmes for future teachers. Has there been any sound discussion on these topics?

Only a few geographers participated in formulating the basic educational programmes (*rámcové vzdělávací programy*, RVP) which also included the role of geography in general education. Most geographers reject the position of geography as part of natural sciences, which is seen as a concept related to the early twentieth century. The imbalance among different subjects within RVP has been widely criticised. The final results, approved by the Ministry of Education, brought a big surprise: the amount of lessons devoted to geography within the basic educational programmes has been reduced, and the outdated approach towards geography education has been largely preserved. The weak rules of the basic educational programmes and inadequate training of many geography teachers resulted in poor definitions of contents and targets.

The ongoing processes of globalisation and transition led to fundamental social and economic changes in the Czech Republic, Europe and the whole world. These changes, which occurred over a short period of time, brought a number of new topics to be discussed:

1. How will current social and economic changes influence one of the basic functions of geography, i.e. securing basic spatial information on the local, regional and global levels? What should such "spatial information" include?
2. Is this "spatial information" really so important that geography (as part of general education) could omit other desired functions (cognitive, applications, etc.)?
3. The interaction between nature and society and its changing nature over time (from local to global) should include synthesis of geographical knowledge as well as knowledge from a number of other sciences (biology, chemistry, history etc.). It is the synthesis – in other words links among findings from different scientific fields – which is most lacking in the current educational programmes. Why did geographers not take this opportunity?
4. How can geography contribute to students' preparation for practical life? By providing the list of the Amazon's tributaries and by detailed information on the population of Madagascar? Or should geography rather provide just basic information on the Earth and teach more about relations among geographical location, climate, way of life, economic level, etc.?
5. What kind of knowledge of regional geography should pupils at primary and secondary schools have? Have any standards been set? Naturally, the amount of regional geographical knowledge will always be different place by place – depending on the students' characters and teachers' requirements. Do teachers really understand what a reasonable amount of regional geographical information entails? Based on our research, the expectations of different teachers cover a very broad span. And, after all, is the above question appropriate with respect to what geography should provide at primary and secondary schools?

In our opinion, geography education at primary and secondary schools should focus on relations between different geographical phenomena on different levels. It should include a selection of countries (regions) where typical geographical problems could be explained. On the other hand, modern geography education should not include just a list of different places, regions and names (capital city, highest mountain, currency, president of the state, etc.).

The above-mentioned strategy, i.e. basic knowledge of respective regions including core-periphery relations and typical geographical phenomena, has been adopted by the two co-authors of this article, who produced one of the available geography textbooks for secondary schools (Bičík et al. 2010; Anděl and Bičík 2015).

12.4 Available Sources: Critical Assessment

Surprisingly, the division of the planet Earth into major geographical regions has been studied by just a few scientists. Most authors just use "major geographical regions" as part of their research methods or databases. Many take these supranational regions as a given and do not care how they were formed. The publications on cultural geography (Goncalves and Sanchez 2014) or economics (Bramall 2008) can serve as examples. The Scopus database includes roughly 50 entries dealing with major geographical regions (September 2015), of which only 15 % study real supranational regions. Among these are publications that examine just one selected region (sub-Saharan Africa, Middle East or Eastern Europe – Smetkowski 2013) or show historical events using GIS (Thornton 2012).

Different concepts of the division of the world into major regions also have been suggested by economists, linguists, ethnologists and political scientists. Most of these concepts, however, do not include much methodology and often reflect just the authors' ideas. The global division of the world has been recently thoroughly studied by Hampl (2010; see also Novotný 2003). Hampl's approach is based on elaborate methodology that allows the importance of groups of nation states and the changes of this importance over time to be compared. Polonský (2012) focuses specifically on major geographical regions and distinguishes among four alternatives: major geographical regions appropriate for scientific studies, textbooks and atlases, practical purposes (regional bodies and global actors) and for different overviews (publications for general public, databases, etc.).

Many publications dealing with the geographical division of the world into supranational regions often lack a methodological basis that would justify such a division. This is the case of many textbooks (see Jackson and Hudman 1990 or Hobbs and Dolan 2009). Also the division of the world used by the United Nations for statistical purposes (Standard Country or Area Codes for Statistical Use 2012, UN) includes little methodology. The United Nations division works with M.49 area codes dividing the world into supranational regions and smaller geopolitical entities. To a great extent, these supranational regions correspond to five continents. Supranational regions are further subdivided into smaller units on the basis of cardinal points (Europe divided into Northern, Southern, Western and Eastern Europe; Americas divided into Northern, Central and South America; etc.). Czechia is part of Eastern Europe here – contrary to common sense – together with the whole territory of Russia (Fig. 12.1).

There are just a handful of scientists that have seriously striven to produce a reasonable geographical division of the world. While creating supranational regions, different criteria have been used: political-economic (Häufler 1985), socio-economic (Morris 1972), social-cultural (De Blij and Muller 1988; Huntington 1996) or the level of technical and economic development (Cole 1996).

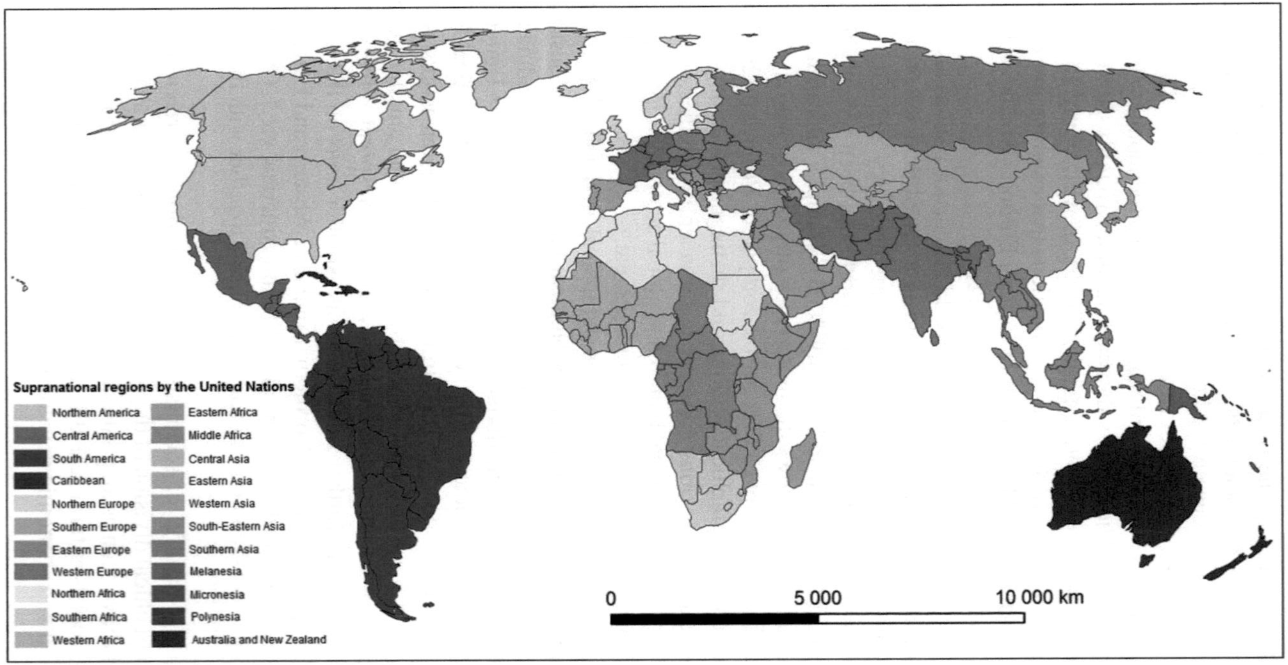

Fig. 12.1 Supranational regions as defined by the United Nations, UN (Source: UNData (2014) (Modified (prepared) by L. Bobr))

12.5 The World's Major Regions: Methodical Remarks

The attempt to construct a reasonable geographical division of the world is a complex task, as any kind of regional division of complex entities cannot meet the reality in full. The author's subjective point of view is inevitably present and it is always going to be "the best bad choice".

The existing divisions into supranational regions are often assessed from one standpoint only, usually the economic one. More aspects should be taken into consideration, however, such as politics and international relations, social and cultural conditions and environmental factors (Hampl 2010). The importance of social and economic aspects was stressed by Morris (1972) and Häufler (1985). The Häufler's approach, however, includes political and economic clichés (capitalist, socialist and third worlds). Social and cultural predispositions were emphasised by de Blij and Muller (1988) and by Huntington (1996). The latter author divided the world into different "civilisations".

Of special interest is the approach of Cole (1996), who divided the world into 12 units according to the level of technological and economic development. Cole also took into consideration raw materials, practical use of modern technologies and capital reserves available for domestic and foreign investment. So far, these are tangible things that can somehow be measured; our considerations intentionally do not include intangible concepts that cannot be reflected in statistics like happiness, hospitality or effectiveness of international aid.

The key methodological task is to find an aggregate measure that would at least roughly reflect social and economic aspects as well as the political relations and environmental conditions of respective regions. It seems that one of the best approaches is to divide the world into major regions by the so-called civilisations that include a certain degree of integrity and present broad cultural entities (Huntington 1996). Approaches that include a sound analysis of the world system (Wallerstein 1979, 1984 etc.) and underlining the core-periphery relations should be taken into consideration, too. De Blij and Muller (1988) successfully adopted a three-tier model of division of the world (see Sect. 12.6).

12.6 The World's Major Regions: Selection of Concepts

Supranational regions as defined by de Blij and Muller (1988) present one of the key concepts in the English written sources. The authors applied a multi-tier division of the world. The largest and most complex regions are called *cultural realms* and share common or similar *cultural traditions*. China and Latin America are examples of these cultural realms. Cultural realms are divided into *cultural regions* – in the case of Latin America, for instance, represented by Brazil or Argentina. Eastern Europe and Southern Europe are also examples of cultural regions. Each cultural region is further divided into *cultural subregions*. For example, Canada constitutes one cultural region, and Québec forms one of the various cultural subregions within Canada (Fig. 12.2).

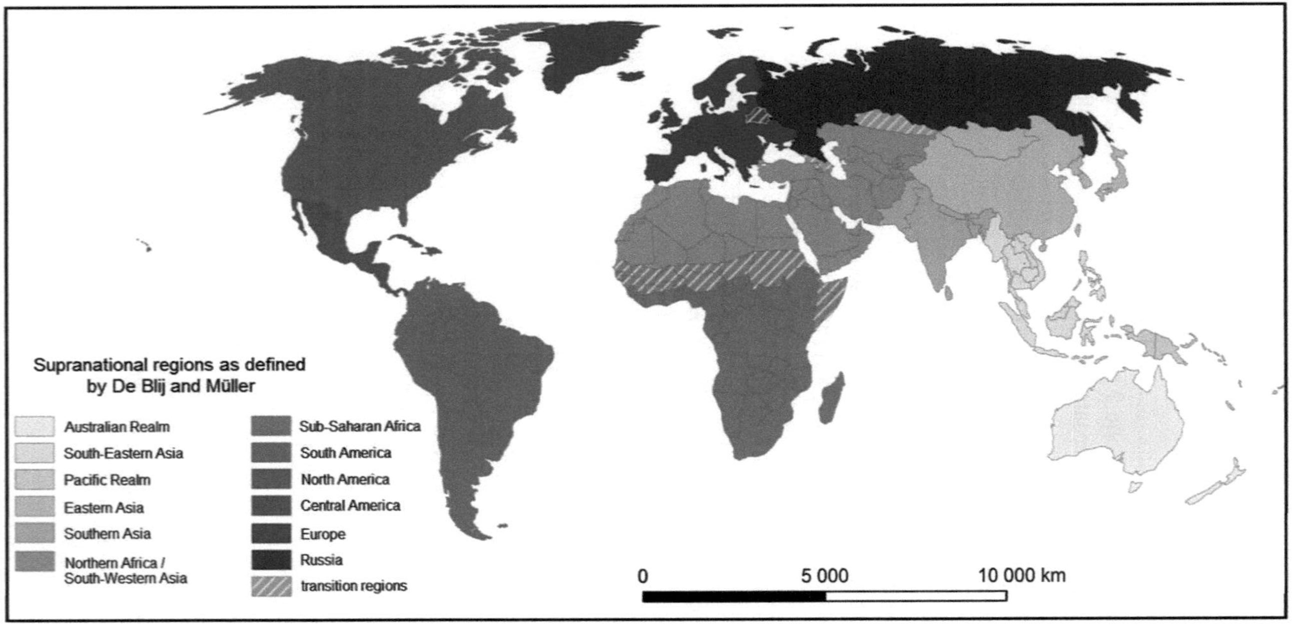

Fig. 12.2 Supranational regions as defined by de Blij and Muller (1988)
(Source: De Blij and Muller (1988) – adapted by Bičík and Anděl and prepared by L. Bobr)

The authors' approach towards cultural realms (the upper tier) is a broad one, taking into account economics, urbanisation and politics as well as physical and historical geography. De Blij and Muller apply geographical synthesis based on the methods used in human geography. As a result, the planet is divided into 12 *geographical realms* or supranational regions. Of these, ten supranational regions are "developed" and seven are "underdeveloped".

The concept of supranational regions defined by de Blij and Muller has developed over time (already sixth edition of the textbook). It serves well the main purpose, i.e. geography courses at secondary schools and universities. In the Czech context, however, this division into supranational regions needs a certain adjustment. First, in our opinion, the supranational regions should be relatively homogeneous. Second, they should reflect the core-periphery relations among more developed (core) areas and less developed (peripheral) regions. Last but not the least, more attention should be devoted to nearby regions, and students should receive more information compared to more distant regions.

The textbook intended for secondary schools produced by Bičík et al. (2010) includes a different division of the world into supranational regions. The authors applied geographical synthesis and used the experience from the above-mentioned publications. Bičík et al. divided the world into ten major regions, each of which is relatively homogeneous on the grounds of economic performance and cultural integrity (Fig. 12.3). Four of these supranational regions are developed (core regions), and five are less developed (peripheral). The remaining one, Eastern Asia, was further divided into the core subregion (Japan, Taiwan, South Korea 5A) and less developed subregion (China, Mongolia, North Korea 5B).

The core regions include Europe (developed and post-Communist parts, excluding the former USSR 1A), Northern Eurasia (the post-Soviet realm 3A), North America (2A) and Australia and Pacific (4A). Among the less developed regions are sub-Saharan Africa (10B), South-Western Asia and Northern Africa (Islamic world 8B), Southern Asia (7B), South-Eastern Asia (6B) and Latin America (9B).

The following table (Table 12.1) outlines different supranational regions as defined by different authors using different concepts.

12.7 Mental Maps of the World's Major Regions

The second part of this chapter is based on a survey focused on how students from different age groups perceive the current division of the world into supranational regions (Figs. 12.4, 12.5, and 12.6).

The results are presented in the form of "average mental maps". Mental maps may be created in different ways: one can talk about comparative mental maps, preferential, negatively preferential or mental maps of the virtual world. Comparative mental maps reflect reality and can be assessed using conventional tools. For example, such maps can reveal the degree of knowledge of a particular field, in this case knowledge of major geographical regions. The map is created in such a way that

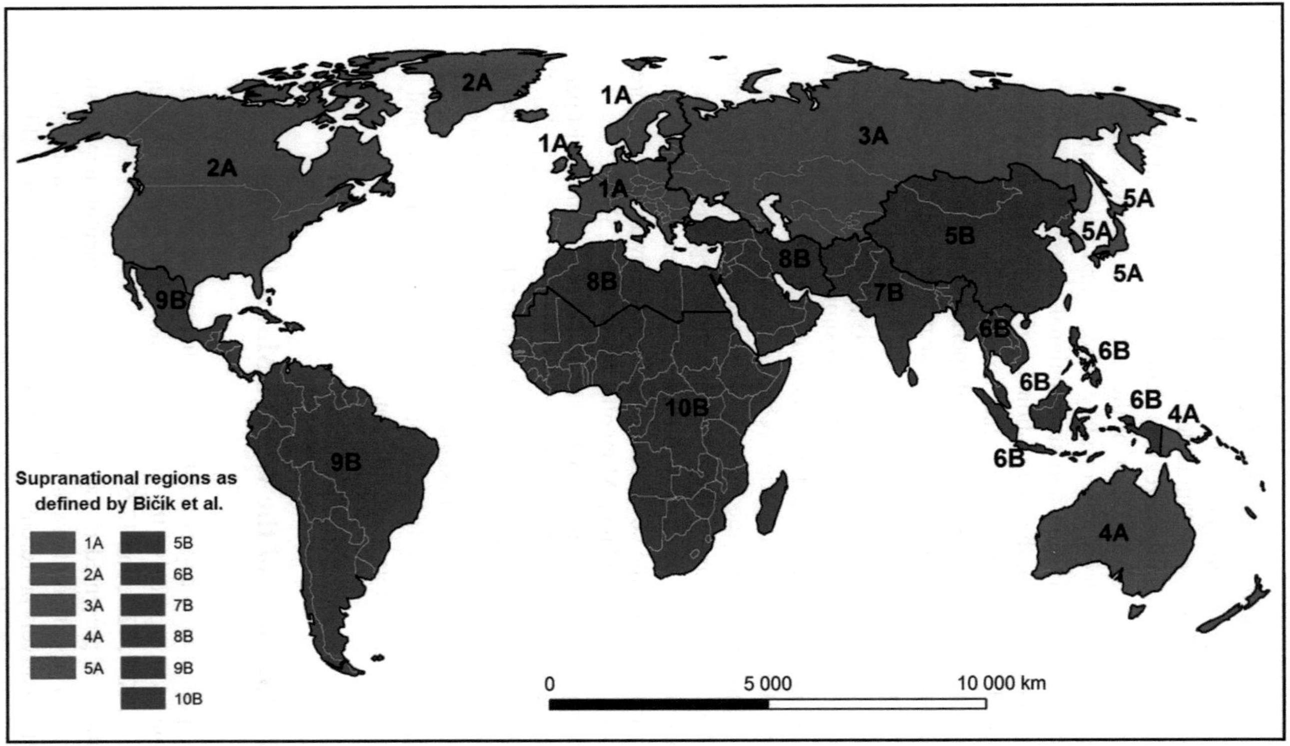

Fig. 12.3 Supranational regions as defined by Bičík et al. (2010) – see the text above (Source: Bičík et al. (2010) – prepared by L. Bobr)

Table 12.1 Supranational regions as defined by different authors

Author	Morris	Häufler	De Blij – Muller	Clawson	Huntington Civilisations	Cole
Year	1972	1985	1997	2007	1996	1996
Europe	Western and Central	Capitalist	Europe	Europe	[d]Western	"Western" Post-Communist
	Southern	Socialist			[c]Orthodox	
	Eastern					
Russia	USSR	USSR	Russia	[a]Russia	[c]Russia	Post-Soviet realm
Asia	[b]Central	[b]South-Western	[b]South-Western	[a]Central	[b]Islamic	South-Western + Southern
	Eastern			Eurasia	Indian	
	Southern	Eastern	Eastern	[b]Middle East	Buddhist	South-Eastern Developing Eastern
	South-Eastern	Southern	Southern	Asia	Chinese	Developed Eastern
	Central and Eastern	China, North Korea, Japan	China, North Korea, Japan		Japanese	
Africa	[b]Central Eastern Sub-Saharan	[b]South-Western Sub-Saharan	South-Western Sub-Saharan	[b]Middle East Eastern Sub-Saharan	[b]Islamic	Northern + Sub-Saharan
					African	
Northern America	Canada	Northern	Northern	USA	[d]Western	Northern
	USA-West			Canada		
	USA-East					
Latin America	Central South	Central South	Central South	Latin America	Latin	Latin
Australia	Australia	Australia	Australia Pacific	Australia	[d]Western	Australia and Pacific
Regions in total	15	13	12	8	8	12

Notes: [a]*political situation is similar, and cultural different (two continents);* [b]*the respective region spans two continents;* [c]*cultural influence of Orthodox Church into some parts of Europe;* [d]*Euro-American civilisation in three continents (Western). For the sake of compatibility, some data were simplified. Source: see References in the end of the article*

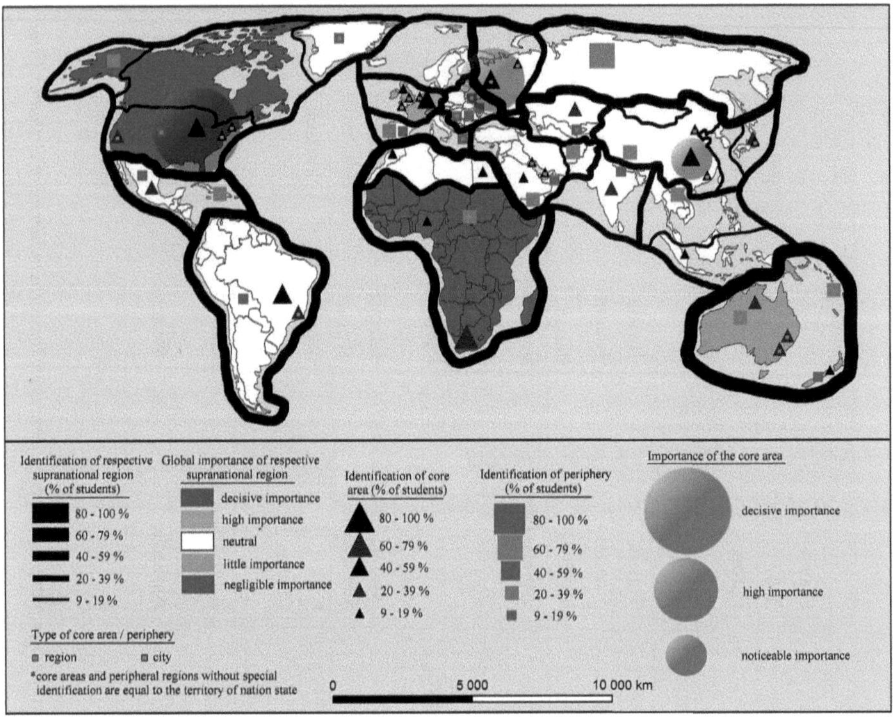

Fig. 12.4 *Average mental map as created by geography students, Charles University, Prague (aged 23–26).* Identification of respective supranational region (percentage of students) 80–100 % and so on
(Note: core areas and peripheral regions without special identification are equal to the territory of nation state. Source: Zavadská (2015))

information (here: world major regions) is drawn into the map. Consequently, the map is based on estimates and do not correspond to reality in full. Differences from the reality can be easily assessed (Siwek 2011).

The creation of the "average mental maps" was again based on a survey. Respondents were encouraged to draw information into blank maps using the methodology produced by Polonský and Novotný (2010) and Polonský (2012). Students from Prague (Charles University), Ústí nad Labem (Jan Evangelista Purkyně University), Moscow (Lomonosov University) and the University of Maribor participated in the survey. Among the Czech students, their specialisation was taken into account (geography vs. non-geography students) as well as age group (19–22 and 23–26 years). Altogether, eight groups of students were examined (Zavadská 2015).

The preceding Sect. 12.5 includes an overview of available scientific sources dealing with the division of the world into supranational regions. These sources were used by Bičík et al. (2010) and also by Anděl and Bičík (2015) in a recently published article. In our opinion, the above concepts should serve as a basis for

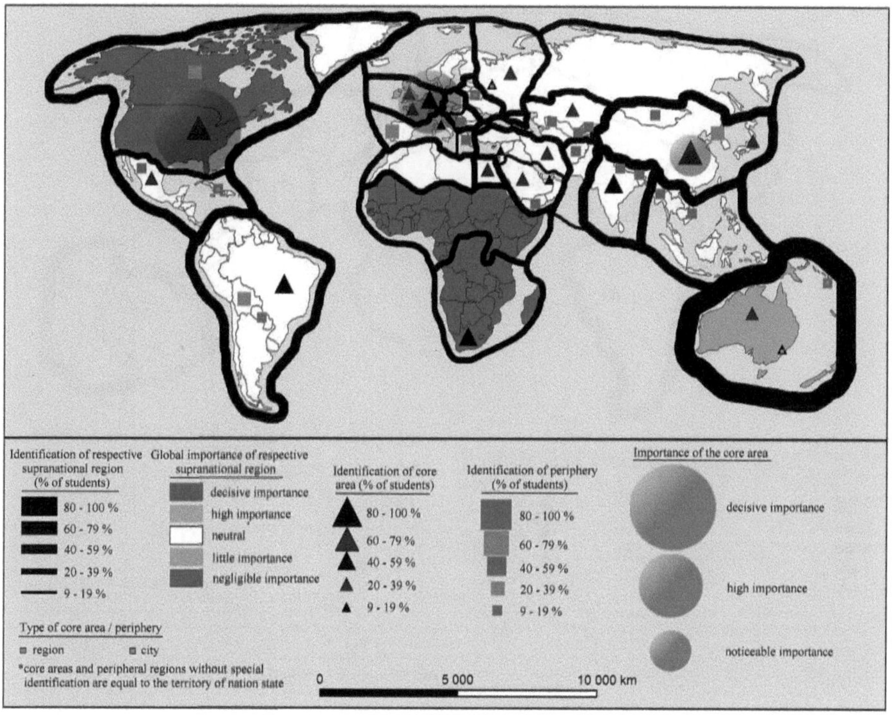

Fig. 12.5 Average mental map as created by geography students, MGU, Moscow, Russia (Note: core areas and peripheral regions without special identification are equal to the territory of nation state. Source: Zavadská (2015))

regional geography courses at secondary schools. Regional geography is supposed to provide basic information on nature and society, including their mutual interaction, at different regional levels. We tried to find out how students from different backgrounds (age groups, domicile) perceive this role of regional geography.

The mental maps clearly show the respective major geographical regions as perceived by different students' groups (eight groups – see methodical remarks) as well as the perception of global importance and location of core and peripheral areas. As an example, mental maps created by students of Charles University, Prague (Fig. 12.4); MGU, Moscow (Fig. 12.5); and the University of Maribor (Fig. 12.6) are shown.

First, let us examine how precisely respective supranational regions were identified by the students. Surprisingly, Australia (in some cases including Pacific) and South America were identified with the highest accuracy. The supranational region of Africa was also clearly identified by most students; many further subdivided Africa into smaller regional units.

The identification of North America was rather ambiguous. Some students included Greenland or even Central America. Similarly, the precise identification of Europe posed problems. Some correctly perceive the European part of Russia as

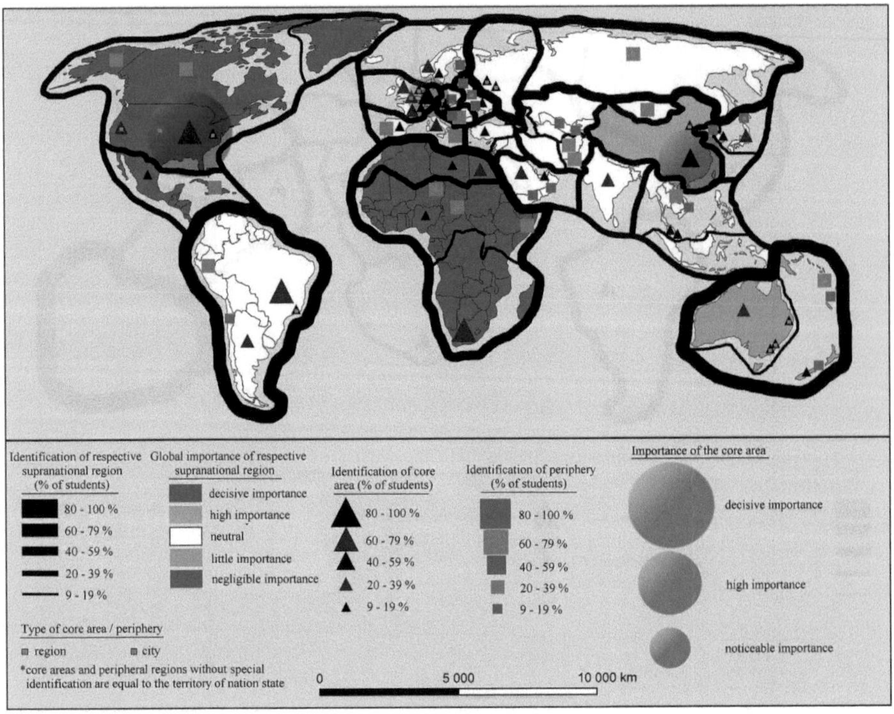

Fig. 12.6 Average mental map as created by geography students, University of Maribor, Slovenia (Note: core areas and peripheral regions without special identification are equal to the territory of nation state. Source: Zavadská (2015))

being part of Europe; other identified Russia as an independent supranational entity or considered Russia as part of Asia.

In the case of Asia, big differences exist in the perception of how Asia should be subdivided into smaller units. In our opinion, the subdivision of Asia as a continent should be as follows: the Asian part of Russia (part of supranational region of Russia), South-Western Asia (including, however, also the countries of Northern Africa), Central Asia (part of former USSR including Kazakhstan), Southern Asia, South-Eastern Asia and Eastern Asia. The latter could be further subdivided into developed Eastern Asia (Japan, South Korea, Taiwan) and developing Eastern Asia (China, Mongolia, North Korea). The biggest differences between our approach and students' perceptions were in the case of Russia. Most students do not perceive Russia as a major geographical region.

When it comes to global core areas, the importance and degree of influence have been examined. All respondents perceive the USA as being the single most important core area. Some identify this "global core area no. 1" more precisely as the US West Coast, while the East of USA is seen as the second most important area. Europe was indicated as the world's second most important core area twice; in one case, Europe ranked third. China ranked second three times as regards global influence. Japan scored well, too (second and third position once, respectively). Russia ranked third in one case only.

Respondents were also asked about the influence of respective supranational regions on world affairs. North America ranked first among an overwhelming majority of respondents (78 %), followed by Europe (50 %), European part of Russia (only 25 % of respondents) and Russia and China (20 %).

12.8 Conclusions

The division of the world into supranational regions is examined from the theoretical standpoint (the process of delimitation and its results) and also with respect to curriculum (investigation among geography students). Understanding major geographical regions is essential for one of the basic functions of geography: proper orientation in the current world under conditions of ongoing globalisation. The results show, however, that many secondary school students and even a number of university students are unable to cope with this task. Our survey showed that many students did not succeed in filling in the questionnaire properly or did not do so completely. Some 40 % of Czech students returned it in such a state that any kind of evaluation was impossible. This is quite surprising, especially when one takes into account that major world regions must have been presented in some way in geography courses at secondary schools and also at universities.

Seen from the broader perspective, the territory of Czechia has been fully exposed to international competition and global trends since the early 1990s, i.e. for approximately 25 years. In light of this, one would assume that major geographical regions are granted appropriate attention in the curriculum. As geography lessons should provide basic orientation in the world, a sound knowledge of major geographical regions is a key element in geography education. In our opinion, this topic (supranational regions) should receive far more attention, especially as regards future geography teachers. A more comprehensive survey testing knowledge of supranational regions at Czech secondary schools and universities is planned in the near future.

The survey showed that there are big differences regarding the perception of major geographical regions. The same applies to the global importance (influence) of respective regions. Most respondents correctly assume that North America is currently the key major geographical region, although definitions of "North America" vary. The perception of Europe and its global role shows great differences; most respondents acknowledge Europe's important global role, but the significance of European core areas seems to be underestimated, probably due to the lack of a clearly defined single core European area.

Many respondents perceive China as an important global entity, although some assume that Japan has the leading role in Eastern Asia. Diverse opinions emerged also in the case of Russia. Many respondents tend to underestimate Russia's global importance and do not even perceive the post-Soviet sphere as a single supranational region (though these countries share a good deal of common history and economic relations).

In general, non-geography students tend to perceive the division of the world along continental lines, i.e. on the basis of major physical features. The mental maps produced by undergraduate geography students already show signs of division of the world into supranational regions, not only on the basis of physical geography factors. There are also differences when one compares students from different age groups and nationalities. Surprisingly, Russian students perceived the global role of Russia as less important than their Czech and Slovak counterparts; the same applies to the identification of Russia as one of the global core areas (Fig. 12.5).

Most Slovak students, unlike other respondents, perceived their own country as being part of Central Europe and underlined the global importance of China. Regarding Czech students, a much better knowledge of major geographical regions had been expected. These poor results are worrisome and have occurred in spite of the fact that the division of the world into supranational regions is discussed in many geographical textbooks already at the secondary school level. Clearly, many secondary school students are not meeting key competences, which may result in inadequate level of knowledge at universities (Zavadská 2015).

The division of the world into supranational regions will always pose a number of problems. Mental maps of major geographical regions much depend on individual preferences and on the abilities to process these preferences. The division into continents, though commonly accepted, seems inappropriate for a number of purposes (Polonský 2012), as it does not reflect economic, social and cultural circumstances. Our survey, based on small groups of respondents, suggests that different mental maps of supranational regions exist depending on the character of university background. However, the number of respondents was limited, and consequently the survey does not allow us to present sound results about differences among mental maps based on nationality.

Division of the world into supranational regions (compared to division into continents) allows unbiased insights into different parts of the world. As an example, the supranational region "Islamic world" (Fig. 12.3) is much more homogeneous than continents like Africa or even Asia. The latter shows the highest degree of internal differences regarding physical features, economic development, cultural environment, population, etc. Thus, the above-described supranational regions serve as a better tool that allows us to understand and analyse the changing character of the current world.

The creation of supranational regions lacks any logical order (methods) and always includes subjective points of view. Supranational regions are often used in scientific geographical publications as well as in non-geographical circles when explaining the territorial distribution of different phenomena (see Bramall, Thornton). For the sake of better understanding of the current world, it is essential to produce a clear and logical methodological background of supranational regions. Also important is the ability to distinguish among crucial and less important major geographical regions, including their internal differences and core-periphery relations. Using the above-mentioned approach would help to produce more accurate results and to apply the knowledge in practical life.

References

Anděl, J., & Bičík, I. (2015). Geografické makroregiony světa [World major geographical regions]. *Geografické rozhledy, 3*(14–15), 2–5.

Bičík, I., Anděl, J., Havlíček, T., et al. (2010). *Makroregiony světa* [World major regions]. Praha: Nakladatelství České geografické společnosti.

Bramall, C. (2008). Development of local financial systems in mainland China. *Eurasian Geography and Economics, 49*(2), 160–179.

Clawson, D. L., Johnson, D. L., Haarmann, V., & Johson, M. L. (2007). *World regional geography*. London: Pearson.

Cole, J. (1996). *Geography of the world major regions*. New York: Routledge.

De Blij, H. J., & Muller, P. O. (1988). *Geography (regions and concepts)*. New York: Wiley.

Goncalves, B., & Sanchez, D. (2014). Crowdsourcing dialect characterization through Twitter. *PLoS ONE 9*(11). doi:http://doi.org/10.1371/journal.pone.0112074.

Hampl, M. (2010). Regionální diferenciace společnosti: Obecné typy vývojových procesů [Regional differences of the society: General types of evolutionary processes]. *Geografie, 115*(1), 1–20.

Häufler, V. (1985). K socioekonomické typologii zemí a geografické regionalizaci Země [Social and economic typology of different countries, geographical regions of the world]. *Sborník ČSGS, 90*(2), 135–143.

Hobbs, J. J., & Dolan, A. (2009). *World regional geography*. Belmont: Brooks/Cole.

Huntington, S. (1996). *The clash of civilizations and the remaking of world order*. New York: Simon and Schuster.

Jackson, R. H., & Hudman, L. E. (1990). *World regional geography*. New York: Wiley.

Morris, J. W. (1972). *World geography*. New York: McGraw-Hill Book Company.

Novotný, J. (2003). Sociogeografická diferenciace současného světa [Sociogeographical differences in the current world]. *Geografie – Sborník ČGS Praha, 108*(1), 14–35.

Polonský, F. (2012). *Makroregionalne struktury sveta; reprezentácie, percepcie a objektivizácie* [World regional structures: Representation, perception, objectivization]. Dizertační práce [Thesis]. Praha: PřF UK.

Polonský, F., & Novotný, J. (2010). Cognitive mapping of major world regions among Czech geography students. *Journal of Maps, 6*(1), 311–318.

Scopus. (2015). http://www.scopus.com. Accessed 15 Sept 2015.

Siwek, T. (2011). *Percepce geografického prostoru* [Perception of the geographical space]. Praha: Česká geografická společnost.

Smetkowski, M. (2013). Regional disparities in Central and Eastern European countries: Trends, drivers, and prospects. *Europe-Asia Studies, 65*(8), 1529–1554.

Thornton, P. M. (2012). Mapping dynamic events: Popular contention in China over space and time. *Annals of GIS, 18*(1), 31–43.

UNData. (2014). *Standard country or area codes for statistical use*. New York: Statistical Services Branch Statistics Division United Nations.

Wallerstein, I. (1979). *The capitalist world – Economy*. Cambridge, MA: Cambridge University Press.

Wallerstein, E. (1984). *The politics of the world - economy: The states, the movements and the civilizations* (p. 191). Cambridge: Cambridge University Press.

Zavadská, K. (2015). Makroregiony světa pohledem různých věkových skupin studentů Česka a Slovinska [World major regions as seen by different age groups of Czech and Slovenian students]. Diplomová práce [Thesis], Ústí nad Labem: Univerzita J. E. Purkyně.

Chapter 13
New Political Geography

Barbara Baarová and Vladimír Baar

13.1 Introduction

Political geography has been expanding its sphere of scientific interest for several decades. The traditional topic of states and their relations have spread to a wide range of other topics, from the macro-regional dimension of international organisations to the processes of globalisation, as well as from the micro-regional national agenda to the local level. Key areas of political geography are no longer only international relations, but also problems associated with localisation conflicts (urban conflicts), the distribution of wealth (redistributive system, welfare system), local administrative organisations, state and local governments and their funding as well as the geography of elections. In addition to those key areas, political geography includes the study of geopolitics, international and border conflicts, integration processes and transborder cooperation, democracy and human rights, nationalism and separatism, linguistic and religious policies, wealth and poverty, global issues (nuclear threat, terrorism, mass migration, environmental risks, depletion of mineral resources, problems of sustainable development, demographic problems, etc.), crime (including international crime), socio-pathological phenomena, the politics of everyday life, the problems of women and children and many other topics. And this list is far from complete.

It is obvious that within the educational process, it is necessary to select topics and adjust their selection to the age of students at various school levels. The current trend largely remains focused on the classic topics associated with the state; some other topics do not appear separately, but are part of teaching of broader regional geographic units. Expanding the scope of political-geographical themes naturally

B. Baarová (✉) • V. Baar
Faculty of Science, Department of Human Geography and Regional Development,
University of Ostrava, Chittussiho 10, 710 00 Ostrava, Czech Republic
e-mail: barbara.baarova@osu.cz

P. Karvánková et al. (eds.), *Current Topics in Czech and Central European Geography Education*, DOI 10.1007/978-3-319-43614-2_13

led to a change in the didactic approach. Descriptive passages about the economic structure of countries and concentration of economic activities in cities or regions are gradually disappearing from geographic education. It is a logical consequence of the fact that the economy is developing dynamically, large companies cease to exist, there are new ones being established, and what was written in textbooks is ageing rapidly and losing its relevance. Purely political-geographical issues are much more likely to be discussed in the media, especially those of a negative character: conflicts, illegal immigration, terrorism, human rights violations, discrimination against ethnic and religious minorities and many others. To a lesser extent, however, we can also hear positive topics: free elections, international and cross-border cooperation and local problems in regions and municipalities. Therefore, it seems necessary to reassess the current concept of political geography and strengthen teaching of the problems encountered by pupils and students in their daily lives.

After the fall of the Iron Curtain, the Central European post-communist space had to face building a significantly different political system. The old system built on promoting a single opinion on virtually everything was replaced relatively quickly. Even in the education system, the first textbooks that tried to correct the lies in communist textbooks appeared in just a few months. The collapse of the communist bloc, however, took a few years and significantly changed the political map of the world. As it was not only the core of this bloc around the USSR which started collapsing, but it was followed by regimes in other parts of the world, more and more changes gradually occurred. Under the new events, political geography as a discipline began not only to broaden its sphere of scientific interest but also its methodological approaches. While it is not the aim of this paper to specify the new approaches and changes in political geography, it is necessary to recall that the political geography of the communist countries was an absolutely marginal part of geography and was limited to the most basic facts about the state, political borders and other factors that were more or less merged with similarly ideological political science. It is therefore not surprising that even in the training of teachers of geography, political geography was of marginal interest, and teachers were not prepared for the changes that occurred.

13.2 Political Geography in Tertiary Education

The modern concept of political geography appeared at Charles University in the 1990s in the Regional and Political Geography Department and at the University of Lodz in Poland, where in 1981 a separate department was established, now called the Department of Political Geography and Regional Studies. Virtually in all geography departments in Czechia, Poland and Slovakia, there are some lecturers specialised in political geography, but a whole team of political geographers has formed only at the University of Ostrava. In 2000, an interdisciplinary degree course called Political and Cultural Geography was accredited here under the guarantee of the co-author of this chapter. This course also includes political and other social science disciplines. However, it should be noted that political geography has to a different extent become an integral part of Political Science, International Relations and Security Studies in all three countries, and it is also represented significantly in a course called International Territorial Studies.

Within the teacher training course, the political-geographical component in the curricula was also strengthened, but the didactics of political geography has for a long time been lagging behind. Political-geographical topics are difficult to grasp, especially for pupils in primary schools. In secondary schools, students are already more advanced and can handle even complex topics. The more important is the didactic preparation of future teachers of geography, but also of history and social sciences, so that they are able to introduce relatively less attractive political-geographical topics to students in an interesting manner. It is particularly important to coordinate the diversity of their opinions and a common ungroundedness of their political views. If we return to the broad spectrum of political and geographical topics and approaches, i.e. to the relationship of space and politics, it becomes clear that only some of them can be used in the educational process. But it is hardly surprising, as many new issues from other subjects cannot be presented at schools. It is true even in natural sciences, where there are constantly new discoveries and approaches to research.

At present, researching political regions is considered to be the most useful part of analysis of the relation between space and politics, while it is not only the state that is regarded as a political region. The *macro-perspective method* has proved to be the most effective for this kind of research. It enables the researcher to define the area of interest without having to take any pregiven limiting conditions into account. The scale can be chosen freely, and it is not important whether it is smaller or larger, as the existing states are not taken into account. In this way, the problems of relations between developed and less developed regions within states are best studied, as well as relations between states integrated in international groups in a global scale. The analysis of causes and results of dynamics of development of different regions has also become very popular, which has subsequently again brought geopolitics into the spotlight. When compared to political geography, geopolitics focuses on territoriality from a different point of view. Out of all the disciplines of political science, it is closest to international relations, as it is focused mainly on the problems of territorial-political arrangement of the world researched in the global perspective. Geopolitics attempts to explain the rules of international relations by means of geography (Romancov 2002, p. 388–390), but it is very challenging for students and requires special preparation, including on the part of teachers. Student discussions in the form of conferences, where they present and defend opinions of fictitious representatives of states or other entities (e.g. Russia and Ukraine, Israel and Palestine, the four Visegrad countries, etc.) on selected topics, are highly beneficial and develop communication skills.

The fact that the political map has changed dramatically has been highly relevant for the Central European area, whose natural centre is Czechia, during the entire period since the fall of the communist regimes. Nearly three dozen new states have formed, some of which have further split (Federation of Serbia and Montenegro).

Others did not receive international recognition and disappeared (e.g. Chechnya or Serbian Krajina), while others exist in fact but still lack wider international recognition (e.g. Abkhazia and South Ossetia) or are totally without international recognition (Transnistria and Nagorno-Karabakh Arcach). The split of Czechoslovakia happened peacefully, while the disintegration of Yugoslavia triggered a bloody conflict that was felt even in Czechia in the form of a higher influx of migrants from the warring states. The collapse of the USSR caused more localised conflicts, however, with a long-term potential of their revival, which could also affect other places (eastern Ukraine). Geography of conflicts, as a part of conflictology, i.e. a science which is primarily focused on the creation, typologies and development of conflicts and conflict resolution, is one of the ways to integrate conflicts and solutions into teaching. The problem for teachers, students and pupils is of course the fact that a textbook that would be relevant is hard to find. At most, the textbooks contain a remark to the effect that there was an ongoing conflict at the time the book was published, which is useless as the textbooks are used continuously for a decade.

13.3 The Status of Political Geography in the Czech Educational Programme for Primary Schools

Although political geography is an important part of human geography, if you enter the keywords "political geography" in the most well-known methodical portal in Czechia, the FEP, no record appears, not even among digital learning materials. The reason is that this branch of science is scattered across various areas of education; many topics are taught more widely in history and civics (i.e. the foundations of the social sciences), but only to a small degree in geography at primary schools. Generally, political geography is barely mentioned, so keywords like polity, democracy and the current political map are found with more success. How are the various outcomes related to political geography incorporated into the Framework Educational Programme in the Czech Republic valid from 1 September 2013? (Kol. 2013)

At primary schools for children aged 6–11 years, we encounter the first outputs after which students gain a basic understanding of the political division of Europe. These are taught in the educational area *Man and His World*, within the topic *The Place We Live In*. At this stage the textbooks are chiefly devoted to the European Union at the expense of basic knowledge about the continent. In the thematic area *People Around Us*, teaching about human coexistence is recommended, which means a curriculum focusing on interpersonal relations, communication, the principles of democracy, trade, businesses, interest groups, political parties, churches, helping the sick and the socially weak and the complex topic of a common "European house" should be included (Kol. 2013). For pupils aged 12–15, there are other educational fields with issues related to political geography prepared within the

Framework Educational Programme. This is mainly the educational area *Man and Society*, which aims at

> ensuring that pupils learn the historical, social and cultural aspects of human life in all their diversity, variability and mutual relations. It familiarises students with the development of society and important social phenomena and processes, which are reflected in everyday life and influence the formation of the social climate. It focuses on creating positive civic attitudes, developing awareness of belonging to the European civilisation and culture and supports the adoption of values on which contemporary Europe is being built, including collective defence. An important part of education in this area is the prevention of racist, xenophobic and extremist attitudes, education towards tolerance and respect for human rights, gender equality and education for respect for natural and cultural environment and the preservation of artistic and cultural values (Kol. 2013).

This text accurately describes the knowledge and skills that are important for teaching political geography, but the problem is that the area of Man and Society explicitly covers only two subjects: history and citizenship education. Apart from these, the curriculum in the fifth to ninth grades of primary school in these subjects focuses on European integration (its essence, meaning and benefits) as well as on international economic, political and security cooperation between states and on important international organisations (Council of Europe, NATO, the UN and others) and globalisation – its manifestations; advantages and disadvantages; major global issues, including war and terrorism; and their possible solutions. So actually, most topics of political geography are not taught in the subject of geography, which is according to the FEP primarily classified in the educational area Man and Nature. Besides the natural geographical aspects, it also deals with the issue of world macro-regions, whose characteristics are discussed from the viewpoint of natural and socio-economic relations, with an emphasis on relationships and contexts (nature, climate, settlement, language, religious areas, cultural areas). Complex natural and social and of course cultural, political and economic problems, and the options for their solutions, are to be taught on model regions.

In Slovakia and Poland, the content of the educational areas is similar. There is little difference in the classification of geography in the framework education system in Slovakia, where it is included in the educational area Man and Society (SVPG 2015). But in Poland, it is a part of the area Man and Nature, similar to Czechia (Podstawa 2011).

For secondary school students, the situation is almost the same. Within the framework of FEP, topics related to political geography are reserved for history and social sciences, where more time is reserved for this issue than in geography.

It follows that the university education of student teachers in terms of political geography should be focused more on the combination with history or social sciences. In these two fields, teacher training should therefore be substantially strengthened on politically geographic issues. Unfortunately, future teachers of political geography can use their knowledge and skills better in interest groups or optional subjects, because at the current time allocation for geography, they cannot manage to discuss important aspects of political geography and geopolitics during lessons. This opens space for nontraditional forms of education, especially for project teaching, which directly offers cooperation through the intersection of topics.

13.4 Insight into the Curriculum of Political Geography in the Textbooks for Primary Schools

In the 1990s, the need to create new textbooks for homeland studies and geography for primary and secondary schools was seen by many geographical institutions in Czechia, Slovakia and Poland. The number of titles of these books written in these three countries from 1990 to 2000 for reasons of political, economic and social transformation after 1989 was in the hundreds (Wahla 2000). The differences between the countries are quite large and proportional to their size. In Poland the number accounts for nearly 60 titles, in Czechia it was 33, and it was 10 in Slovakia. The reason for major changes in textbooks reflects not only the creation of two independent states (Czechia and Slovakia in 1993) but also the efforts of the Central European countries to join the European Union. All the books began to react to the changes in the political map after the collapse of the bipolar world, the possibility of a real assessment of the economic level of states and the new teaching methods.

The co-author of the text was the first in Czechia, when he in early 1991 issued a Supplement to the Geography Textbook, which included fundamental political changes occurring in the late 1980s and 1990s (Baar 1991). Given the dynamic development, upgraded second and third editions were released in the following 2 years in nearly half a million copies. Although the first classical textbooks started being issued at that time, the supplement mentioned above was used by teachers throughout the 1990s. During the time Ateliér Milata Publishing House also began to publish methodical manuals for teachers of geography, history and other social sciences offering the teachers truthful information about the history of post-communist countries and the events of the revolutionary transformation period, including changes in political structures within the series named Scholaforum. These guides were published in thousands of copies and were in great demand. Books were also published devoted not only to the new countries in Europe but also, for example, to South Africa, Israel and Palestine, Vietnam, Laos, Cambodia, China, Taiwan and other states.

Many publishing houses published textbooks in Poland and Czechia at such a pace that supply far exceeded demand (unlike in Slovakia, where there was still only a limited number of titles available). Nowadays this trend has stopped, but publishers have significantly changed their marketing strategy. Many of them use the word "new" as a postscript, e.g. Fraus publishes "new-generation" textbooks and the other publishing house is directly called the "New School". Attractive graphic design has become commonplace, but is no longer the main criterion for schools for selecting the purchase of textbooks. The continuity in the content of homeland studies from the first to fourth grades has become more important, as well as whether a guide for teachers, which contains not only guidelines, but mainly photocopiable materials and information sources, is a part of the textbook. Another important selection criterion is the level and cost of workbooks and of course whether the textbooks are prepared in accordance with the Framework Educational Programme.

The entire offer must be supplemented by an e-book, which is based on the printed version, but adds audio recordings, videos, images and curricular links.

The same trend can be observed on the Polish market, but after a closer examination of geography textbooks for the fifth to ninth grades of primary school (in Poland gimnazjum) and secondary schools (liceum, szkoła ponadgimnazjalna), a big difference in the whole concept is revealed. The Czech market is dominated by textbooks that systematically deal with various macro-regions and their important state; these topics are similarly discussed by the Czech Republic's neighbours. Only textbooks designed for the eighth or ninth grade are more focused on social, economic and political geography. Generally they are introduced in the market under attractive names, such as *Contemporary World*, *People and Economy*, *Political Map of the Contemporary World*, etc. The question remains, however, how much these books are used in particular educational programmes in comparison with "classical" regional geography textbooks. Where the offer of textbooks for Czech secondary schools is concerned, it naturally depends on the type of school where the subject is taught (with the exception of physical geography). Thus, there are textbooks targeted especially at business academies (*Regional Aspects of the World Economy*, *Global Aspects of the World Economy*), but these are often used by grammar school teachers as well, especially by those who incline to human geography. The latest publication, *Macro Regions of the World*, is dedicated to grammar schools. It is also based on economic advancement of the world's macro-regions and dedicated to the most important states within them.

In Poland, the situation is different. All the textbooks in primary schools teach only model states across macro-regions and explain on their example feature natural, social and political phenomena and processes. Unlike the Czech and Slovak textbooks, in terms of regional geography, they focus much more extensively on Poland's neighbours. At secondary schools, textbooks are invariably focused on current changes, phenomena and processes in the economic, political and cultural spheres, which means comprehensively on human geography in the world and in Poland. The structure of all textbooks provided with an endorsement of the Ministry of Education is a reflection of the educational programme, which in Poland differs significantly from the Czech one.

An analysis of textbooks of the three Slavonic Central European states shows that in Polish textbooks more space is devoted to the aspects of political geography than in the Czech and Slovak ones. However, in practice this is not so, due to the very low number of hours at the fifth to ninth grades of primary school and secondary school. It is significantly different at primary schools. Weekly allocation for 3 years at the fifth to ninth grades of primary school in Poland is 4 h, in Czechia it is 7–8 h for 4 years, and in Slovakia it is 5 h for 5 years (Table 13.1).

For pupils of the eighth or ninth grade of primary school or grammar school, the essential information from political geography is compiled in textbooks from different publishers, and as the chapters do not differ in many respects, we can take as an example the textbook by the Czech Geographical Society – *The Current World* (Herink and Valenta 2010). The chapter Political Map of the Contemporary World

Table 13.1 Overview of lessons of geography at the second level[a] of primary schools

	Age of pupils	Name of the level	Year	Number of grades	Lessons per week (45 min) of geography at this level
Czechia	12–15	2. stupeň ZŠ	6–9	4	7–8
Poland	13–15	Gimnazjum	7–9	3	4
Slovakia	11–15	2. stupeň ZŠ	5–9	5	5

Note: [a]Primary schools in this region are divided in two levels – the first level starting from the age of 6. The conception of the second level varies according to the country

deals with the history of states, examples and definitions of independent states, dependent territories and changes of the political map of the state since 1900. Learning these topics is appropriate for students who already have a basic knowledge of regional geography. Another chapter deals with the location, size and population of states and introduces the interesting topic of microstates, types of state borders and their permeability (closed borders versus the Schengen Treaty), ways of delimitation of boundaries and conflicts that accompany them. It includes an interesting description of situations where boundaries separate individual cities, such as Cieszyn, Zgorzelec, Gubin or Berlin. In the chapter on state establishment and administrative division, the textbook outlines the structure of the administration in the USA in a diagram. Information about ways of governance is included in the newly written textbooks of the 1990s. Until then an ideologically distorted view focused on the difference between the "socialist and capitalist state" dominated education in the socialist sector.

A suitable annex to the topic is a chart comparing electoral systems in democratic and totalitarian regimes. The economic level of states is associated with the political systems through thematic maps and tables with GDP per inhabitant in purchasing power parity; the problems with the terms "rich North" and "poor South" are explained. Pupils learn about international political organisations and associations, where the UN, NATO and ASEAN serve as examples. The political geography curriculum is completed with a text on the outbreaks of unrest in the world today. Pupils will learn about the reasons for conflicts, examine a map of current conflicts and read a description of selected types of armed conflicts, but also receive information on the participation of Czech soldiers in military missions. A textbook published by the Nová škola (New School) publishing house (Chalupa and Hübelová 2010) is enriched by English and German translations of key terms. A textbook called the *Political Map of the Contemporary World* published by Fortuna (Baar 2005) covers almost identical topics, only elaborated differently. They differ in details only by the range of text and attachments. The reason is obvious: for the textbook to be included in the official list of textbooks of the Ministry of Education, the authors have to follow the Framework Educational Programme and the topics therein.

As mentioned in the introduction about the political geography within the Czech Framework Educational Programme, more opportunities for teaching this topic are provided by history and civics. In Poland, the situation is similar. The foundations of social sciences are taught under the name *wiedza o społeczeństwie* (civic education), and in Slovakia they are called *občianská nauka.*

13.5 Materials, Resources and Methods of Teaching the Topics of Political and Cultural Geography

Some of the topics of political and cultural geography appear already in the teaching of geography at the first grade of primary school. They are not included in textbooks, but many new methodological materials inspire teachers to present interesting information in the fourth and fifth grades of primary school. By analysing these materials, we can conclude that their demands are suitable for the second grade of primary school or for secondary school. So why are these relatively challenging topics already discussed in the first grade? The explanation is simple in principle: fifth grade classes have a greater percentage of talented and intelligent students, who later leave for 8-year secondary schools. Pupils at the first stage also show greater activity and curiosity, which is very important when implementing creative teaching ideas. This chapter provides examples of compiled material resources from the Czech and Polish environment. No new or innovative teaching methods are introduced, but methods will be mentioned that are relatively new in Czech education, but still underutilised in political geography. Only teachers who are inquisitive, proactive and open teach in this manner. They take part in the methodological workshops offered not only by universities (most often through projects) but also by other organisations and companies involved with education, for example, regional training centres. The best-known organisations in Czechia include the educational programmes Varianty (ČvT 2013b); Cenia (2012); Rezekvítek (2008); Národní institut pro další vzdělávání (National Institute for Further Education) (NIDV 2006) and A-Z Schola (2014). Methodological workshops are also organised by the publishing houses themselves to present all the benefits of their teaching materials.

One of the newly chosen teaching methods for teaching political geography in Czechia is Philosophy for Children, long known in other countries, which is based on sensitive, forthcoming, critical and creative thinking. These four listed essential characteristics seem to many as obvious and understandable, but it is not easy to teach students and pupils to think, and considering the complex contemporary issues in political geography, it is even harder. Philosophy for Children is not widely taught in Czechia, so we will try to explain its advantages in education. Sensitive students and teachers show a genuine interest in the ideas of others, thus developing empathy and respect. Because an atmosphere of mutual attention and listening is created in this way, participants must argue with caution and concentrate on a sensitive way of speaking. Open thinking is characterised by the ability to build on the thoughts of others, which helps to develop new directions of exploration. Naturally, the ability of communication and cooperation is deepened, because an openly thinking individual encourages other participants of the discussion or, on the other hand, reconciles with them. Another important characteristic of Philosophy for Children is critical thought, which is to lead to appropriate judgement, recognition of assumptions and consequences. This is the most difficult part of Philosophy for Children, so teachers should cultivate the ability of correct critical thinking, which involves disrupting prejudices, analysing various aspects and searching for meaning, causes and differences. But teaching political geography itself is subject to a biased approach of each of us. Different political opinions, beliefs or value systems are precisely the criterion that determines the complexity of the teaching of critical thinking. Equally important is creative thinking that is to lead to new ideas, to imagine new possibilities and alternative explanations, and to the ability to search for context or criteria. The concept of Philosophy for Children is very beneficial for strengthening skills which Czech pupils and students do not have. This is mainly the ability to argue, listen and cooperate (Varianty 2015). Teachers should focus on guidance phrases during the discussion: I agree, because …. I want to ask you …. I'd like to come back to …. Does a person have the right to …? We recommend the methodology and courses within the Varianty (2015) educational programme by People in Need for teachers of geography to learn to use the methods of Philosophy for Children.

Another beneficial material in the Czech Republic is a journal for the modern educational programme *The World of Today* (Dnešní svět). It belongs among the activities and programmes supported by the Depositum Bonum Foundation, which are divided into methodical training programmes and motivational-educational programmes (this includes the geography knowledge contest Eurorebus). The structure of each issue offers a variety of thematic maps, current graphs and tables, a general introduction to the topic and mainly a topic developed according to the individual continents, including Czechia. It includes a regular column called Use in Teaching, which is enriched with worksheets according to the level of the school. There is also news and a glossary in which the terms are translated, thus enabling easier orientation in the topic in foreign language texts (English, German, French, Spanish). A particularly good example for teaching political geography is issue number 4, named The State (Dnešní svět 2014/2015). In this issue, there is much more information on the state than in the textbooks for primary and secondary schools. The basic characteristics of states, national sovereignty, international recognition, the function of the state and theories of the state are clearly discussed here. Students will become familiar with the map of the State Power Efficiency Index (FSI), state power, state authorities, elections, types of state in the twentieth century, etc. The worksheet provides, e.g. a task in which a group of students allocate matching subindicators to the list of 12 basic indicators that are manifested by a negative assessment of the state according to FSI. Another worksheet also encourages discussion which assesses the importance of FSI indicators for tourism, foreign investors or the media. Worksheets contain the recommended solutions and mainly specific Internet resources and QR codes.

In light of the Central European region, it would be appropriate to discuss migration and the life of compatriots abroad, not only among Czechs, but also Poles and

Slovaks. Cross-cutting issues in this case are Thinking in a European and Global Context, Multicultural Education and Media Education. The topic of Czech expatriates in the world was elaborated by Mgr. Klára Smolíková for one of the publications of the series *Good School* (Thinking in a European and Global Context, Raabe). Appropriate teaching methods are independent and group work with worksheets, presentation of the results, separate search and processing of information, problem identification and problem solving. The purpose of educational material about Czech expatriates is to draw attention to the real fate of Czechs abroad, outlining the motives for emigration, characterisation of emigration waves and mapping out major areas. The fact that in the timeline of Czech emigration, religious, economic, military or political reasons alternate, allows students and pupils to understand today's strong wave of migration in Europe.

Other high-quality materials to the problem of migration are offered by Varianty. cz. First, we shall mention a blended learning course for teachers called the "labyrinth of migration" that deepens knowledge on this issue and offers tips on how to explain the causes and impacts of migration and asylum issues to the pupils and students. It further acquaints the participants with the testimony of those who have come to live in Czechia. Information and media "Migration Programme" has been dealing with the issues of migration in Czechia since 2002. One of the goals is that each migrant is seen as a unique personality with his own life story. In the framework of this programme, media trainings for students, journalists and migrants are organised (ČvT 2013a). In recent years, activities connected with lectures or seminars with prominent figures increased significantly at Czech secondary schools, but also at primary schools. Today's youth also need positive moral models, and they tend to receive information from well-known figures or eminent experts positively. Students do not come to the universities, but academics, travellers, ambassadors, MEPs and other experts are invited directly to schools under the projects. Presidential visits to schools and subsequent discussion could perhaps also be included in teaching political geography.

As mentioned above, in Poland the teaching of political geography is also implemented mostly in the foundations of social sciences (*wiedza o społeczeństwie*). Many interesting ideas for the course are available at the well-prepared and constantly updated educational platform called Centrum Edukacji Obywatelskiej, within the chapter named The Whole World in the Classroom (CEO 2015).

Teachers can find modern methodological materials here, including filled-in worksheets designed for history, geography and civics. Topics such as Colonialism from a Different Perspective, Immigrants in Poland, Africa in Sepia, Distant Countries – Nearby Problems Where Does Inequality Come from?, and others will surely be interesting for pupils and students (Witkowski 2011).

In terms of teaching methods, dramatisation prevails. This entails the visual presentation of an event or story, as well as a discussion, which is initiated by authentic stories and further stimulated by appropriate questions on the part of the teacher, because discussion assumes a certain level of knowledge related to the problem addressed by all parties involved. The assumption, however, that Polish youth growing up in a society with such a significant percentage of Polish nationals living abroad approach migration with more empathy than Czech and Slovak youth is not confirmed in practice. In the introduction of the report on the condition of Polish culture and Polish nationals abroad from 2012 (Raport 2012), the Polish Minister of Foreign Affairs, Radosław Sikorski, states that since the relaxation of borders in the European Union, almost two million Poles have used the opportunity to migrate abroad for work.

At present, approximately 15 million Poles live abroad, where in Poland itself it is 37 million. However, education on migrations is much needed in Poland. An interesting teaching material called Immigrants in Poland (CEO 2015) can be adapted to the Czech and Slovak conditions as well. Students work in groups with worksheets containing authentic stories of migrants (a Vietnamese, a Chinese, a Ukrainian, a Nigerian). Based on the text, students reply to questions such as where did the immigrant come from, why, what is his living standard in Poland and what difficulties he has encountered there. They then write on the blackboard in two columns the reasons for migration and the problems that foreigners in Poland encounter. After that they are asked to complete the following sentence: "The situation of migrants is as follows…, because…". Their last task is to write in pairs the ideas and suggestions that could improve the situation of immigrants in Poland.

Other Polish materials on migration are available on the Sfery educational portal (Przyroda 2015). Although these multimedia materials for teachers have been created since 2012, when today's issue of migration to Europe was unknown, the topics prepared are extremely relevant now under the increased pressure of migrants on Europe. Among purely natural scientific topics or regionally geographic, we can find a range of topics called Science in the Media, which are intended for geography, alongside other sciences. And the names of particular topics already suggest why they are relevant: policy of restrictions against immigrants, illegal immigrants, terrorism and migration, the consequences of migration, the proportion of migrants in the population and Greece – operation against illegal migration.

It is clear that education is not focused only on migration; other interesting material comes from the methodology The Whole World in the Classroom mentioned above and is called The Story of Eva. This highlights the problem of the lack of water in South Sudan, showing the negative impact of this situation on local life. In the beginning, countries with water shortage are listed and the area of Sahel is delimited. The aim of the work is to understand and explain the need of rational use of water and to describe the relation between water supplies and appropriate forms of economic development. The Story of Eva is very evocative for students, because Eva is their age and her story is plausible. After the story is introduced, the class is divided into two groups – the first finishes telling the story or draws a cartoon using keywords: new well in the village, education, health, development, happiness and human rights. The other group has different keywords: non-functional well, drought, disease, sorrow, etc. Their work is then presented and other issues are discussed.

In Czechia, similar materials can be found at the website of the methodological portal – the inspiration and experience of teachers (MP 2015). The Struggle for Water (author Mgr. Petr Vinš) (Boj 2008) is the name of a lesson based on a cartoon video from the project Paragraf 11/55, which deals with increasing legal awareness of youth. The video contains a fictional story of a kingdom ruled by an old despotic king. After he dies, he is succeeded to the throne by a young king. After coming to power, he proposes new laws that should make life in the country easier. A strong earthquake had damaged most sources of drinking water. The only source that remained was located on the property of the king's mean uncle. He took advantage of the water shortage for his own enrichment. Current laws have failed to punish him. The task of the pupils is to imagine themselves in the situation of the young king and modify the law in order to prevent situations described in the story. Additionally, the students must propose three current laws which they revoke or modify and, on the contrary, suggest three new laws.

13.6 Mutual Picture of Neighbours: Political-Geographical Topic in Interdisciplinary Relations

Political and cultural geography in the eyes of today's pupils and students is equated mainly with relations between states, religious and ethnic differences, the political map of the world and international organisations.

To a much lesser extent, the existence of neighbouring states, common borders and cross-border cooperation are perceived as political-geographical issues. The topic Our Neighbours is nevertheless important for the inhabitants of any state. But do the students ask questions such as:

- Where do neighbouring nations come from?
- Do we have anything in common?
- Have we always had the same borders with each other?
- Do we know them much better than other states and nations?
- Do we visit them more often?
- How do they see us? (Baar 2001)

For many inhabitants of our country living in the border regions, these issues are irrelevant, because they speak two languages, they regard knowledge of people and nature abroad as commonplace and they participate in transborder cooperation, especially in terms of culture but also economics. Do they know, however, the other neighbours of their own state?

Analysis of the framework educational programmes of Czechia, Poland and Slovakia ensures that nearby states are given sufficient attention. So what should secondary school students know about each of the three Slavic states?

Linguistically and by character, *Slovaks* are undoubtedly the closest to Czechs. Our predecessors already lived with them within Great Moravia, the borders of which spread across today's Czech regions and western Slovakia. People then still spoke the same Slavic language, and even today there are little differences between the Czech and Slovak languages. Since the arrival of the Magyars, who in the ninth century contributed to the downfall of Great Moravia, however, today's Slovakian territory became part of the Hungarian state, and the political, cultural and economic development of Slovaks occurred in different cultural environments. While in the western part of the former empire, the Czech state was formed; the eastern part was exposed to the influence of the Hungarian language, a language very different, for nearly a millennium. Apparently this is why Slovaks have retained their language, but only in relatively isolated mountainous areas. In the lowlands, where large numbers of Hungarians settled, the Slavic language gradually disappeared. The different political developments had a major impact on the fact that Czechs and Slovaks eventually formed two distinct nations. It is thus not surprising that during the decomposition of the Habsburg Empire in the twentieth century, the representatives of Czechs and Slovaks agreed to form a joint state called Czechoslovakia, which was officially announced on 28 October 1918.

Czechoslovakia contributed significantly to the mutual recognition of the linguistically related nations. They are so close that most people gradually stopped to realise whether it was the Czech or Slovak language which was being spoken on the radio or TV. Czechs have no problem reading Slovak books and magazines, and Slovaks read the Czech ones. People migrated and settled in the territory of the other friendly nation and linguistically adapted to the majority. Numerous mixed marriages were created; creation of joint sports teams and musical ensembles became commonplace. Only occasionally were there misunderstandings. Czechs and Slovaks got on with each other, and their mutual sympathy remained intact even after the division of Czechoslovakia into two independent states, Czechia and Slovakia, which occurred on 1 January 1993. The rapid weakening and collapse of the communist regimes in Central and Eastern Europe have contributed greatly to this event. While Czechs perceived Czechoslovakia as a continuation of their own state formed already on the ruins of Great Moravia and expanded in 1918 by the Slovak territory, for some Slovaks a statehood shared with the Czechs was not adequate.

However, Czechs and Slovaks still clearly like each other – football or hockey matches can serve as proof. When the opponent is someone else, most Czechs support the Slovak team and most Slovaks the Czech one. It can thus be seen that the designation "brotherly nations" remained true even after the division of Czechoslovakia.

Poland is our second Slavic neighbour. Even though the Polish language is not as close to Czech as Slovak, it is still quite comprehensible to the Czech people. This is especially true in the border areas of the country's northeast. This is where a smaller part of historical Silesia can be found, which is also contained in the title of the Moravian-Silesian Region. Historical Silesia stretches along almost the entire Polish border area with Czechia and for a long time was a part of the Czech state, although it was inhabited by a large German minority. The Czech kings of the Přemyslid dynasty occasionally ruled in Poland, and Polish Piasts sometimes reached to the Czech throne. The Czech state, however, was focused mainly towards the west, while Poles expanded to the east, so that mutual political contacts gradually weakened. Along with the Czech crown, the Habsburgs gained all of Silesia, and with the weakening of Poland in the second half of the eighteenth century, they added other areas with Polish population. Although the greater part of Silesia was acquired by German Prussia in 1742, Czechs along with some Poles lived under the large multinational Habsburg Empire until 1918. The emergence of two independent states brought about the problem of defining a common border in the Silesian Cieszyn region. Czechoslovakia acquired its smaller part with a predominantly Polish population, which the Poles perceived as an injustice. Czechoslovakia had then needed to gain this territory in order to secure a railway connection with Slovakia, but Poland never resigned itself to the loss of Cieszyn in 1938 and took advantage of the weakening of our state by the German occupation of the Czech border areas and joined Czech Cieszyn Silesia to Poland with the support of Germany. The subsequent German war against Poland and its elimination showed Poles that the hostile act against Czechoslovakia was not a good investment in Czech-Polish relations. After the war, the borders returned to their status prior to 1938, and on the 70th anniversary, Poland apologised to the Czechs and Polish President Kaczynski described the annexation of Cieszyn as a sin (Baar and Houdková 2010).

Czech students should also know in detail their German and Austrian neighbours. Polish students should know Germany, Lithuania, Russia, Belarus and Ukraine, and Slovak students Ukraine, Hungary and Austria. What will they learn in the context of the subject of geography?

At Czech schools, pupils are introduced to the regional geography of neighbouring states already between the first and fourth grades. However, in the fifth to ninth grades, one lesson is devoted to each of the states, where their knowledge is not extended, but merely repeated. At secondary schools (i.e. at grammar schools), they will be able to enumerate several cities and describe tourist attractions. The Polish educational programme also remembers neighbouring countries and deals with them in textbooks in the range of one lesson. If geography for the seventh to ninth grades is taught only 1 h a week at Polish primary schools and students are to become familiar with model states, phenomena and processes throughout the regional geography of the world, then the text on neighbouring states can be considered very rich. In Slovak textbooks, similar space is devoted to teaching about neighbouring states as in the Czech ones.

How strong then is mutual awareness really? Experts try to answer this question at conferences, where geographers from universities, authors of geography textbook and publishers from all three states meet. These meetings are beneficial in terms of examining the books and exchange of experience. A second platform for these meetings is the mutual image of each state, not only in education but also in society. Geographers thus are not the only ones who deal with this problem; naturally historians and linguists discuss it as well.

An analysis of school history textbooks focusing on the mutual image of Polish and Czech key events and important personalities has shown that the space reserved for the Middle Ages and the twentieth century is sufficient (Gracová et al. 2014). The main difference is the degree of emotionality in the interpretation of disputed historical moments in Polish-Czech relations, which is much more marked in the Polish textbooks. In the Czech textbooks, minimal attention is devoted to the Polish minority in the Czech lands. This fact, of course, occurs in geography textbooks as well and not only there but also in the media, because the ties of compatriots with Poles living beyond their borders, or the Polish diaspora (i.e. Polonia), are much stronger than in the Czech population. The importance of "Polonia" in connection with the attitude of the society towards migration will be mentioned further in the text. The fact that the history of art is missing entirely in the textbooks of both countries shows a certain laxity in the attitude of the authors of the textbooks to their closest neighbours. In the context of school education, it is expected that the most objective image of the neighbouring country is obtained through an interdisciplinary approach of teachers who are knowledgeable in several fields so that they are able to fulfil the curriculum and teach about transborder cooperation and linguistic, cultural and historical proximity. This is ensured by transborder cooperation between schools and other institutions that provide educational and cultural activities in individual regions. Textbooks cannot substitute some of the educational materials of a complementary nature, regional textbooks and other events providing the bulk of information related to the regional topic.

13.7 The Conflict in the Balkans in Czech, Slovak and Polish Schools

The model example of the Balkan conflict, which is the best-known problem area closest to the Central European space, is a living political-geographic problem. On this example, the roots of the conflict can be demonstrated, its course and post-conflict solution, the question of the states recognised as existing de jure or de facto as well as the impact of this problem on international relations. Within this example, the disintegration of a state formation (Yugoslavia), border disputes (e.g. between Serbia and Croatia), ethnic and religious conflict (Bosnia and Herzegovina) and separatist and irredentist processes (Kosovo, efforts to unify Serbs in one state) can be discussed in detail. Related to this topic is also the violation of state sovereignty, the involvement of international organisations in conflicts, forced migrations and relocation of residents, xenophobia and genocide, multiculturalism, the process of international recognition of new states and geopolitical disputes of powers, but also the cartographic aspects of the conflict (displayed textbooks, atlases, media). For educational purposes, the Balkan conflict is therefore appropriate, as it has already passed through all the stages and has to a certain extent been resolved.

In Tables 13.2 and 13.3, some of the geography textbooks for the second grade of primary schools and secondary schools, which were issued between 2003 and 2012, are listed. They are used as representative samples to demonstrate how the subject matter of the Balkan conflict is handled in textbooks for pupils and students in primary and secondary schools in Czechia, Poland and Slovakia.

Table 13.2 Summary of information about the conflict in the Balkans in the secondary school textbooks in Czechia, Poland and the Slovakia

The author and the name of the textbook	Year and place of issue	Designation	Topic	Content
V. Baar *Hospodářský zeměpis, regionální aspekty světového hospodářství (Economic Geography, Regional Aspects of the Global Economy)*	2010 Prague	Business academies	Southeast Europe	The reasons for the disintegration of Yugoslavia, comparison of the economic level of individual republics of the former Yugoslavia in 1991, the political organisation of the Republic of Bosnia and Herzegovina, the current national structure in individual states, Kosovo (grounds for the declaration of independence), information about the civil war, the history of the Balkans since the Roman Empire until the twenty-first century
J. Anděl, I. Bičík, T. Havlíček *Makroregiony světa, regionální geografie pro gymnázia (Macro-regions of the World, Regional Geography for Secondary School)*	2010 Prague	Grammar schools	Southeast Europe	The history of the Balkans, a comparison of the economic level of individual republics of the former Yugoslavia in 1991, brief information about the civil war
Stark, G. Wnuk *Teraz Geografia (Geography Now)*	2012 Torun	Professional schools	Changes to the current world map	A list of newly created states of former Yugoslavia with the year of origin, including Kosovo
D. Makowska *Geografia 2 (Geography 2)*	2004 Warsaw	Grammar schools, professional schools	Balkan conflict	Religious structure of individual nations, the borders of Yugoslavia in 1992 and after the breakup, the history of the Balkans from the Ottoman Empire to the twenty-first century, the current national and language structure, the intervention of international forces, information about the civil war

(continued)

Table 13.2 (continued)

The author and the name of the textbook	Year and place of issue	Designation	Topic	Content
J. Korba, J. Mordawski, W. Wiecki *Geografia 2, geografia i czlowiek* (*Geography 2, Geography and People*)	2008 Gdynia	Grammar schools	Balkan problem	Religious structure, a brief history of the conflict since 1914. A detailed description of the situation in Kosovo, the declaration of independence
J. Stasiak, Z. Zaniewicz *Geografia, vademecum maturalne* (*Geography, Handbook for High School Graduation*)	2011 Gdynia	Preparation for high school graduation	Armed conflicts worldwide	The causes of the conflict between Serbs and Albanians, Kosovo's independence
F. Plit, W. Osuch M. Sielatycki, J. Wrona *Geografia – czlowiek i jego dzialalnosc* (*Geography – Man and His Activities*)	2003 Warsaw	Grammar schools	Ethnic conflicts on the Balkan Peninsula	The causes and consequences of the civil war, history since 1914, the UN and NATO intervention, map of ethnic structures in Bosnia at the end of the twentieth century, waves of migration
B. Lenartowicz, E. Wilczynska, M. Wojcik *Geografia na czasie* (*Geography at the Time*)	2008 Warsaw	Professional schools	Armed conflicts and terrorism worldwide	A description of the civil war, religious structure, the unilateral declaration of Kosovo, international NATO forces
A. Gajdoš *Hospodárska geografia* (*Economic Geography*)	2006 Bratislava	Business academies	Southeast Europe	States which have emerged after the breakup of Yugoslavia, a brief remark about the genocide in Kosovo, the ethnic composition of individual states

A curriculum analysis of the Balkan conflict has shown that in many textbooks there is no significant difference between publications designed for primary and secondary schools in terms of the amount and complexity of information. The main difference lies in the area in which the subject matter is incorporated. In some of them, the text appears within the regional geography of Southeast Europe; in others it is a part of teaching about conflicts and global issues. In one of the textbooks

Table 13.3 Summary of information about the conflict in the Balkans in primary school textbooks in Czechia, Poland and the Slovakia

The author and the name of the textbook	Year and place of issue	Designation	Topic	Content
D. Hübelová, P. Chalupa *Zeměpis, Evropa* (*Geography, Europe*)	2009 Brno	Eighth grade	Southeast Europe	The names of new states, the area was part of the Roman Empire, information about the civil war, the declaration of independence of Kosovo, peacekeeping – UN intervention troops, including the Czech Republic, religious structure, nationality structure
D. Hübelová, P. Chalupa *Zeměpis – Lidé a hospodářství* (*Geography – People and Economy*)	2010 Brno	Ninth grade	Development of the political map of the world	Religious structure, the course of the civil war, the new states, the independence of Kosovo
J. Demek, I. Mališ *Zeměpis světadílů* (*Geography of Continents*)	2008 Prague	Seventh to eighth grade	Southeast Europe	States which have emerged after the breakup of Yugoslavia, including Kosovo
P. Chalupa, J. Demek, J. Rux *Lidé a hospodářství* (*People and Agriculture*)	2009 Prague	Eighth to ninth grade	Outbreaks of ethnic, political and religious conflicts	The disintegration of Yugoslavia, the newly created states with the date of foundation, the independence of Kosovo
J. Brinke, V. Baar, V. Kašpar, M. Pollaková *Zeměpis Ameriky, Asie a Evropy* (*Geography of America, Asia and Europe*)	2005 Prague	Sixth to seventh grade	Southeast Europe	The history of the Balkans in the twentieth century; ethnic, religious and linguistic composition; the republics of the former Yugoslavia; the newly created states; the military forces of the UN

(continued)

Table 13.3 (continued)

The author and the name of the textbook	Year and place of issue	Designation	Topic	Content
J. Herink, V. Valenta a kol. *Současný svět* (*Contemporary World*)	2010 Prague	Eighth to ninth grade	Political map of the modern world	The disintegration of Yugoslavia, the characteristics of the civil war, Kosovo (the problem), the declaration of independence, the military forces of the UN
M. Zaťková, J. Brezovská, A. Durmisová *Zeměpis 6* (*Geography 6*)	2008 Bratislava	Sixth grade	Southeast Europe	Brief information about the history of the Balkans, the ethnic composition of Yugoslavia in 1991, characteristics of individual states after the collapse of Yugoslavia, Kosovo's independence, about the civil war in each country separately
R. Uliszak *Puls Ziemi* (*Pulse of the Earth*)	2012 Warsaw	Seventh to ninth grade	The most important regions of conflict and their consequences	The reasons of the war in Kosovo, the description of the disputed region
B. Dobrosik, A. Hibszer, J. Soja *Puls Ziemi* (*Pulse of the Earth*)	2013 Warsaw	Eighth grade	Political division of Europe	All the states of the former Yugoslavia, including Kosovo (photo of the declaration of independence in Pristina)
J. Mordawski *Planeta, Eurazja* (*Planet, Eurasia*)	2003 Straszyn	Seventh to ninth grade	Regional armed conflicts	Causes of the civil war, a time delimitation, the peace treaty of Dayton

(Plit et al. 2003, p. 275), the topic called Ethnic Conflicts in the Balkans is prepared as a seminar (*warsztat geograficzny*), which is dedicated to this conflict to a greater extent than others. Questions and tasks are included in the text and complemented by maps, and the chapter truly serves as a model for understanding conflict formation and resolution. A fundamental factographic difference in the textbooks is the recognition of Kosovo's independence. The Serbian province of Kosovo unilaterally declared independence at the parliament session in Pristina on 17 February 2008. The Czech government agreed to the establishment of diplomatic relations with the Republic of Kosovo on 21 May 2008, thereby recognising Kosovo as an independent state (Vláda 2008). The Polish government did this much sooner, on 26 February 2008, expected to approach other European countries, like the Czechs (Portal 2008). Slovakia is among the countries that refuses to recognise the Republic of Kosovo's independence, along with Russia, China, Venezuela and Azerbaijan, or the EU member states of Spain, Greece, Cyprus and Romania. The reason for this non-recognition is fear of separatist tendencies of minorities living in their own territories. Thus Slovak textbooks appear to contain only simple information about the unilateral declaration of independence, but an independent Kosovo is not recorded in maps or atlases.

Sample study materials for teachers at primary and secondary schools for teaching the Kosovo conflict, including methodological notes and worksheets, are provided in the book *Geopolitics and Geoculture in Selected Regions* (Baarová 2015). This support includes texts describing the history of Kosovo from the first millennium in two versions, both from the perspective of Serbs and Albanians. The text is especially dedicated to the history of the twentieth century, with an emphasis on the existence of "Great" Albania, and the fact that Kosovo and Metohija became part of the newly federated Yugoslavia. It is not just a description of the events, but a compendium of information on individual populations, their emotional relationship to the new state and discrimination of certain nations. Students will learn about the activity of Slobodan Milosevic, the initial declaration of Kosovo as an independent state in 1991 and the Serb response.

A more detailed description of the civil war does not focus only on the data, but tries to explain the reasons for the actions of individual politicians and the response of international organisations and great powers such as Russia and China. More space than in textbooks is also dedicated to the events in Pristina in 2008, which are accompanied by a map of the states according to their opinions on Kosovo's declaration of independence. Links to information about the KFOR mission (MO 2014) or to the latest report on the condition of the political and economic scene in Kosovo are complemented by news from the region (EC 2014).

Regarding the level of the pupils, the following types of questions and tasks for primary schools can be asked:

- Why has Slovakia not recognised the independence of Kosovo? Do Hungarians in Slovakia have the right to autonomy?
- How did Milosevic's regime contribute to the worsening of the conflict? Was Milosevic punished?
- Find the most important countries which recognise the independence of Kosovo on the map or on websites.
- Find out what currencies are valid in the countries of the former Yugoslavia. Out of those opting for the euro, are all of them members of the European Monetary Union? Give reasons.
- What symbols do the states strike on the euro coins? (search the Internet)

For secondary school students:

- What is the difference between the Serbian and Kosovo Albanian version of history?
- On what basis can international organisations engage in a conflict?
- In your opinion, was the 12-year mission of the Czech Army in the Balkans justified?
- What distinguishes countries that did not recognise Kosovo? Do they have anything in common?

13.8 Conclusion

Analysis of the status of political geography in the education system shows that it is a discipline with an irreplaceable role in primary and secondary education. Given the close interconnection of political geography and politics, it is not surprising that a number – what might even say many – of classical concepts and topics related to the state as the basic political unit of the international system are taught not only in geography but logically also in history and general social science. Teaching time for political-geographical topics is very low and the possibility of its expansion is virtually zero. Space for teaching these topics can only be found in specialised workshops. And as political geography is, along with scientific disciplines, constantly evolving, there are many ways to inspire.

References

Anděl, J., Bičík, I., & Havlíček, T. (2010). *Makroregiony světa, regionální geografie pro gymnázia* [World major regions, regional geography for grammar schools]. Praha: ČGS.

A-Z Schola DVPP. (2014). Další vzdělávání pedagogů [Further education teacher]. http://www.azschola.cz/dalsi-vzdelavani-pedagogu.html. Accessed 12 Sept 2015.

Baar, V. (1991). *Doplněk k učebnici Zeměpis 6, Evropa* [Textbook geographic 6 supplement, Europe]. Praha: Scientia SPN.

Baar, V. (2001). *Národy na prahu 21. století: Emancipace nebo nacionalismus* [Nations on the threshold of the 21st century: Emancipation or nationalism?]. Ostrava: Ostravská univerzita.

Baar, V. (2005). *Politická mapa dnešního světa* [Current political map of the world]. Praha: Fortuna.

Baar, V. (2010). *Hospodářský zeměpis, regionální aspekty světového hospodářství* [Economic geography, regional aspects of the world economy]. ČGS: Učebnice pro obchodní akademie a jiné střední školy. Praha.

Baar, V., & Houdková, Z. (2010). *Výchova k myšlení v globálních a evropských souvislostech* [Education to thinking in European and global contexts]. Praha: NČGS.

Baarová, B. (2015). *Geopolitika a geokultura ve vybraných regionech pro ZŠ* [Geopolitics and geoculture in selected regions for elementary schools]. Ostrava: Ostravská univerzita.

Boj o vodu [The Fight for Water]. (2008). Metodický portál RVP. http://rvp.cz/detail/id/dum-d1041. Accessed 12 Sep 2015.

Brinke, J., Baar, V., Kašpar, V., & Pollaková, M. (2005). *Zeměpis Ameriky, Asie a Evropy pro 6. a 7. ročník základní školy* [Geography of America, Asia and Europe for 6th and 7th grade of elementary school]. Praha: Fortuna.

Cenia – česká informační agentura životního prostředí. (2012). Rámcové vzdělávací programy [Framework educational programmes]. http://www1.cenia.cz/www/evvo/skolni. Accessed 16 Sep 2015.

CEO – Centrum Edukacji Obywatelskiej. (2015). http://www.ceo.org.pl/pl/koss. Accessed 15 Sept 2015.

Chalupa, P., & Hübelová, D. (2010) *Zeměpis. Lidé a hospodářství* [Geography. People and economy]. *Učebnice pro 9. ročník základní školy*. Nakladatelství Nová škola Brno.

Chalupa, P., Demek, J., & Jaromír Rux, J. (2009). *Zeměpis. Lidé a hospodářství pro základní školy* [Geography. People and economy for elementary schools]. Praha: SPN.

ČvT – Člověk v tísni. (2013a). Migrace [Migration]. http://www.clovekvtisni.cz/cs/migrace. Accessed 12 Sept 2015.

ČvT – Člověk v tísni. (2013b). Varianty [Variants]. http://www.clovekvtisni.cz/cs/clanky/varianty?gclid=CNbu7qW5l8gCFQbkwgod7MEPow. Accessed 13 Sept 2015.

Demek, J., & Mališ, I. (2008). *Zeměpis světadílů pro základní školy* [Geography of continents for elementary schools]. Praha: SPN.

Dobrosik, B., Hibszer, A., & Józef Soja, J. (2013). *Puls Ziemi. Podręcznik do geografii dla klasy drugiej gimnazjum* [Pulse of the earth. Guide of geography for the second class of the grammar school]. Warszawa: Nowa Era.

EC – European Commission. (2014). *Kosovo*Progress report*. http://ec.europa.eu/enlargement/pdf/key_documents/2014/20141008-kosovo-progress-report_en.pdf. Accessed 10 Sept 2015.

Gajdoš, A. (2006). *Hospodárska geográfia – pre obchodné akadémie* [Economic geography – For business academy]. Bratislava: Orbis Pictus Istropolitana.

Gracová, B., Labischová, D., & Szymeczek, J. (Eds.). (2014). *Vzájemný obraz souseda v polských a českých školních učebnicích* [Reciprocal image of a neighbor in the Polish and Czech school textbooks]. Ostrava: Ostravská univerzita.

Herink, J., Valenta, V., et al. (2010). *Současný svět. Základy společenského, hospodářského a politického zeměpisu, lidé a příroda – životní prostředí* [Contemporary world. Basics of social, economic and political geography, people and nature – Environment]. *Učebnice pro 8. a 9. ročník ZŠ*. Praha: Nakladatelství České geografické společnosti s.r.o.

Hübelová, D., & Chalupa, P. (2009). *Zeměpis, Evropa* [Geography, Europe]. *Učebnice pro 8. ročník základní školy*. Brno: Nová škola.

Kol. (2013). *Rámcový vzdělávací program pro základní vzdělávání /RVP ZV/* [Framework Education Programme for Basic Education FEP BE]. Praha: Ministerstvo školství, mládeže a tělovýchovy, VÚP. http://www.msmt.cz/vzdelavani/zakladni-vzdelavani/upraveny-ramcovy-vzdelavaci-program-pro-zakladni-vzdelavani. http://www.vuppraha.cz/wp-content/uploads/2009/12/RVP_ZV_EN_final.pdf. Accessed 15 Sept 2015.

Korba, J., Mordawski, J., & Wiecki, W. (2008). *Geografia 2, geografia i człowiek. Podręcznik dla liceum ogólnokształcącego* [Geography 2, geography and people. Textbook for grammar schools]. Gdynia: Operon.

Lenartowicz, B., Wilczyńska, E., & Wójcik, M. (2008). *Geografia na czasie. Podręcznik dla szkół ponadgimnazjalnych* [Geography at the time. Textbook of geography for higher secondary schools]. Warszawa: Wydawnictwo szkolne.

Makowska, D. (2004). *Geografia 2, podręcznik dla liceum ogólnokształcącego, liceum profilowanego i technikum* [Geography 2, textbook for higher secondary schools]. Warszawa: Wydawnictwa Szkolne i Pedagogiczne.

MO – Ministerstvo obrany České republiky. (2014). *Zahraniční mise. Aktuální mise. Kosovo* [The Ministry of Defence of the Czech Republic. Foreign Missions. Current Missions. Kosovo]. http://www.mise.army.cz/aktualni-mise/kosovo/kosovo-kfor-15559/. Accessed 10 Sept 2015

Mordawski, J. (2003). *Planeta, Eurazja, wybrane zagadnienia geografii świata* [The planet, Eurasia, selected issues of the world geography]. Guide to geography Straszyn. Wydawnictwo Rożak.

MP – Metodický portál Rámcového vzdělávacího programu pro základní školy [Methodological Portal of the Framework Education Programme for Basic Education FEP BE]. (2015). http://dum.rvp.cz/index.html. Accessed 12 Sept 2015.

NIDV – Národní institut pro další vzdělávání [National Institute for Further Education]. (2006). http://www.nidv.cz/cs/. Accessed 18 Sept 2015.

Plit, F., Osuch, W., Sielatycki, M., & Wrona, J. (2003). *Geografia – człowiek i jego działalność. Podręcznik dla liceum ogólnokształcącego* [Geography – Man and his activities. Manual for the grammar school]. Warszawa: Wydawnictwa Szkolne i Pedagogiczne.

Podstawa programowa przedmiotu geografia [Curriculum of the subject of geography]. (2011). http://men.gov.pl/wp-content/uploads/2011/02/5e.pdf. Accessed 15 Sept 2015.

Portal spraw zagranicznych. (2008). Polska/Rząd uznał niepodległość Kosowa [Poland/ Government recognized the independence of Kosovo]. http://www.psz.pl/92-polska/polska-rzad-uznal-niepodleglosc-kosowa. Accessed 16 Sept 2015.

Przyroda w liceum [Nature in Lyceum]. (2015). http://www.sfera.lublin.pl/sobieski/co-nas-wyroznia/projekt-przyroda-w-liceum.html. Accessed 15 Sept 2015.

Raport o sytuacji Polonii i Polaków za granicą [Report on the Situation of the Polish Diaspora and Poles Living Abroad]. (2012). Ministerstwo Spraw Zagranicznych. https://www.msz.gov.pl/resource/b8b3993a-2df7-408b-a4c4-20b7ef465d34:JCR. Accessed 4 Sept 2015.

Rezekvítek. (2008). Sdružení pro ekologickou výchovu a ochranu přírody [Association for Environmental Education and Nature Protection] http://www.rezekvitek.cz/?idm=113. Accessed 10 Sept 2015.

Romancov, M. (2002). Politická geografie a geopolitika [Political geography and geopolitics]. In L. Cabada & M. Kubát (Eds.), *Úvod do studia politické vědy* (pp. 388–444). Praha: Eurolex Bohemia.

Stark, A., & Wnuk, G. (2012). *Teraz geografia. Podręcznik dla szkoly ponadgimnazjalnej* [Geography now. Textbook of geography for higher secondary schools]. Toruń: SOP.

Stasiak, J., & Zaniewicz, Z. (2011). *Geografia. Vademecum maturalne* [Geography. Handbook for high school graduation]. Gdynia: Operon.

SVPG – Štátny vzdelávací program pre gymnáziá [State Educational Programme for Grammar Schools]. (2015). http://www.minedu.sk/data/att/7900.pdf. Accessed 10 Sept 2015.

Uliszak, R. (2012). *Puls Ziemi. Podręcznik do geografii dla gimnazjum* [Pulse of the earth. Textbook of geography for grammar schools]. Warszawa: Nowa era.

Varianty [Variants]. (2015). www.varianty.cz. Accessed 13 Sept 2015.

Vláda České republiky. (2008). Česká republika uznala nezávislost Kosova [The Czech Republic has recognized the independence of Kosovo] http://www.vlada.cz/cz/media-centrum/aktualne/ceska-republika-uznala-nezavislost-kosova-35465/. Accessed 15 Sept 2015

Wahla, A. (Ed.). (2000). *Učebnice geografie 90. let* [Geography textbooks in the 90s]. Sborník z mezinárodní konference. Ostrava: Ostravská univerzita.

Witkowski, J. (2011). *Cały świat w klasie* [The whole world in the classroom]. Warszawa: Centrum Edukacji Obywatelskiej.

Zaťková, M., Brezovská, J., & Durmisová, A. (2008). *Zemepis 6, učebnica pre 6. ročník základnej školy* [Geography 6, textbook for sixth grade of elementary schools]. Bratislava: Poľana.

Chapter 14
Changes in Demographic Behaviour: Possible Use of Its Findings in Didactic Practice

Jozef Mládek, Marcela Káčerová, Dagmar Popjaková, and Michal Vančura

14.1 Introduction

Several scientific disciplines have been examining changes in demographic behaviour. In particular social practice is greatly involved with the changes. Since the 1960s, changes in demographic behaviour of the population of Western and Northern Europe have become evident; later they were also identified in Central, Eastern and Southern Europe. The changes related to the second demographic transition are strongly manifested in the following three areas: reproduction behaviour, population ageing and family behaviour. In terms of reproduction behaviour, the birth rate, fertility rate and degree of reproduction have decreased. At the same time, the population is ageing, i.e. the number and proportion of senior citizens is on the rise. Family behaviour is also undergoing big changes. People get married much later, there are more single people, cohabitation is popular, families are getting smaller, and the divorce rate is increasing (Lapierre-Adamcyk and Charvet 2000).

The study results of demographic processes and their current changes are irreplaceable when learning about and managing all social areas. This also concerns the preparation of zoning maps at all scales. The analysis of demographic behaviour is used by state administration authorities and local government when dealing with education, healthcare, transport, trade issues, etc.

To monitor the causes and manifestations of different demographic behaviours, to look for the schemes and patterns and to be aware and be able to identify examples

J. Mládek (✉) • D. Popjaková • M. Vančura
Faculty of Education, Department of Geography, University of South Bohemia in České Budějovice, Jeronýmova 10, 371 15 České Budějovice, Czech Republic
e-mail: mladek@pf.jcu.cz

M. Káčerová
Faculty of Natural Sciences, Department of Human Geography and Demography, Comenius University in Bratislava, Ilkovičova 6, 842 15 Bratislava, Slovak Republic

P. Karvánková et al. (eds.), *Current Topics in Czech and Central European Geography Education*, DOI 10.1007/978-3-319-43614-2_14

from their own surroundings can also be an important, interesting and attractive topic for pupils and students at all types of schools.

This paper analyses the changes of demographic behaviour of the European population, with a special emphasis on the populations of Czechia and Slovakia. Based on detailed analyses of the population processes and structures, the validity of some generally accepted patterns is evaluated. This paper also aims at helping geography teachers to make demographic topics more attractive, namely by guiding pupils to actively participate in research and teaching them to question, learn, evaluate and present demographic findings.

14.2 Theoretical Context

One of the main tasks and objectives of scientific learning is to discover and formulate rules, patterns and laws which are generally valid and guiding for relevant scientific disciplines, based on the application of correct methods and study techniques. Compared to the natural sciences, it is far more difficult, and less frequent, for social sciences to attain such objectives. This is due to the complexity of the subject matter, i.e. a society with its multicausal relationships between its individual components and the biosphere. Often, time dynamics and spatial differentiation of social phenomena are pointed out which restricts the validity of the sought patterns.

We come across similar problems in the cognitive efforts of demography. Frequently, the results of demographic analysis have a restricted time and spatial validity. Generally accepted regularities and patterns have a different position, such as the theory of demographic revolution, demographic transition and demographic cycle.

14.2.1 Demographic Revolution

The term demographic revolution (transition, cycle) was introduced by Landry (1934), and later by Notestein (1944), to name a set of demographic changes connected to the transition from high death and birth rates to low death and birth rates. In general, a demographic revolution also refers to a stage in population development when the character of demographic reproduction changes radically, in particular due to lower death rate and consequently birth rate. In the narrow sense, it is the transition from one balanced state to another one, whilst at the start and end of the stage, the population shows a low population growth. During the transition the number of inhabitants increases dramatically (Vance 1952; Mackenroth 1953; Pavlík et al. 1986; Mládek 1992; Kirk 1996; van de Kaa 1987).

A more detailed analysis allows the demographic revolution to be divided into several stages. According to the theory of the demographic cycle, high crude death and birth rates are typical for the first stage; however, both rates show significant fluctuations. The increase is low and determined mostly by the death rate. In the second stage, the crude death rate decreases; however, the birth rate remains at approximately the same level. In the third stage, the birth rate starts dropping; however, thanks to a high difference between the two processes, the natural increment is high, and the size of the population grows fast, which is interpreted as a population explosion. In the fourth stage, the crude death rate stabilises, and the birth rate accelerates. The population growth decreases. In the last stage, the crude death and birth rates are stabilised at a relatively low level, the natural increment is low, and the population stagnates; however, the total population is higher (Mládek 1992). The crude death and birth rate at the beginning of the demographic revolution is approximately 40–50 per 1000. Based on the progress of initial stages, we speak of the French, English or Mexican-Japanese model (Pavlík 1964). The bigger the delay between the start of the decrease in the death and birth rates, the higher the population growth. Originally, it was assumed that both processes will stabilise and the natural increment will be kept at a minimum level (Fig. 14.14 – Annex 1).

14.2.2 Second Demographic Transition (Revolution)

Changes in demographic behaviour, recorded by Nordic countries in the second half of the twentieth century and demonstrated, in some modifications, in the countries of Southern Europe, can be considered as one of the most significant changes in the history of population. The significance of the changes is not in their intensity, but mostly in their complexity. They have influenced nearly all areas of demographic behaviour, in particular the development of population processes and consequently also the changes in some population structures. Therefore, they were, by right, called revolutionary, and the periods when these typical changes took place are called *the second demographic revolution* (SDR) *or second demographic transition* (SDT) (van de Kaa 1980, 1987, 1996, 1998, 1999; Lesthaeghe 1983, 1991, Lesthaege and van de Kaa 1986; Birg 1996; Mládek 1998, 1999; Pastor 1997, 1998, 2002).

SDT refers to a development of population when the demographic behaviour and the population value system have radically changed. It is a stage when individualism and personal freedom are overvalued, the role of marriage and family is weakened, natality and fertility are reduced to a level which cannot guarantee population recovery and the population is rapidly ageing.

By comparing the characteristic features of the FDR and SDT, we can see major differences in the demographic behaviour of population. For the FDR, altruism is typical, demonstrated by actions for the benefit of others. The focus on family and posterity is dominant. The process of industrialisation and urbanisation influences the decrease in fertility. Secularisation decreases the influence of religion on family size and leads to family planning.

The predominant feature of the new demographic behaviour is individualism, emphasising personal freedom in all aspects of life. The essential condition for such behaviour is a certain degree of education and motivation to develop one's talents.

Also a higher degree of own economic security is expected, which guarantees economic and social independence. Starting a family and having children can significantly influence economic and social independence. The society is gradually secularised, weakening traditional demographic behaviour. A very typical albeit very simplified feature of the SDT is a major decrease in the dynamics of population, which is demonstrated through a marked decrease in the reproduction rate. The crude birth rate is decreased below the level of a relatively stable crude death rate, resulting in a natural population decline.

The changes in demographic behaviour during the SDT are demonstrated in three areas: *reproduction behaviour, ageing of population and family behaviour.* Mainly, such division enables learning more about the aforementioned changes; however, the phenomena from all three areas are in fact mutually conditional, often intermingling.

14.3 Changes in Reproduction Behaviour

The death rate, or its decrease, is considered a crucial process in the FDR. This is followed by a decline in birth rates, and its speed significantly influences the natural increase of certain populations. During the SDT the decisive process is natality or fertility (Table 14.3 – Annex 2). The death rate, and its crude rate (Fig. 14.1), is relatively stable (this does not apply absolutely, since there are variables such as specific death rate based on age and sex and the causes of death, Meslé and Vallin 2002).

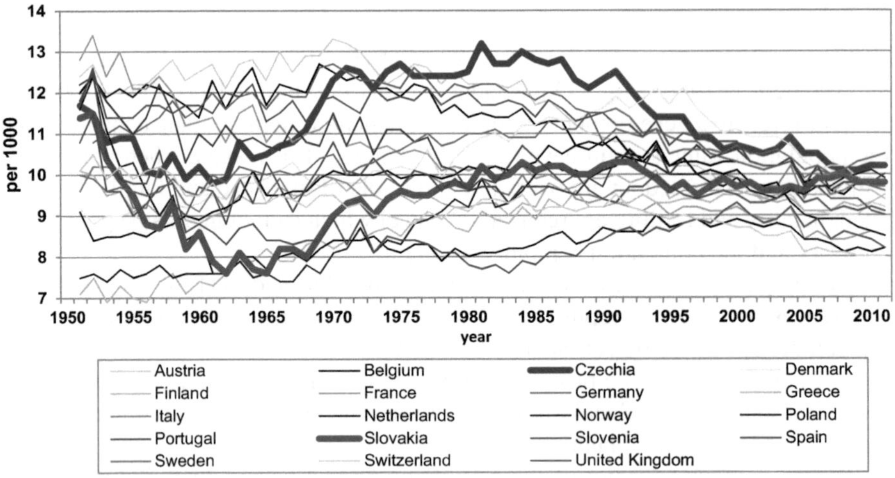

Fig. 14.1 Crude death rate development in selected European countries (1950–2010) (Source: Eurostat 2015a)

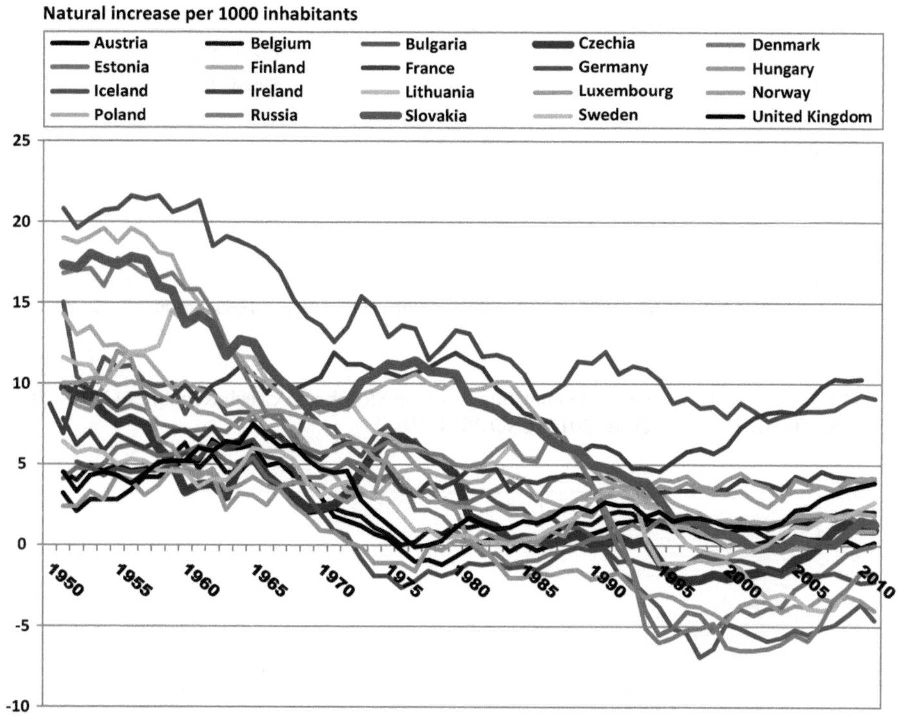

Fig. 14.2 Natural increase per 1000 inhabitants in selected European countries (1950–2010) (Source: Eurostat 2015a)

A general view indicates two time periods of decrease in reproduction indicators. The first period was the 1960s and 1970s, and the second period was the 1990s and the beginning of the next century. Until 1972, the population of socioeconomically advanced European countries (but also of the USA, Canada, Japan, Australia, etc.) showed a natural population growth. Figure 14.2 clearly shows the overall trend of decrease over a period of time by individual regions. Many populations show a natural decline (Bulgaria, Hungary, Germany, Romania, Russia, Italy, Lithuania, Latvia), and many were in decline for several years. Both the Czech and Slovak populations have shown a significant decline in natural population growth in the last 50 years, and for several years they have even recorded a natural population decline (Fig. 14.3).

The total fertility indicator has a good ability to characterise changes in reproduction behaviour of the population. A drop below the level of 2.0 (two births per woman of a childbearing age) indicates insufficient reproduction. Since the 1960s the total fertility rate has gradually dropped below this level in all countries (Fig. 14.4). A general view shows two time periods of decline in reproduction indicators.

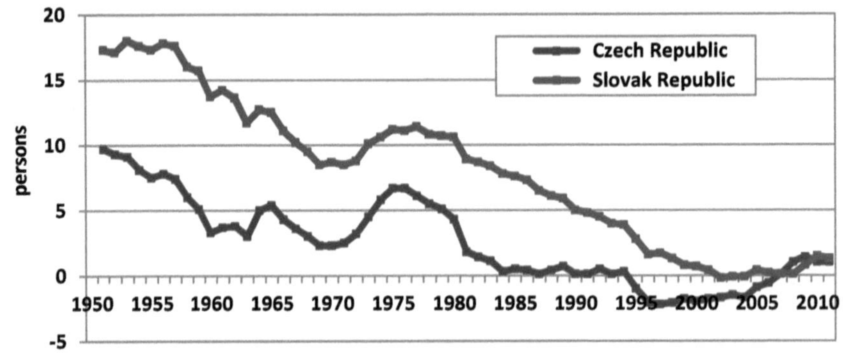

Fig. 14.3 Natural increase per 1000 inhabitants in the Slovak and Czech Republics (1950–2010) (Source: SO SR – Infostat 2001, 2012; CSO 2014, 2015)

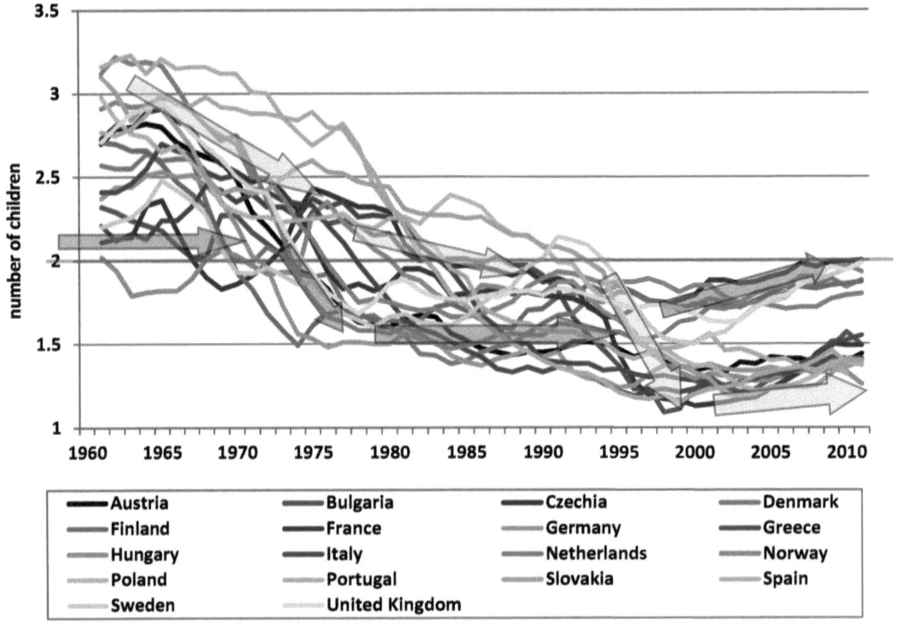

Fig. 14.4 Total fertility rate in selected European countries (1960–2010) (Source: Demographic yearbook 2012; Ined 2015)

The first decline (1960s and 1970s) was recorded in Western European countries (Germany, Netherlands, Sweden and others). This resulted from natural changes of social structures in these countries. In the 1990s, the decline in the total fertility rate in Southern Europe was much faster, and the values reached very low values (Poland 1.2, Slovakia 1.2, Greece 1.3, Spain 1.3, Czechia 1.2, Hungary 1.3, etc.). These changes occurred under the strong pressure of social and, in particular, political-economic changes in these countries.

Two development periods of reproduction behaviour can be identified in the second demographic transition. The first could be called destructive, when the total fertility rate in Western and Northern European countries dropped below 1.5–1.7. In the second compensation period, the fertility rate gradually rose to 1.8–2.0 (Denmark1.9, Netherlands 1.8, Finland1.9, United Kingdom 2.0, France 2.0). A similar development can be observed in the second group of countries (Central and Southern Europe), even though the rise of the total fertility rate is slower (Bulgaria 1.5, Czechia 1.5, Slovakia 1.4, Spain 1.4, Greece 1.5, etc.). A more pronounced demonstration of compensation can be expected.

14.4 Population Ageing

New peculiarities in the behaviour of the population, i.e. demonstration of the second demographic transition, provide a theoretical basis for the process of population ageing. To define and measure the process of population ageing means to express changes in the age structure according to the fall in the number or ratio of children in the population (bottom-up process of ageing) and the increase in the number or proportion of senior citizens (top-down process of ageing) (Káčerová 2005; Káčerová and Ondačková 2015a, b). The process of ageing, such as forming the age structure of the population, can be considered a demographic phenomenon with a relatively high degree of complexity. The development of many fundamental population processes, such as natality, mortality and migration moves, is reflected in the age structure. On the other hand, the age structure of each population can significantly influence the development of other population phenomena and processes (in addition to the aforementioned, also the marriage rate, workforce potential).

Europe is the oldest continent, but it is not homogeneous. It is essential to emphasise the spatial differentiation of ageing in Western European countries. As stated above, the time shift of the second demographic transition, i.e. the change of reproduction and family behaviour, was first demonstrated in Western European countries and later in Southern Europe. By comparison, in Eastern and Central Europe, such behaviour was prevented by totalitarian political systems that were controlling the reproduction behaviour of its inhabitants by anti-natality measures. This timing difference caused a different start and intensity of the process of population ageing.

The bottom-up ageing process, expressed as a proportion of population aged 0–14, showed there were more children in Eastern European countries in the second half of the twentieth century and less in Western European countries (Káčerová et al. 2014). In 2015 the spatial polarity of ageing from below has not been absolutely maintained. The lowest representation of the proportion of population aged 0–14 is in Germany (12.8 %), Bosnia and Herzegovina, Italy and Portugal. The population of Germany was one of the first where in the 1960s the fertility rate dropped, which resulted in an intensive start of bottom-up ageing. Later, the 1972 Abortion Act contributed to this fact (Coleman 1993). Figure 14.5 helps to identify

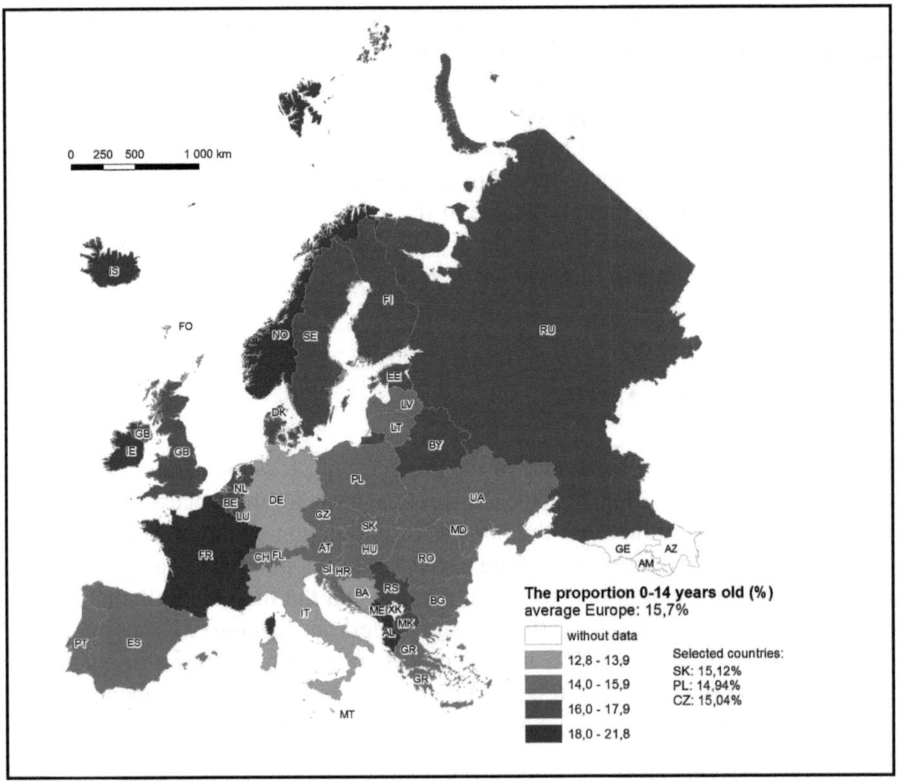

Fig. 14.5 Proportion of 0–14-year-old population in Europe (2015) (Source: Eurostat 2015a, own processing)

similarly low or below average representation of children in Central, Eastern and Southern Europe: Bulgaria, Greece, Czechia, Slovakia, etc. Radical changes in reproduction behaviour occurred in these countries at the end of the 1980s and beginning of the 1990s. Most of them underwent both economic and social transformations, and exactly in this period, the intensive bottom-up process of ageing starts. On the other side of the spectrum, maximum values of the percentage of children, are shown in the population of Ireland (thanks to the important migration gains), Iceland or Albania. Also the Benelux countries belong to countries with an above-average representation of children.

The European process of *ageing* mentioned above shows the proportion of 65+ populations. The value of its variation span of 12.6 percentage points shows a much more pronounced diversification of this process, compared to the process of bottom-up ageing (Fig. 14.15 – Annex 3). In general, it can be stated that Nordic and Western Europe countries have a higher degree of top-down ageing and Eastern Europe and Balkan countries show a lower degree of top-down ageing (Fig. 14.6). More than one fifth of the 65+ populations were registered in Germany, Italy, Greece, Portugal, Finland and Bulgaria in 2015. "This group of countries is heterogonous in terms of

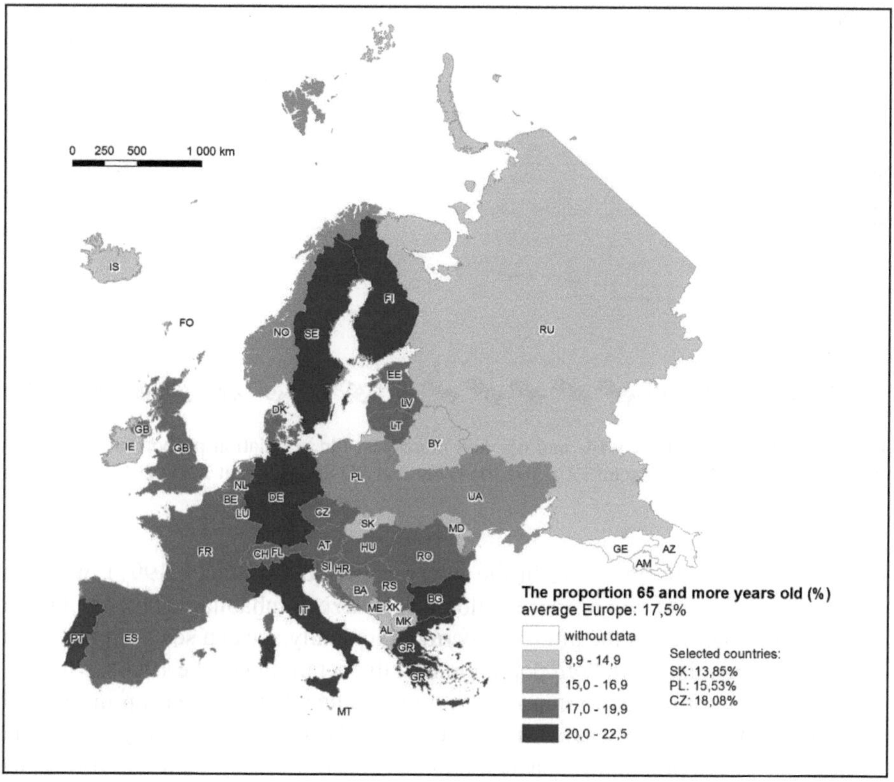

Fig. 14.6 Proportion of 65- and more year-old population in Europe (2015) (Source: Eurostat 2015a, own processing)

the start of the population ageing. The German and Italy population belongs amongst the traditional leaders of European ageing. The same high values are shown in Nordic countries (Sweden, Finland), and South Europe (Portugal, Greece), where the decisive factor is high median life expectancy and a long-term low total fertility rate" (Káčerová and Ondačková 2015b). Moldavia, Macedonia, Albania and Ireland, amongst others, are some of the countries with the lowest representation of senior citizens. The populations of Macedonia and Moldavia have poor mortality rates. Although the median life expectancy for men has increased in Central Europe over a course of 10 or 20 years, it still lags behind the EU average (Hoff 2008).

The process of ageing in Central Europe deserves special attention. An example of different degrees and timing of the process can be seen in the populations of Czechia and Slovakia. Although the indicators of ageing identify a similar development of population ageing (Fig. 14.7 – nearly parallel curves), the time ahead of the Czech population is evident, looking at the indicators. This time ahead became even more prominent in the 1930s. Until the beginning of the Second World War, Czechia belonged to the Western type of the Hajnal line (Rabušic 2001) of family behaviour, whilst Slovakia belonged to the Eastern model. This led to the

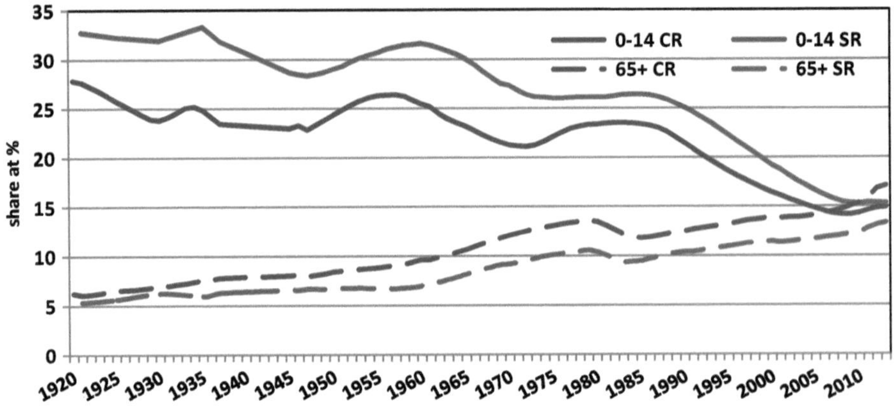

Fig. 14.7 Development of 0–14- and 65- and more year-old population proportion in Slovakia (SR) and Czechia (CR) between 1920 and 2014 (Source: SO SR – Infostat 1922–2014; CSO 2015)

preference of having one child in the interwar period (Roubíček 1997), whilst in Slovakia families had two or more children. The Czech inhabitants belonged to the older population even in the after-war period. Gradually, we can see a slight growth of the senior population in both populations at the beginning of the 1960s, which, to some extent, is connected to improved mortality rates. Due to the drop in the birth rate during WWI, there was a slight decline in the 65+ population group in the 1980s. The Czech population reached a higher level of ageing than the Slovak one in all periods. At the same time, the level of ageing was getting closer in both populations.

14.5 Family Behaviour

The third area of population changes in the SDT reflects changes in family behaviour. These are processes which are directly or indirectly connected to family formation, or its collapse, and its function in reproduction processes. During these changes a common behaviour feature of the population is used in the second demographic transition.

Changes in family behaviour are mostly demonstrated as follows:

- Higher average marriage age and the related decrease of the number of weddings
- Introduction of effective hormonal contraception and related sexual freedom
- Higher divorce rate
- Lower number of single people and a higher number of divorced people getting remarried
- Increase in the rate of cohabitation
- Higher extramarital fertility
- Smaller households and an increase in one-person households
- Growing trend of incomplete families

The idea basis explaining changes in family behaviour is provided by the *theory of civilisation waves*. It also presents the idea basis for the interpretation of the most general global development trends in society. We will be mostly interested in the findings and ideas concerning the development of family behaviour, family forms and family functions. The issues of civilisation waves theory and family behaviour were largely presented in the works of Toffler (1980, 1990, 1992), Toffler and Tofflerová (1996a), Naisbitt (1982), Naisbitt and Aburdenová (1992) and Fukuyama (2005).

The First Wave Civilisation is characterised by the origin of agriculture and soil cultivation. This created a spatial stability of the population and enabled the foundation of settlements: mostly of villages and later on, hand in hand with the development of trade, of towns. The era of the first wave was quite long (8000 BC to 1650–1750); however, some of its characteristic features are still evident in the less developed regions of the world today (Toffler and Tofflerová 1996b).

The form, functioning of the family or households can be derived from the method of householdry. Numerous groups of inhabitants – *multigenerational households* – lived together as one production management unit. The main objective of these communities was to ensure essential sources of livelihood, provide protection against the enemy and reproduce in the difficult conditions caused by complex family and generational relationships.

In the Middle Ages in Central Europe, an extended family (household) was formed, sometimes in the form of a large family unit. The family consisted of blood-related family members. It is characterised as a production, consumer and residential community owning property. The rights and obligations of individual members were determined on the basis of age and gender.

The second wave civilisation was linked to the development of science, in particular technical science. A major driving force behind the development was factory production, mass production system, mass consumption, mass education and mass media. Production was concentrated, which demanded a lot of workforce and resulted in an intensive process of urbanisation.

During the development of the second wave civilisation, the family tie to the agricultural land weakened, and the tie to industry became crucial. This improved the mobility of the inhabitants and families. Some of the original family functions were transferred to specialised institutions – school education for children and healthcare to hospitals. A new model of a standard family became *the nuclear two-generation family*. The father looked after economic matters and the mother looked after the household and child-rearing. This became the leading model in developed countries, and it is still the most common model in the world.

The third wave civilisation (information age, electronic epoch, cosmic age, post-industrial society, science and technical revolution) or the *information revolution* is based on the development and distribution of findings and information which is crucial for the growth of productivity at work and social development. Gradually, computers and computer technologies have become involved in production processes, education, information and innovation which strongly influences financial services, healthcare, military protection and company management. Mass production, a typical feature of the second wave civilisation, has been replaced with highly specialised, small-series production. *Demassification* has an impact on all sorts of areas of society. The homogeneity of the society in the second wave is gradually replaced by the disparity of the society in the third wave (Toffler and Tofflerová 1996a, b; Hudeček 2006; Rankov 2006; Mládek 2009; DeGenova et al. 2010).

The universal model of the nuclear family is also subject to the process of demassification when other family forms develop. Most often, one-child families, no-child families, single households, informal cohabitation, etc. become the most common family forms (Hradil 1995). The position of women has changed radically, both in terms of the economy and in terms of their life missions. Women with higher education seek to apply their life ambitions also outside their family interests. Their ambition to be economically independent, to increase their education and to have a career often leads to delaying marriage and in particular delaying the birth of their first child. In consequence, the number of marriages falls, but also the natality and fertility rates.

Fukuyama (2005) has a far more dramatic view of the development trends in the second half of the twentieth century. In his book *The Great Disruption*, he expressed his critical attitude and serious considerations. The transfer from an industrial to an information society is accompanied by a number of negative features which are reflected in our social and moral lives. Even during reproduction behaviour, Fukuyama points out the consequences of *The Great Disruption*, such as the fall of birth rates when not even simple reproduction is secured. Such unfavourable development is demonstrated by a lower number of marriages, a higher rate of divorces and a higher number of extramarital births. The growth of individualism weakens all forms of authority (family, neighbourhood, state).

14.5.1 The Third Wave Civilisation and the Population Family Behaviour in Europe, Slovakia and Czechia

14.5.1.1 Decrease of Nuptiality, Increase of Average Marriage Age and Models of Family

The intensity of marriage rates, expressed by the crude rate, decreased by half, from the original 8–9 % to 4–5 % (Fig. 14.8). Similarly to reproduction indicators, the decrease occurred in two stages. In the 1970s and 1980s, the decrease occurred in Western Europe (Sweden, Germany, France, Switzerland, Belgium, Norway). In the 1990s a similar decrease was experienced by the countries of Central and Eastern Europe that were undergoing transformation (Bulgaria, Hungary, Poland). Similarly to the other group of states, the populations of Slovakia and Czechia responded to the changes in the first years of the transformation period (1991–1994) by a significant drop in the number of marriages. The drop in the number of marriages is often explained by delaying marriage until one is older. The traditional distribution of the marriage rate by age in Czechia and Slovakia was, in most cases, between the ages

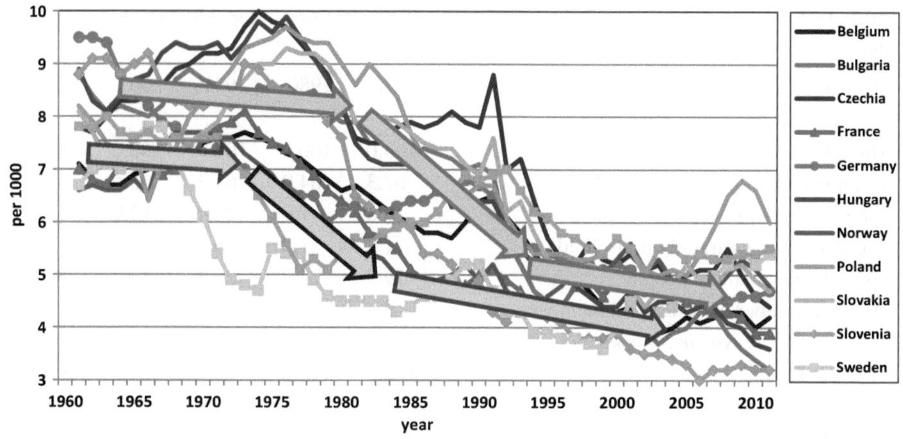

Fig. 14.8 Crude marriage rate in selected European countries (1960–2011) (Source: Eurostat 2015b)

of 20 and 24 for both genders. For men, the maximum age started to move to 25–29 at the end of the 1990s, as it eventually did for women, but not until very recently. The specific data by age also confirms a drop, mainly in the case of younger age categories, and a slight increase for older age categories.

14.5.1.2 Size and Structure of Families and Households

The development of 1970–2011 shows a fast decrease in the size of Slovak and Czech households. The average number of household members decreased in Slovakia from 3.37 to 2.61 persons in 1970–2011, and in the same period, it decreased from 2.89 to 2.34 persons in Czechia (Table 14.1). Most inhabitants of Czechia and Slovakia live in complete families (husband and wife, with or without children), although this ratio is constantly getting lower. Whilst in 1970 complete families represented 78.5 % of all households (Slovakia), i.e. 67.3 % (Czechia), by 2011 this figure dropped to 52.4 %, i.e. 47.9 % (strongly influenced by the increase of one-person households). The ratio of incomplete families doubled.

14.5.1.3 Single Households (Singles)

The increased number of one-person households (flourishing "singles" culture) has become a typical feature of family behaviour of the population in all developed countries. Their permanent or temporary life programme does not include getting married or starting a family. This lifestyle corresponds to a typical feature of social development, which is individualism, and it is not aided by the fact that it is socially accepted, and the cultural and housing conditions are favourable.

Table 14.1 Household types in the Czech Republic (CR) and Slovak Republic (SR) between 1970 and 2011

Indicators	1970	1980	1991	2001	2011
Households[a] by type in Czechia					
Household total	100.0	100.0	100.0	100.0	100.0
Complete families	67.3	64.9	61.3	53.6	47.9
Incomplete families	7.8	7.9	10.4	12.9	13.0
More member nonfamily households	1.2	1.4	0.4	2.0	4.9
Households of single people	19.9	23.7	26.3	30.3	32.5
Average number of household members	2.9	2.7	2.6	2.4	2.3
Households[b] by type in Slovakia					
Household total	100.0	100.0	100.0	100.0	100.0
Complete families	78.5	70.6	67.4	56.4	52.4
Incomplete families	8.6	8.2	10.4	11.9	16.1
More member nonfamily households	1	1.4	0.4	1.7	2.1
Households of single people	12	19.8	21.8	30.0	29.5
Average number of household members	3.4	3.0	2.9	2.6	2.6

Source: SO SR (1970, 1980, 1991, 2011) and CSO (1970, 1980, 1991, 2011)
Note: [a]Private households, [b]census households

This is the lifestyle of up to 1/3 of households. In 1970, the share of single households represented 12 % out of all households, and by 2011 this share increased to 36.1 %. In the same period in Czechia, these types of households increased from 19.9 to 32.5 % (Fig. 14.9, Table 14.2). This form of lifestyle is mainly preferred by the younger generation of single people. The older generation often attains "single household" status due to divorce or the death of their partner. Due to longer life expectancy, the number of individuals who live in single households will inevitably increase.

14.5.1.4 Cohabitations

During the last 50 years, informal partnerships and cohabitation have massively spread in most industrial countries and became a standard form of partnership which significantly influenced the importance and function of the family as a social institution.

Universal approaches to education, the growing participation of women in the job market, an emphasis on professional quality and personal fulfilment in the public sphere, better lifestyle and growing availability of efficient contraception contributed to cohabitation in the second half of the twentieth century (Mládek and Širočková 2004a, b).

In many countries, cohabitation has to a large extent replaced marriage in recent decades. The rule "first live together, then get married, or never get married" started spreading in Western Europe in the 1960s. This attitude became common in Czechia and Slovakia after the political changes of 1989.

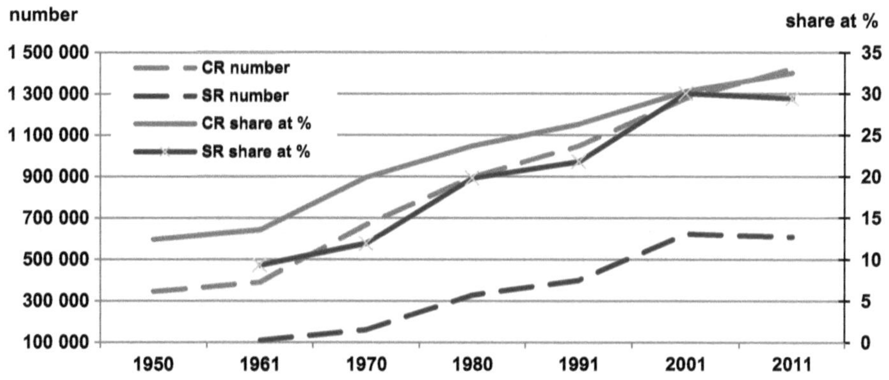

Fig. 14.9 Singles in the Czech Republic (CR) and Slovak Republic (SR) between 1950 and 2011 (Source: SO SR 1961, 1970, 1980, 1991, 2001, 2011; CSO 1961, 1970, 1980, 1991, 2001, 2011)

The standard way of starting a family, i.e. getting married without previous cohabitation with children born after getting married, is less common. This model started to be replaced with a model "cohabitation – wedding – birth of a child" in most societies and sometimes even with a model "cohabitation – one or two children – marriage", or sometimes the couple does not get married at all (Sobotka and Toulemon 2008). During the 1980 census, there were 31,200 cohabitations recorded in Slovakia and 89,423 de facto marriages[1] in Czechia. The growing figures show that these forms of cohabitation have become widespread, i.e. by 2011 the figures showed 89,308 cohabitations in Slovakia (2.86 times as many) and in Czechia 234,346 (2.62 times as many). The lower intensity of cohabitation in Slovakia and certain regional differences in both republics have survived (Fig. 14.10).

In international comparison, both populations belong to a group of European countries with a very low intensity of cohabitation (2–4 % – person in a consensual union per 100 total population). The highest degree of cohabitation (Fig. 14.11) is in Western Europe (9–14 % – United Kingdom, Switzerland, Netherlands, Denmark, France, Norway, Estonia, Sweden).

14.5.1.5 Extramarital Birth

The data on extramarital fertility is often used when full and reliable data about the cohabitations is missing in the intercensal period (Fig. 14.16 – Annex 4).

The number of extramarital children has traditionally been very low in Slovakia and in Czechia. In 1950, 5.5 % of extramarital children were born in Slovakia and 6.5 % in Czechia, and by 1990 this ratio increased to 7.6 % in Slovakia and 8.6 % in Czechia. In the 1990s, this indicator started to grow significantly, and in 2010 it was

[1] The terminology is not uniform. The terms cohabitation, de facto marriages and unmarried couples are used and can have certain methodological differences.

Table 14.2 Single households in Czechia and Slovakia (1961 and 2011)

	1961	2011
Czechia	390 thousand persons	1422 thousand persons
Slovakia	110 thousand persons	607 thousand persons

Source: SO SR (1961, 2011) and CSO (1961, 2011)

33 % in Slovakia and 40.3 % in Czechia (Fig. 14.12). The graph shows that the total number of children born and the number of marital children born were much lower in the 1990s. Only the number and percentage of extramarital children show a growing tendency.

14.5.1.6 Divorce Increase

The divorce is one of the characteristic processes of family behaviour and represents a family breakdown. For the divorce rate, an increased number of indicators and a certain spatial differentiation are typical. It is caused by a number of factors, amongst others, by the degree of economic development, national and religious structure of population, urbanisation ratio, age structure and others.

Using the crude divorce rate (year 2011), the populations of Eastern Europe traditionally show a high divorce rate – in per 1000 (Russia 4.7, Belarus 3.8, Lithuania 3.6, Latvia 3.5, Moldavia 3.1). Some of the populations of Western Europe also have a high divorce rate (Denmark 2.6, Belgium 2.5, Portugal 2.5, Finland 2.5, Sweden 2.5, Switzerland 2.3). On the other hand, conservative behaviour is demonstrated by populations with a lower divorce rate (Romania 1.8, Poland 1.7, Iceland 1.6, Slovenia 1.1, Italy 0.9, Ireland 0.6).

From the European point of view, Czechia (2.5) and Slovakia (2.1) have close to the average divorce rate – per 1000 (EU 28 shows 2.8). The development of the divorce rate and the typical growth of indicators are far more significant. Both populations have viewed family as something conservative which represents a stable society. From the 1950s, the number of divorces increased due to a gradual liberalisation of the divorce legislation. The crude divorce rate (in per 1000) increased in Czechia from 1.1 in 1950 to 2.5 in 2011 and similarly in Slovakia from 0.5 to 2.1. The number of divorces in Czechia increased from 10,261 in 1950 to 28,113 in 2011 and similarly in Slovakia from 1800 to 11,102 divorces. As recently as 1990, there were only 11 divorces per 100 marriages in Czechia and 22 divorces per 100 marriages in Slovakia. In 2011, the divorce index increased to 62 % divorces in Czechia and 44 % in Slovakia. This is also due to the steep decline in the number of weddings in this period (Fig. 14.13).

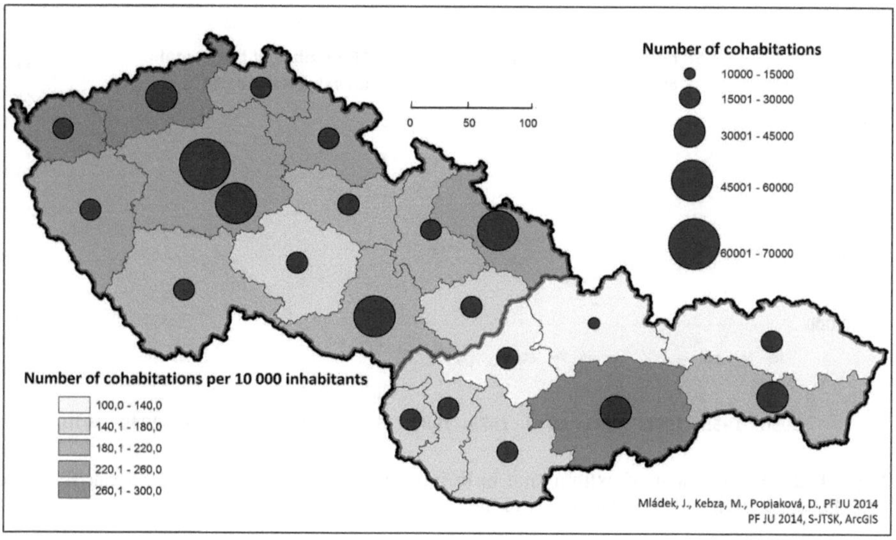

Fig. 14.10 Cohabitations in the Czech and Slovak Republics in 2011 (Source: SO SR 2011; CSO 2011, own processing)

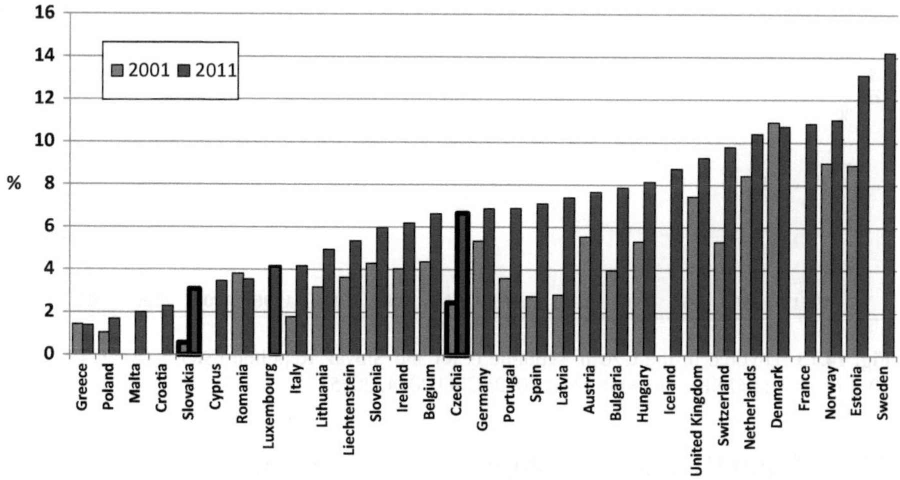

Fig. 14.11 Persons in a consensual union per 100 total population in Europe (Source: SO SR 2011; CSO 2011; Eurostat 2015b)

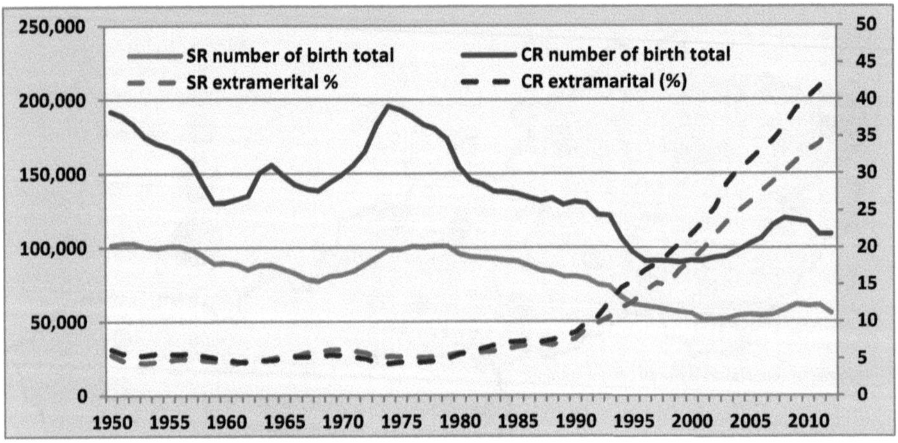

Fig. 14.12 Development of extramarital birth in Czechia Republic (CR) and Slovak Republic (SR) between 1950 and 2011 (Source: SO SR – Infostat 2001, 2012; CSO 2014, 2015)

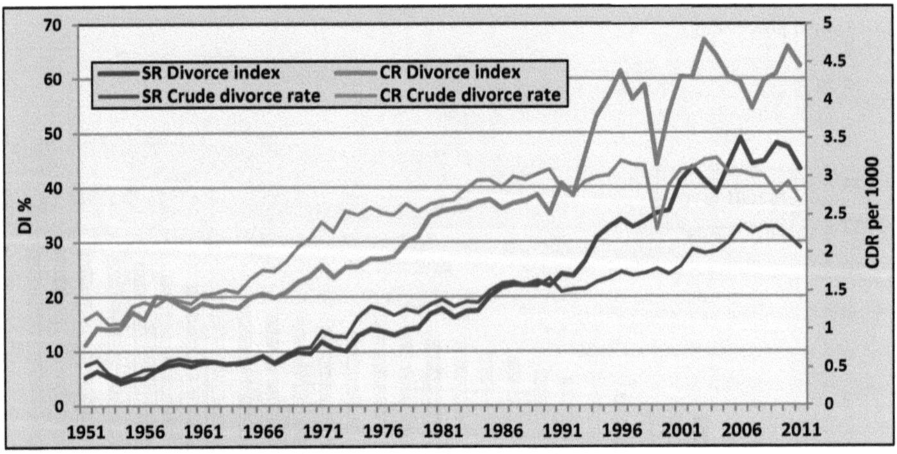

Fig. 14.13 Divorce in the Czech Republic (CR) and Slovak Republic (SR) between 1951 and 2011 (Source: SO SR – Infostat 2001, 2012; CSO 2014, 2015)

14.6 Final Remarks on Population Behaviour During the Second Demographic Transition

Population is the most dynamic feature from all geographic spheres in the country. It is also the only feature in the entire landscape system that can influence or directly manage the others. It has an irreplaceable position in the country, in particular in its social area. The dynamic development also concerns all three population processes covered by our paper whose dynamic aptly represents the period of the second demographic transition (SDT).

The decrease in the birth and fertility rates taking place in European countries at different times and with different intensities is often regarded as the decisive process. The development of the death rate is becoming relatively stable, and the development of fertility is decisive for the total reproduction of the population. The drop in the fertility rate in Czechia, and in other countries in Central Europe, strongly influenced the transformation of the social and political systems in these countries.

Population ageing is regarded as a certain regularity in the development of the population (the population can show signs of rejuvenation in shorter intervals). Changes in age structure usually influence the decline in reproduction (decreasing the share of younger age categories) and increase the share of seniors, extending the median life expectancy as a result of improved lifestyle of the population. Ageing can influence migration of the population by differentiated representation of age categories of migrants. Population ageing is connected to a number of social consequences that require special care and security.

A family is regarded as one of the "eternal" social institutions. In some modifications a family has existed in all historical periods of societal development, in all cultures. It is not a static institution, its form and size changes, and so does the relationship structure between its members and its various functions, including the change of the reproduction function. The clear dominance of a nuclear family is gradually disappearing in the countries showing features of the third wave civilisation. There are more families without children and more incomplete families. Nonfamily households started appearing in the household structure. The share of single households is significantly increasing. These development trends can be considered part of the demassification process of societal development (politics, science, education, housing, business enterprise, etc.).

The issue of population development, in particular the dynamics of individual processes and structures, represents one of the key society changes. Due to this it should be incorporated into educational processes. A whole spectrum of modern constructive educational methods and techniques can be used.

14.7 Outline of Didactic Exercises on Demographic Topics in the Teaching Geography

It is advisable to include the topic of changes in demographic behaviour during the second demographic transition (SDT) into geography lessons and to practise these topics in a simple manner. Pupils from year 6 onwards are able to perceive these changes, using examples from their families, and to identify with them. The topics are closely connected and develop the curriculum on population that forms part of the social geography curriculum. This topic can also be discussed during geography lessons when learning about regional geography and specific problems of the selected regions.

To be more specific, the monitored demographic topic has been included in the Framework Educational Programme for Primary Education (Kol 2013) in the valid curriculum document for the Czech Republic under an educational area entitled Humans and Nature taught in geography. Population problems also partly overlap with the expected outcomes from other topical areas, such as the Social and Economic Environment, World Regions and the Czech Republic.

Demographic topics are taught in school educational programmes in years 8 and 9; however, they are also discussed during regional geography of selected continents in year 6 and year 7.

From a didactical point of view, this subchapter represents a framework for selected topics in population geography which can be analysed with pupils in geography lessons. The selected topics can be developed by teachers based on current needs. Teachers can adapt the topics for the relevant age groups for years 8 and 9 at primary schools and for years 10 and 11 at grammar schools. These topics can also be used in geography lessons at secondary schools and universities. The topics are presented in the form of exercises which serve as a platform for further didactical work using various didactical methods, depending on demands put on pupils and students at the specific level of education. The invention of teachers is crucial in terms of how they will interpret demographic exercises to their pupils and students.

Four thematic frameworks "Educational training – practice, exercise" were selected which characterise the changes in demographic behaviour in SDT. Each of them is presented in the form of simple exercises. Exercises comprise the wording of the tasks, usually accompanied by a graph, a table and additional questions. Questions to the answers are provided in the first six parts of this chapter.

The first framework compares a demographic revolution and the second demographic transition. The second exercise focuses on practising changes in reproduction behaviour and a drop in birth rate. The third block looks at the issue of population ageing. Changes in family behaviour are addressed in the fourth framework (see Annexes 1, 2, 3, and 4).

Annexes

Annex 1: Educational Training – Practice, Exercise

Demographic Revolution

Dramatic changes in the development of world population are described by the term "demographic revolution". The revolution concerns reproduction and population changes and the dynamics of population recovery arising from changes between the degree of natality and mortality. We speak about the first and second demographic revolutions. The time course of these demographic transitions is connected to the social-economic development of the society.

Demographic changes in developed countries started earlier. The first demographic revolution (FDR) ran parallel with industrialisation, i.e. at the end of the eighteenth century in some advanced countries. In Czechia it was most intense at the end of the nineteenth century, especially at the turn of the nineteenth and twentieth centuries.

The second demographic transition (SDT) was identified in the 1970s in advanced countries. In the Czech population, it was recorded from the mid-1980s onwards and in the Slovak population as late as from the beginning of the twenty-first century. In addition to natality and mortality, other demographic characteristics also have an impact on population recovery, such as marriage, divorce, etc.

Exercise

What are the reasons of the decrease in mortality during the FDR?
Which countries are currently going through the FDR?
When and in which countries did the SDT start?
What are the typical features of the SDT?

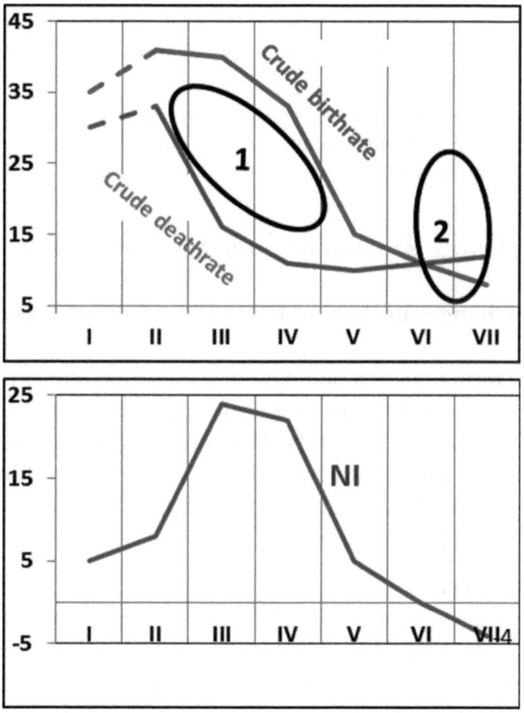

Fig. 14.14 First (1) and second (2) demographic transition phases (Note: *NI* natural increase (number of persons per 1000 inhabitants))

Table 14.3 Development of crude birth rate in selected European states

Birth rate[a]	1950	1955	1960	1965	1970	1975	1980	1985	1990	1995	2000	2005	2010
Austria	15.5	15.6	17.9	17.9	15.0	12.4	12.0	11.6	11.8	11.2	9.8	9.5	9.4
Belgium	16.5	16.8	17.0	16.4	14.6	12.2	12.7	11.6	12.4	11.3	11.2	11.3	11.7
Bulgaria	25.2	20.1	17.8	15.3	16.3	16.6	14.5	13.3	12.1	8.6	9.0	9.2	10.0
Czechia	21.4	17.9	13.5	15.1	15.1	19.1	15.0	13.1	12.6	9.3	8.9	10.0	11.1
Denmark	18.6	17.3	16.6	18.0	14.4	14.2	11.2	10.5	12.3	13.3	12.6	11.9	11.4
France	20.6	18.6	17.9	17.8	16.7	14.1	14.9	13.9	13.4	12.6	13.1	12.7	12.7
Germany	–	15.6	17.3	17.4	13.4	9.9	11.1	10.5	11.4	9.4	9.3	8.3	8.3
Hungary	20.9	21.4	14.7	13.1	14.7	18.4	13.9	12.3	12.1	10.9	9.6	9.7	9.0
Ireland	21.4	21.1	21.5	22.1	21.8	21.1	21.7	17.6	15.1	13.5	14.4	14.8	16.5
Netherlands	22.7	21.3	20.8	19.9	18.3	13.0	12.8	12.3	13.2	12.3	13.0	11.5	11.1
Norway	19.1	18.5	17.3	17.8	16.7	14.1	12.5	12.3	14.4	13.8	13.2	12.3	12.6
Poland	30.7	29.2	22.6	17.4	16.8	19.0	19.6	18.3	14.4	11.2	9.9	9.5	10.8
Romania	–	25.6	19.1	14.6	21.1	19.7	18.0	15.8	13.6	10.4	10.5	10.2	9.9
Slovakia	28.8	26.7	22.1	19.3	17.8	20.6	19.1	17.5	15.1	11.5	10.2	10.1	11.1
Slovenia	24.5	20.9	17.6	18.5	15.9	16.6	15.7	13.2	11.2	9.5	9.1	9.1	10.6
Spain	20.1	20.4	21.5	20.9	19.4	18.8	15.3	11.9	10.3	9.2	9.9	10.7	10.4
Sweden	16.5	14.8	13.7	15.9	13.7	12.7	11.7	11.8	14.5	11.7	10.2	11.2	12.3
United Kingdom	16.2	15.5	17.5	18.3	16.2	12.4	13.4	13.2	13.9	12.5	11.5	12.0	13.0

Source: Demographic yearbook (2012) and Ined (2015)
[a]Crude birth rate – number of born children per 1000 inhabitants. It is usually computed for one calendar year

Annex 2: Educational Training – Practice, Exercise

Changes in Reproduction Behaviour of Population, Decrease of Natality

During the first demographic transition, the main demographic process is mortality. On the other hand, the decisive process of the second demographic transition is the birth rate and mainly the decrease in birth rates. This means that over a period of time, fewer children are born. The indicator of the evaluation of the decrease in birth rates is the crude birth rate. Monitor the decrease in birth rates based on the following indicators in the table below.

Exercise

Make a graph of the development of crude birth rates in selected countries in Europe in 1950–2010 (in Excel).

Characterise the overall trend of the development of birth rates in Europe.

What are the differences in the fall of birth rates in Western and Central Europe?

Compare the number of children in the families of your parents, grandparents and great grandparents or in the families of your close relatives.

Compare the development of birth rates in your country with the development of birth rates in selected European countries.

1950				
Age category	males	females	total	men negative
0-4	25 768	24 661	50 430	-25 768
5 – 9	21 520	20 935	42 455	-21 520
10 - 14	25 330	24 991	50 321	-25 330
15 - 19	23 070	23 029	46 099	-23 070
20 - 24	23 754	25 074	48 827	-23 754
25 - 29	19 227	23 016	42 243	-19 227
30 - 34	14 052	17 114	31 166	-14 052
35 - 39	18 121	22 577	40 697	-18 121
40 - 44	17 892	21 445	39 338	-17 892
45 - 49	15 999	19 536	35 534	-15 999
50 - 54	12 985	16 528	29 513	-12 985
55 - 59	10 406	13 920	24 326	-10 406
60 - 64	8 979	12 399	21 378	-8 979
65 - 69	7 153	10 130	17 283	-7 153
70 - 74	5 371	7 846	13 217	-5 371
75 - 79	3 269	5 156	8 425	-3 269
80-84	2 083	3 953	6 036	-2 083
spolu	254 979	292 308	547 287	

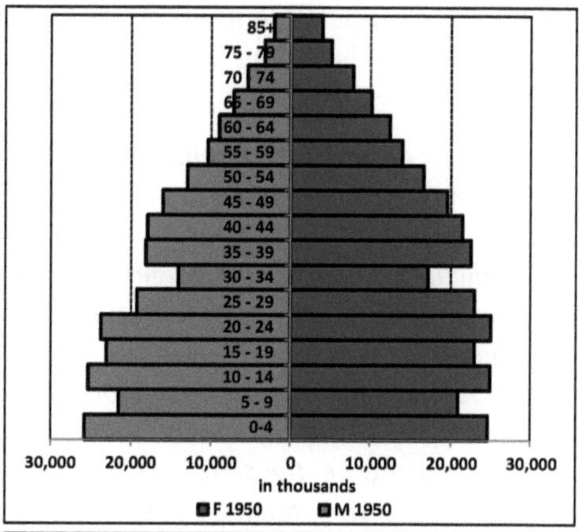

2015				
Age category	males	females	total	men negative
0-4	20 524	19 427	39 951	-20 524
5 – 9	20 198	19 196	39 393	-20 198
10 - 14	18 914	17 982	36 896	-18 914
15 - 19	19 212	18 283	37 495	-19 212
20 - 24	22 262	21 358	43 620	-22 262
25 - 29	25 905	25 171	51 076	-25 905
30 - 34	26 281	25 824	52 105	-26 281
35 - 39	25 737	25 783	51 520	-25 737
40 - 44	25 845	26 140	51 985	-25 845
45 - 49	25 768	26 380	52 147	-25 768
50 - 54	26 544	27 928	54 472	-26 544
55 - 59	24 422	26 847	51 269	-24 422
60 - 64	21 596	25 115	46 711	-21 596
65 - 69	17 482	20 793	38 275	-17 482
70 - 74	12 718	16 628	29 346	-12 718
75 - 79	11 011	16 590	27 601	-11 011
85+	11 748	22 831	34 579	-11 748
spolu	356 166	382 275	738 441	

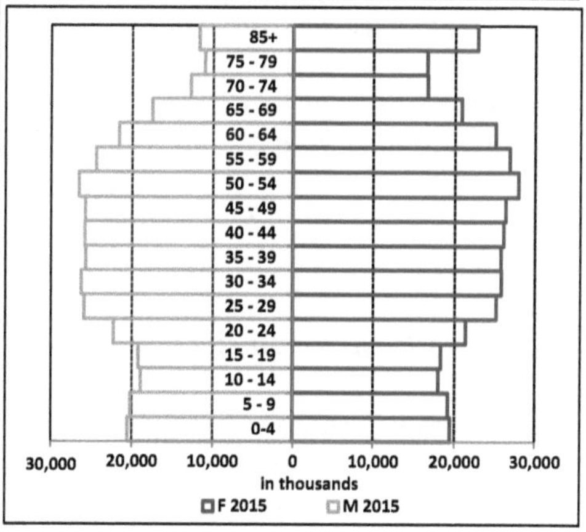

Fig. 14.15 Age structure of European population (1950 and 2015) (Source: UN 2015)

Annex 3: Educational Training – Practice, Exercise

Population Ageing

Each population is undergoing changes in the age structure. We are always interested in the age structure of a whole group of inhabitants, not only the individual age of an individual person. Whilst the population can age or get younger, individual age develops only towards ageing.

Globally, the European dimension of population ageing is the most dynamic one. It was here where the changes in behaviour during the SDT showed most radically in the second half of the twentieth century. These changes had a major impact on the age structure of Europe, which has changed from a progressive type (1950) to a regressive one (2015).

The changing shape of age pyramids of the European population shows several fundamental facts. It is evident that there is an enormous rise in the total number of the European population from 549 to 738 million. It is also evident that the base of the age pyramid is getting narrower and the top is getting wider.

Exercise

The process of below-up ageing resulted in a drop in the percentage of children. By how many percent did the percentage of children in the population of Europe drop?

The growth of the productive age group of the European inhabitants is significant. How did the number and proportion of inhabitants of the age group 20–64 change?

The increase in the number of inhabitants 65 years and over is extreme, which indicates a process of top-down ageing. By how many percent did the proportion of seniors grow, i.e. the share of the "old" population of 65 + ?

Evaluate the development of this age category also in terms of gender structure.

What are the reasons for bottom-up and top-down ageing of the population of Europe?

Try to prepare an age pyramid for Europe in Excel. When constructing a graph (two combined and rotated bar charts), use data in negative values in the last column.

Annex 4: Educational Training – Practice, Exercise

Family Behaviour, Extramarital Birth

Natality differentiates by legitimacy between marital and extramarital children. It is one of the significant indicators of family behaviour of the population. The number of extramarital children in Western Europe is one of the characteristic features of changes in family behaviour during the SDT which started appearing in the 1970s (Fig. 14.16). In this period, the ratio of extramarital children increased to 30–40 % and in the next years to 40–50 % (Finland, France, Iceland, Netherlands, Norway, Sweden, United Kingdom). The population of Central Europe recorded an increase in this indicator 40 years later.

Exercise

What are the reasons for the higher extramarital birth rate in general?

Which countries have the highest extramarital birth rate?

What is the rate in your country?

Evaluate the development of the percentage of extramarital children in Central Europe compared to other European countries using the graph, first orally and then in writing.

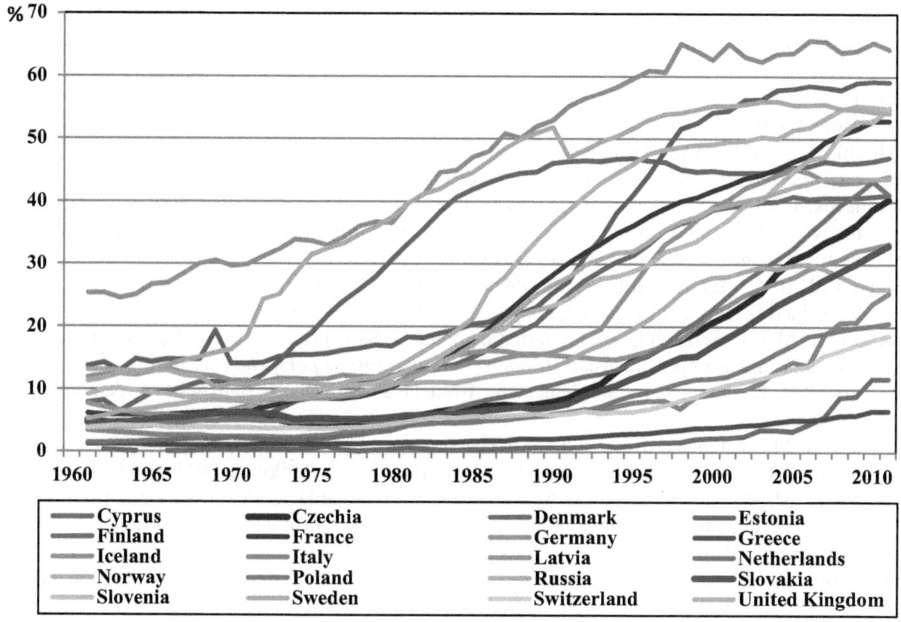

Fig. 14.16 Development of extramarital birth (Source: Ined (2015))

References

Birg, H. (1996). *Die Weltbevölkerung. Dynamik und Gefahren*. Becksche Reihe Wissen, Nr. 2050. München: C. H. Beck.

Coleman, D. A. (1993). Contrasting age structures of Western Europe and of Eastern Europe and the former Soviet Union: Demographic curiosity or labor resource? *Population and Development Review, 19*(3), 523–555.

CSO – Czech Statistical Office [Český statistický úřad]. (1961, 1970, 1980, 1991, 2001, 2011). *Sčítání obyvatel, domů a bytů* [Population and housing census]. Praha.

CSO – Czech Statistical Office [Český statistický úřad]. (2014). *Demografická příručka 2013* [Demographic handbook 2013]. Praha.

CSO – Czech Statistical Office [Český statistický úřad]. (2015). Population – Yearly time series. https://www.czso.cz/csu/czso/population_hd. Accessed 4 Aug 2015.

DeGenova, M. K., Rice, F. P., Stinnett, N., & Stinnett, N. (2010). *Intimate relationships, marriage and family*. New York: McGraw-Hill Education.

Demographic yearbook. (2012). *Regular Issues 2011–2012*. http://unstats.un.org/unsd/demographic/products/dyb/2010_round.htm. Accessed 5 Aug 2015.

Eurostat. (2015a). *Population change – Demographic balance and crude rates at national level*. https://open-data.europa.eu/en/data/dataset/3s22pBCm8CcIRnLqWTUUg. Accessed 3 June 2015.

Eurostat. (2015b). *Population and social conditions. Demography and migration*. http://ec.europa.eu/eurostat/web/population-demography-migration-projections/statistics-illustrated. Accessed 1 June 2015.

Fukuyama, F. (2005). *Veľký rozvrat. Ľudská prirodzenosť a opätovné nastolenie spoločenského poriadku* [The great disruption. Human nature and the reconstitution of social order]. Bratislava: Agora.
Hoff, A. (2008). Population ageing in Central and Eastern Europe as an outcome of the socio-economic transition to capitalism. *Social Darbas, 7*(2), 14–25.
Hradil, S. (1995). Die "Single-Gesellschaft" ["The single society"]. In *Perspektiven und Orientierungen*. München: Schriftenreihe des Bundeskanzleramtes.
Hudeček, T. (2006). Spoločnosť tretej vlny v diele H. a A. Tofflerovcov [The third wave society in the work of H. and A. Toffler]. In R. Ištok (Ed.), *Transformácia politicko-priestorových systémov a systémov demokracie*. Prešov: Prešovská Univerzita.
Ined. (2015). *Institut national d'études démographiques – Developed countries database*. http://www.ined.fr/en/pop_figures/bdd_conjoncture/. Accessed 1 Aug 2015.
Káčerová, M. (2005). Demografické starnutie populácie Slovenska a Európy [Population ageing of Slovakia and Europe]. In *Naša demografia, súčasnosť a perspektívy*. Zborník z 10. demografickej konferencie v Smoleniciach. Bratislava: Slovenská štatistická a demografická spoločnosť.
Káčerová, M., & Ondačková, J. (2015a). The process of population ageing in countries of the Visegrad Group (V4). *Erdkunde, 69*(1), 49–68. doi:10.3112/erdkunde.2015.01.04.
Káčerová, M., & Ondačková, J. (2015b). Proces starnutia populácie Slovenska v európskom kontexte [Ageing process of Slovakia population in European context]. *Slovenská štatistika a demografia, 25*(3), 44–58.
Káčerová, M., Ondačková, J., & Mládek, J. (2014). Time-space differences of population ageing in Europe. *Hungarian Geographical Bulletin, 63*(2), 49–68.
Kirk, D. (1996). Demographic transition theory. *Population Studies, 50*(2), 361–387.
Kol. (2013). *Rámcový vzdělávací program pro základní školy RVP ZV* [Framework Education Programme for Basic Education FEP BE]. Praha: VÚP.
Landry, A. (1934). *La révolution démographique. Études et essais sur les problèmes de la population* [Demographic revolution. Studies and essays on the problems of population]. Paris: INED.
Lapierre-Adamcyk, E., & Charvet, C. (2000). Nichteheliche Lebensgemeinschaft und Ehe: Bestandsaufnahme der demographischen Arbeiten [Non-marital cohabitation and marriage: Inventory of demographic works]. *Zeitschrift für Bevölkerungswissenschaft. Jg., 25*(1), 203–210.
Lesthaeghe, R. (1983). A century of demographic and cultural change in Western Europe: An exploration of underlying dimensions. *Population and Development Review, 9*(3), 411–435.
Lesthaeghe, R., & van de Kaa, D. J. (1986). Two demographic transition? In D. J. van de Kaa & R. Lesthaeghe (Eds.), *Population: Growth and decline*. Deventer: Van Loghum Slaterus.
Lesthaeghe, R. (1991). Moral control, secularization and reproduction in Belgium (1600–1900). In Societé Belge de Démographie (Eds.), *Historiens et Populations. Liber Amicorum Etienne Hélin* (pp. 259–279). Louvain-la-Neuve: Editions Academia.
Mackenroth, G. (1953). *Bevölkerungslehre – Theorie, Soziologie und Statistik der Bevölkerung* [Population teaching – Theory, sociology and statistics of the population]. Berlin: Springer.
Meslé, F., & Vallin, J. (2002). Mortality in Europe: the divergence between East and West. *Population (English Edition), 57*(1), 157–197.
Mládek, J. (1992). *Základy geografie obyvatelstva* [Foundation of population geography]. Bratislava: SPN.
Mládek, J. (1998). Druhý demografický prechod a Slovensko [Second demographic transition and Slovakia]. *Acta Facultatis Studiorum Humanitatis et Naturae Universitatis Prešoviensis, Folia Geographica*, 2(30), 42–52.
Mládek, J. (1999). Population development in Slovakia in the European context. *Acta Facultatis Rerum Naturalium Universitatis Comenianae, Geographica*, Supplementum, 2/1, 59–71.
Mládek, J. (2009). Teória civilizačných vĺn, postavenie a budúcnosť rodiny na Slovensku [Theory of civilization waves, family state and future in Slovakia]. In B. Bleha (Ed.), *Populačný vývoj Slovenska na prelome tisícročí*. Bratislava: Geografika.

Mládek, J., & Širočková, J. (2004a). Kohabitácie jako jedna z foriem partnerského spolužitia obyvateľstva Slovenska [Cohabitation like one form of spouses coexistence in Slovakia]. *Sociológia*, *36*(5), 423–454.

Mládek, J., & Širočková, J. (2004b). Premeny demografického, najmä rodinného správania obyvateľstva na Slovensku [Demographic, especially familiar behaviour population metamorphoses in Slovakia]. *Geografické informácie*, (8), 17–26.

Naisbit, J. (1982). *Megatrends. Ten new directions transforming our lives*. New York: Warner Books.

Naisbitt, J., & Aburdenová, P. (1992). *Megatrendy 2000* [Megatrends 2000]. *Desať nových smerov na deväťdesiate roky* [Ten new directions for the 1990s]. Bratislava: Bradlo.

Notestein, F. W., et al. (1944). *The future population of Europe and the Soviet Union: Population projections 1940–1970*. Geneva: League of Nation.

Pavlík, Z. (1964). *Nástin populačního vývoje světa* [Study of the world population development]. Praha: Československá akademie věd.

Pavlík, Z., Rychtaříková, J., & Šubrtová, A. (1986). *Základy demografie* [Foundations of demography]. Praha: Academia.

Pastor, K. (1997). Súčasný populačný vývoj na Slovensku a demografické teorie. *Slovenská štatistika a demografia*, *7*(4), 45–58.

Pastor, K. (1998). *Druhá demografická revolúcia* [Second demographic transition]. Slavnostná konferencia 30 rokov Slovenskej štatistickej a demografickej spoločnosti, Bratislava 26. 3. 1998. Zborník (pp. 34–36). Bratislava: SŠDS.

Pastor, K. (2002). *Rodina a rodinná politika v druhej demografickej revolúcii* [Family and family policy in the second demographic transition]. In rodina v ohrození – výzva pre sociálne vedy. Zborník k 10. výročiu obnovenia Trnavskej Univerzity v Trnave (pp. 30–41). Trnava: Katedra sociológie Fakulty humanistiky Trnavskej univerzity.

Rabušic, L. (2001). *Kde ty všechny děti jsou? Porodnost v sociologické perspektivě* [Where are all the children? Natality in sociological perspective]. Praha: Sociologické nakladatelství.

Rankov, P. (2006). *Informačná spoločnosť – perspektívy, problémy, paradoxy* [Information society – Perspectives, problems, paradox]. Levice: LCA Publishers Group.

Roubíček, V. (1997). *Úvod do demografie* [Introduction to demography]. Praha: Codex Bohemia.

Sobotka, T., & Toulemon, L. (2008). Changing family and partnership behaviour: Common trends and persistent diversity across Europe. In T. Frejka, T. Sobotka, J. M. Hoem, & L. Toulemon (Eds.), *Childbearing trends and policies in Europe* (Demographic research, pp. 85–138). http://www.demographic-research.org/Volumes/Vol19/6/. Accessed 25 Aug 2015.

SO SR – Statistical Office of SR [Štatistický úrad SR] – Infostat, Demographic Research Centre. (1922–2014). *Vekové zloženie obyvateľstva Slovenskej republiky 1921–2013* [Age structure of the population of the Slovak Republic 1921–2013]. Bratislava.

SO SR – Statistical Office of SR [Štatistický úrad SR] – Infostat, Demographic Research Centre. (2001, 2012). *Stav a pohyb obyvateľstva v Slovenskej republike* [State and movement of the population in Slovakia]. Bratislava.

SO SR – Statistical Office of SR [Štatistický úrad SR]. (1961, 1970, 1980, 1991, 2001, 2011). *Sčítanie obyvateľov, domov a bytov* [Population and housing census] Bratislava.

Toffler, A. (1980). *The third wave*. New York: Bantam Books.

Toffler, A. (1990). *Powershift: Knowledge, wealth and violence at the edge of the 21st century*. New York: Bantman Books.

Toffler, A. (1992). *Šok z budúcnosti* [Future shock]. Praha: Práce.

Toffler, A., & Tofferová, H. (1996a). *Utváranie novej civilizácie. Politika tretej vlny* [Creating a new civilization. The politics of the third wave]. Bratislava: Open windows.

Toffler, A., & Tofflerová, H. (1996b). *Nová civilizace. Třetí vlna a jej důsledky* [New civilization. Third wave and its consequences]. Praha: Dokořán.

UN – United Nations. (2015). World population prospects: The 2015 revision, CD ROM Edition. http://esa.un.org/unpd/wpp/Excel-Data/population.htm. Accessed 4 Aug 2015.

Van de Kaa, D. J. (1980). Recent trends in fertility in Western Europe. In R. W. Hiorns (Ed.), *Demographic patterns in developed societies* (pp. 55–83). London: Taylor and Francis.

Van De Kaa, D. J. (1987). Europe's second demographic transition. *Population Bulletin, 45*(1), 1–57.

Van de Kaa, D. J. (1996). Anchored narratives: The story and findings of half a century of research into the determinants of fertility. *Population Studies, 50*(3), 389–432.

Van de Kaa, D. J. (1998). *Postmodern fertility preferences: From changing value orientation to new behaviour* (Working Papers in Demography. No 74, pp. 1–56). Canberra: The Australian National University.

Van de Kaa, D. J. (1999). Europe and its population: The long view. In European populations: Unity in diversity. European Population conference, Hague 1999 (pp. 1–50). Dordrecht: Academic Publisher.

Vance, R. B. (1952). Is theory for demographers? *Social Forces, 31*(1), 9–13.

Resumé

AKTUÁLNÍ VÝZVY VE VÝUCE ZEMĚPISU V ČESKU A STŘEDNÍ EVROPĚ

Geografické myslenie v Česku, na Slovensku a tiež v Poľsku zaznamenalo v posledných desaťročiach výrazné premeny. Výsledkom je obohatenie poznatkovej sféry vo všetkých geografických disciplínach. Podieľa sa na tom, okrem iného, rozvoj metód a techník vedeckého poznávania. Núka sa potreba transféru nového myslenia a nových bohatých poznatkov do výuky na všetkých typoch školského vzdelávania. Pre odborovú didaktiku geografie, ktorá je situovaná medzi vednými odbormi geografia a všeobecná didaktika, ako veda o výchove a vzdelávaní, tak vyvstávajú nové úlohy. V snahe splniť ich, didaktika geografie, nachádza nové, efektivne metódy vyučovania. Tieto by mali vychádzať z niekoľkých paradigiem – a) prepojenie vedy a vzdelávania, b) redukcia nadmerného objemu učiva a stanovenia reálneho kompendia geografického vzdelávania, c) priťahovanie detí a budenie ich záujmu aj o geografické poznatky v dnešnom dynamicky vyvíjajúcom sa svete.

Předkládaná publikace si klade za cíl zabývat se aktuálními výzvami souvisejícími s výukou geografie především na 2. stupni základních škol a nižším stupni osmiletých gymnázií (týkající se žáků ve věku 12–16 let). Představuje nejenom platformu poukazující na soudobé problémy zeměpisné výuky, ale především se snaží navrhnout možné změny, akceptovat a zapojit do výuky moderní trendy a reagovat na impulzy, vedoucí ke zvýšení zájmu žáků a studentů o zeměpisnou tematiku.

Publikace přináší aplikace vybraných moderních metodických postupů ve výukové praxi zeměpisu. Mezi ně patří např. badatelsky orientované vyučování, globální rozvojové vzdělávání, silně ukotvená výuka a projektová výuka, které jsou podpořené kvalitním využitím informačních a komunikačních technologií. Uplatnění inovativních metod ve výuce je potřeba stavět na bázi vytvoření obrazu historického vývoje a politického pozadí kurikulárních změn, které proběhly či probíhají v rámci sledovaného středoevropského „školského"prostoru. Publikace je zaměřena na změny, které proběhli v didaktice geografie v Česku, na Slovensku a v

P. Karvánková et al. (eds.), *Current Topics in Czech and Central European Geography Education*, DOI 10.1007/978-3-319-43614-2

Polsku. Obecně didaktika geografie v těchto zemích střední Evropy byla ovlivněna sociálně-politickými podmínkami jejich vývoje jako socialistických krajin. V rámci celospolečenské transformace, která zde probíhala po demokratických změnách v roce 1989, bylo nezbytné měnit vzdělávací systém. Především šlo o zavádění moderních výukových metod a technik ve výuce, které jsou již dobře propracované a delší dobu realizované ve vyspělých evropských zemích.

V první části publikace je tedy pozornost věnována stavu kurikulárních dokumentů v Česku, na Slovensku a v Polsku. Jsme si plně vědomi skutečnosti, že se pojetí i obsahy geografického vzdělávání v jednotlivých zemích od sebe liší v kontextu vyplývajícího z geografického poznání, což může být dáno jeho vývojem, jazykem a kulturou. Jinými slovy, každé kurikulum je ovlivněno svým národním charakterem. Prvá časť publikácie je rozdelená na tri kapitoly, ktoré sa venujú predstaveniu didaktiky geografie v Českej, Slovenskej a Poľskej republike. Každá kapitola prezentuje všeobecný systém vzdelávania v týchto krajinách, a tiež aktuálne otázky, vývoj, problémy, zmeny a trendy vo výučbe zemepisu. Pozornosť je venovaná silným a slabým stránkam súčasnej výučby zemepisu. Záverečné zhrnutie je zamerané na porovnanie vybraných učebných postupov. Tvorci geografického kurikula tu nájdu celý rad námětov, ako realizovať výučbu zameranú na podporu a aktivizáciu procesu výučby.

Zavedení inovativních metod, které vzbuzují v žácích zájem nejenom o danou problematiku, ale jsou rovněž předávány zábavnou a pro žáky snadno akceptovatelnou a přijatelnou moderní formou s využitím široké škály informačních a komunikačních technologií, je předpokladem úspěšné výukové praxe. Základem a rovněž cílem vzdělávání je schopnost uplatnění nabytých zkušeností a dovedností žáků a studentů v praktickém životě. Jedny z nejaktuálnějších metodických přístupů, splňujících uvedená kritéria v přímé praxi a umožňující jejich využití na hodinách zeměpisu představuje část publikace. Prvý príspevok je zameraný na predstavenie možností aplikácie metódy bádateľsky orientovanej výučby (BOV). Prezentuje námety českej praxe k využitiu BOV fyzického zemepisu, ktoré môžu byť vhodnou inšpiráciou pre učiteľov. Konkrétne ponúka príklad bádateľskej aktivity zameranej na pedológiu pod názvom „Půda aneb po čem všem to vlastně chodíme“. Druhá kapitola seznamuje s příklady dobré praxe v oblasti globální výchovy, s cílem ukázat, že práce s aktuálním světovým děním se není třeba obávat. Naopak vyvolání u dětí zájmu o dění v jiných částech světa může být velmi užitečným nástrojem k podpoře aktivního učení a vnímání citlivých témat v jejich okolí. V příspěvku je prezentována ukázková vyučovací hodina Globálního rozvojového vzdělávání (GRV).

V rámci druhé části publikace je rovněž ponechán prostor na prezentaci možností terénní výuky, která je součástí vzdělávacího oboru zeměpis. Terénní výuka patří nejen v České republice, ale i v mnoha dalších zemích k silným výukovým strategiím. Její realizace je však v mnoha ohledech obtížná. Jejím prostřednictvím se v rámci výuky geografie dostáváme k řešení tzv. případových studií. Například v rozhodovacím procesu, v kterém studenti řeší možnosti a důsledky výstavby lyžařského areálu ve městě Brně. Svoje závěry předkládají městskému zastupitelstvu k jejich posouzení. V praxi to teda znamená řešení různých reálných situací, které v krajině nastávají nebo mohou nastat a žáci a studenti hledají cesty a zejména geografické nástroje pro jejich řešení. Konečným cílem je naučit žáky kriticky přemýšlet. Což nejde bez změny přístupu učitelů i žáků. Role učitelů se mění z vševědoucího poskytovatele znalostí na manažera či organizátora a role žáků od tradičního spotřebitele informací k aktivnímu učícímu se člověku.

Hlavným cieľom štvrtej kapitoly je predstaviť rôzne prístupy a stratégie realizovania projektovej výuky (PBL) v rámci českého geografického vzdelávania. V prípadových štúdiách je venovaná pozornosť obvyklým chybám učiteľom pri využívaní PBL. Záverečné zhrnutie zachytáva hlavné rozdiely a prístupy k aplikácii PBL v českej didaktike v kontexte príkladov jeho použitia v medzinárodnom kontexte školskej praxe. Pátá kapitola se věnuje možnostem využití cloudových nástrojů ve výuce zeměpisu (např. Google Disk, YouTube, formuláře, rozšířené vyhledávání a snadná publikace informací, on-line spolupráce při výuce aj.). Zároveň prezentuje praktickou ukázku využití mobilní technologie, sociálních sítí a infografiky ve výuce zeměpisu. Představuje také nástroj Arcgis online pro vytváření map a mapových aplikací, který je skvělým pomocníkem při vytváření interaktivní výuky. V samostatnej časti kapitoly sú prezentované možnosti integrácie satelitných snímok, leteckých fotografií a máp a ich interpretácie na hodinách zemepisu. Otázke schopnosti čítania snímok a máp, tj. kartografickej a vizuálnej gramotnosti českých študentov vo veku 11, 15 a 19 rokov sa venoval prezentovaný výskum. Analyzované boli rozdiely vo výsledkoch spracovania úloh pri jednotlivých typoch podkladov.

Geografie je vědní disciplína, která leží na pomezí přírodních a sociálních věd. Svým tradičním přírodovědně zaměřeným objektem studia má své nezastupitelné místo mezi přírodními vědami i při rozvíjení výuky na školách. Zároveň, svou šířkou a zaměřením studia na krajinu jako místo života člověka, má ambice v nejbližší budoucnosti vytvářet v didaktické oblasti prostor pro výuku také sociálně-ekonomických témat na hodinách zeměpisu. Jeden z vybraných příkladů aplikace metody při výuce demografických témat přináší příspěvek, který je součástí třetí části publikace. V príspevku sú analyzované zmeny demografického správania populácie Európy, pričom osobitný zreteľ je kladený na poznanie vývoja správania obyvateľstva Českej republiky a Slovenskej republiky. Demografické správanie obyvateľstva je akýmsi barometrom stupňa vývoja spoločnosti. Ide o aspekt sociálnych prejavov, ktorý je napriek komplikovaným súvislostiam a väzbám, významný a vhodný na bližšie poznávanie a identifikovanie už deťmi základných škôl.

Obecně je závěrečná část publikace věnována tematice výuky sociální, regionální, politické a kulturní geografie na školách, a to z hlediska uplatnění regionálně geografických znalostí a dovedností v odborné praxi či v každodenním životě žáků a studentů. Příspěvky prezentují již několik desítek let v české i slovenské geografii diskutovanou otázku, spojenou s existencí, podstatou a zaměřením regionální geografie. Zároveň je navrhnuta optimální míra politicko-geografických a kulturně-geografických znalostí pro jednotlivé stupně škol. Vztah k politickým strukturám je nejvíce ovlivněn subjektivními názory každého pedagoga, který je pak předává svým studentům. Příspěvky proto mohou napomoci přípravě budoucích učitelů v rámci moderní výuky politické a kulturní geografie.

Geografie je vzhledem ke své multidisciplinaritě, jež ji nabízí úzkou vazbu na všechna průřezová témata RVP ZV v Česku, více než vhodným předmětem pro aplikaci inovativních metodických postupů. Prostor pro efektivnější způsob výuky zeměpisu na základních a středních školách spatřujeme v možnostech integrace předmětu s průřezovými tématy skrze využití výše zmíněných metodických postupů. Geografie tak získává možnost zatraktivnění výukových hodin.

Úspešnosť plnenia hlavných vzdelávacích úloh v geografii je závislá od mnohých faktorov, objektívnych i subjektívnych. Za jeden z rozhodujúcich možno považovať symetrickú vyváženosť (geografických) vedeckých poznatkov a didaktických metód. Predložená publikácia sa pokúša o prezentáciu nových prístupov vo výuke geografie, ktoré vychádzajú zo skúseností v Česku, na Slovensku a tiež v Poľsku. Jej ambíciou je podporiť ciele Komisie pre geografické vzdelávanie Medzinárodnej geografickej asociácie (IGU CGE), ktoré su obsiahnuté v Mezinárodnej charte o geografickom vzdelávaní a v Mezinárodnej deklarácii o geografickom vzdelávaní v oblasti kultúrnej rozmanitosti. Štruktúra publikácie je nasledovná:

1. kapitola Geografické vzdělávání z českého, slovenského a polského pohledu

I. část Obecný vzdělávací systém a vyučování geografie v České, Slovenské a Polské republice

2. kapitola Geografické curriculum v Českéj republice: výzvy a příležitosti
3. kapitola Súčasné geografické vzdelávanie na Slovensku – premeny a podmienky
4. kapitola Výuka na základních a středních školách v Polsku

II. část Inovatívne metódy vo výuce zeměpisu – cesta jako aktivizovat žáky

5. kapitola Badatelsky orientované vyučování fysického zeměpisu
6. kapitola Globální rozvojové vzdělávání: český přístup při šíření globální povědomí
7. kapitola Případové studie ve výuce zeměpisu jako účinný způsob výuky
8. kapitola Projektově orientované vzdělávání

III. část Informační technologie ve výuce geografie

9. kapitola Využití informačních a komunikačních technologií a internetových zdrojů v přírodovědném vzdělávání
10. kapitola Informační technologie ve výuce zeměpisu z pohledu učitele
11. kapitola Čtení satelitních a leteckých snímek a map – Vývoj kartografické a vizuální gramotnosti

IV. část Aktuální témata politické, regionální a sociální geografie a jejich didaktické aplikace

12. kapitola Makroregiony světa jako část výuky regionální geografie
13. kapitola Nová politická geografie
14. kapitola Zmeny demografického správania v 2. polovici 20. a na začiatku 21. storočia, možnosti využitia poznatkov v didaktickej praxi

Streszczenie

AKTUALNE ZAGADNIENIA EDUKACJI GEOGRAFICZNEJ W CZECHACH I W EUROPIE ŚRODKOWEJ

Myślenie geograficzne w Republice Czeskiej, na Słowacji i w Polsce uległo istotnym zmianom w ostatnich dziesięcioleciach, w wyniku wzbogacenia sektora wiedzy we wszystkich dyscyplinach geograficznych, między innymi poprzez rozwój metod i technik badawczych. Proces ten prowadzi to do konieczności transferu nowych myśli i ciągle uaktualnianej wiedzy, aby uczyć na różnych poziomach kształcenia. Dlatego dla dydaktyki geografii, usytuowanej pomiędzy poszczególnymi dyscyplinami geograficznymi a dydaktyką ogólną i naukami o wychowaniu i edukacją, powstają nowe zadania, a także wyzwania.

W celu zaspokojenia tych wyzwań, dydaktyka geografii poszukuje nowych, skutecznych metod nauczania opartych na wielu paradygmatach: a) łączeniu wiedzy naukowej z edukacją; b) zmniejszaniu nadmiernej ilości programów nauczania oraz ustanowieniu prawdziwego kompendium edukacji geograficznej; c) przyciąganiu dzieci i młodzieży i rozbudzaniu w nich zainteresowania wiedzą geograficzną w dynamicznie zmieniającym się współczesnym świecie.

Celem niniejszej publikacji jest ukazanie aktualnych wyzwań związanych z nauczaniem geografii, głównie w szkołach średnich, dla dzieci w wieku 12–16 lat (w Polsce na poziomie gimnazjum). Prezentują one nie tylko zbiór aktualnych zagadnień w nauczaniu geografii, ale przede wszystkim stara się zaproponować ewentualne zmiany, zaakceptować i zaangażować nowoczesne trendy w nauczaniu, a także reagować na bodźce prowadzące do wzrostu zainteresowania geografią wśród uczniów i studentów.

Publikacja przedstawia wybrane nowoczesne strategie i procedury w nauczaniu geografii. Są to: uczenie się przez doświadczenie, edukacja dla rozwoju globalnego oraz efektywne nauczanie, które są wspierane przez zastosowanie wysokiej jakości technologii informacyjnych i komunikacyjnych. Zastosowanie innowacyjnych

P. Karvánková et al. (eds.), *Current Topics in Czech and Central European Geography Education*, DOI 10.1007/978-3-319-43614-2 metod w nauczaniu musi opierać się na znajomości rozwoju historycznego oraz tła politycznego zmian programowych, które miały miejsce w obserwowanym obszarze edukacyjnym Europy Środkowej. Obszar ten w niniejszej publikacji jest reprezentowana głównie przez znajomość systemu nauczania geografii w Czechach, Słowacji i Polsce. Edukacja geograficzna w centralnej przestrzeni europejskiej była długo pod wpływem ogólnych warunków społeczno-politycznych w byłych krajach komunistycznych. W transformacji politycznej i gospodarczej koniecznym jest również zmiana systemów edukacyjnych w tych krajach. Przede wszystkim istotne jest wprowadzenie nowoczesnych metod i technik nauczania, które zostały już zastosowane w rozwiniętych krajach europejskich. Te znaczące zmiany przyniosły nowe, potransformacyjne doświadczenia w systemach edukacyjnych w przedstawianych krajach.

Dlatego I część publikacji koncentruje się na właściwym programie nauczania. Zawiera on propozycje możliwych inspiracji, wyzwań i zmian poruszanych w dokumentacji szkolnej z krajów rozwiniętych.

Warunkiem udanej praktyki pedagogicznej jest wprowadzenie innowacyjnych metod, które wzbudzą zainteresowanie uczniów, a także które są przekazywane w ciekawy i prosty sposób, także w nowoczesnej formie, z wykorzystaniem szerokiego zakresu technologii informacyjnych i komunikacyjnych. Fundamentem oraz celem edukacji jest możliwość zastosowania w praktyce nabytych przez uczniów doświadczeń i umiejętności. Niektóre z najbardziej aktualnych metodycznych systemów, które spełniają wyżej wymienione kryteria w bezpośredniej praktyce i umożliwiają ich wykorzystanie na lekcjach geografii, zostały przedstawione w rozdziałach II części publikacji. Można w nim również znaleźć prezentację możliwości edukacji w terenie, która należy (nie tylko w Czechach, ale i w wielu innych krajach) do strategii „efektywnego nauczania" (ang. powerful learning). Jednak realizacja tego procesu jest trudna w wielu aspektach. W kontekście nauczania geografii prowadzi to do rozwiązywania tzw. studiów przypadku. W praktyce oznacza to analizowanie różnych prawdziwych lub prawdopodobnych sytuacji, gdzie uczniowie i studenci poszukują sposobów oraz różnych instrumentów geograficznych do rozwiązania problemu. Ostatecznym celem jest nauczenie uczniów myślenia. Cały ten proces nie dojdzie do skutku bez zmiany nastawienia zarówno wśród uczniów, jak i nauczycieli. Rola nauczyciela zmieniała się od wszechwiedzącego „dostawcy wiedzy" do „kierownika" lub „organizatora", a rola ucznia i studenta zmieniała się od tradycyjnego "konsumenta informacji "do aktywnie uczącej się i przyswajającej wiedzę osoby.

Geografia jest dyscypliną, która usytuowana jest na pograniczu nauk przyrodniczych i społecznych. Ma też niezastąpioną rolę wśród przedmiotów szkolnych. Z uwagi na zakres poruszanych zagadnień związanych z miejscem życia dla człowieka, geografia ma szanse w najbliższej przyszłości stworzyć przestrzeń w dziedzinie edukacji – do nauczania zagadnień społeczno-gospodarczych.

Część III, ostatnia, dotyczy kwestii nauczania geografii społecznej, regionalnej, politycznej oraz kulturalnej w szkołach, w zakresie stosowania wiedzy i umiejętności geograficznych w praktyce zawodowej oraz w życiu codziennym uczniów głównie w aspekcie regionalnym. Badania przedstawiają problem, który był już dyskutowany od kilkudziesięciu lat, zarówno w czeskiej jak i słowackiej geografii, i który jest połączony z istnieniem, istotą i treściami geografii regionalnej. Rozdziały trzeciej części próbują przedstawić optymalny zakres wiedzy polityczno-geograficznej i kulturowo-geograficznej na każdym poziomie szkoły. Stosunek do struktur politycznych zależy najbardziej od subiektywnych opinii każdego nauczyciela, który przekazuje je do swoich uczniów. Zgromadzony materiał pomoże w przygotowaniu nauczycieli do nowoczesnego nauczania geografii politycznej i kulturalnej. Ostatni rozdział opisuje wybrany przykład metody podczas nauczania zagadnień demograficznych. Analizuje zmiany zachowań demograficznych w populacji europejskiej, w której szczególny nacisk położony jest na zrozumienie zachowania czeskiej i słowackiej populacji. Zachowania demograficzne populacji są swego rodzaju barometrem poziomu rozwoju społeczeństwa. Zwłaszcza w aspekcie manifestacji społecznych, które mimo skomplikowanych połączeń i więzi, są już znaczące i nadają się do dalszej nauki i identyfikacji przez uczniów szkół średnich.

Ze względu na swoją interdyscyplinarność, geografia oferuje bliskie powiązania ze wszystkimi tematami przekrojowymi w Ramowym programie nauczania na poziomie podstawowym i stanowi odpowiedni obszar do stosowania innowacyjnych procedur metodologicznych. W ten sposób geografia zyskuje możliwość zwiększenia atrakcyjności zajęć dydaktycznych.

Pomyślne rozwiązanie podstawowych wyzwań edukacyjnych w geografii zależy od wielu czynników, zarówno obiektywnych i subiektywnych. Symetryczną równowagę pomiędzy geograficzną wiedzą naukową, a metodami nauczania można uznać za jeden z decydujących czynników. Poniższa publikacja stara się przedstawić nowe podejścia w nauczaniu geografii na podstawie doświadczeń w Czechach, Słowacji i Polsce – na tle wybranych powiązań w Europie Środkowej. Pragnie wspierać cele Komisji Edukacji Geograficznej IGU (IGU CGE), które zawarte są w Międzynarodowej Karcie Edukacji Geograficznej oraz w Deklaracji Międzynarodowej Edukacji Geograficznej dla Różnorodności Kulturowej. Publikacja składa się z trzech głównych części i poszczególnych rozdziałów:

Rozdział 1 Edukacja geografii z czeskiego, słowackiego i polskiego punktu widzenia

I. część Ogólny system edukacji i nauczania geografii w Czechach, Słowacji i Polsce

Rozdział 2 Program nauczania geografii w Czechach: ambitne możliwości

Rozdział koncentruje się na krótkiej ocenie reformy programowej w Czechach. Autorzy są świadomi faktu, że koncepcja i treści edukacji geograficznej w poszczególnych krajach różnią się od siebie ze względu na zróżnicowany rozwój, historię, język i kulturę. Innymi słowami, narodowy charakter ma istotny wpływ na zasoby wiedzy geograficznej i jego przyszły program nauczania. Twórcy programu nauczania geografii w Czechach przedstawiają szereg propozycji dotyczących sposobu wdrażania poszczególnych dyscyplin nauki w celu stymulowania procesu nauczania. Rozdział zawiera ogólny obraz systemu edukacji, aktualnie podejmowanych problemów, wydarzeń, zmian oraz nowych trendów w nauczaniu geografii.

Jednocześnie zwraca się uwagę na mocne i słabe strony obecnego nauczania geografii. Końcowe podsumowanie skupia się na porównaniu wybranych metod nauczania i rozwiązań w szerszym kontekście międzynarodowym, zwłaszcza w Europie Środkowej.

Rozdział 3 Aktualna edukacja geograficzna na Słowacji — konwersja i warunki

Transformację nauczania geografii na Słowacji charakteryzują dwa podstawowe atrybuty: konwersja treści i cechy ilościowe. Geografia (jako przedmiot nauczania) na Słowacji rozwija się od końca lat 80. ubiegłego wieku (era Czechosłowacji) w sposób ewolucyjny. Typowe dla jego rozwoju były stopniowe kroki, dzięki którym w geografii stopniowo eliminowano zakłócenia ideologiczne. Usunięta została gloryfikacja komunizmu i zbyt skomplikowana problematyka i tematy, które były nieodpowiednie dla uczniów w tym wieku. Transformacja z regionalnego podejścia do problematyki zorientowanej edukacji nie została jednak ukończona. Aspekt orientacji i definiowania problemu został włączony do geografii regionalnej i był szczególnie podkreślony, ale nie został zakończony. Zakres zmian całego programu był jeszcze niewystarczająco wysoki. Wiele innych fundamentalnych zmian warunków trzeba było dokonać prawdziwymi. Potrzebują oni środowisko, które ich akceptuje i osób, które wykonują te zmiany. Nauczanie geografii zostało znacznie zredukowane, podobnie jak innych przedmiotów nauczania w programie edukacyjnym szkoły. Nauczanie geografii zostało znacznie zredukowane ze względu na stworzenie nowych miejsc i obszarów nauczania w programie edukacyjnym szkoły. Liczba godzin została zmniejszona z dziewięciu do poziomu trzech godzin tygodniowo (dla uczniów w wieku 11–15 lat), a w szkołach średnich (w sumie pięć lat) do pięciu godzin, a obecnie do sześciu godzin w tygodniu dla wszystkich klas w całym cyklu nauczania. Nauczanie geografii w liczbie jednej godziny w tygodniu jest ogromną stratą dla przedmiotu nauczania.

Rozdział 4 Nauczanie i uczenie się geografii w szkole ponadpodstawowej w Polsce

Polski nowy system edukacji określa trzy poziomy edukacji: w pełnym wymiarze godzin obowiązkowej edukacji trwające 10 lat i obejmujący ostatni rok edukacji przedszkolnej, sześć lat nauki w szkole podstawowej i trzy lata szkoły średniej niższej (gimnazjum). Ponadto dalsze kształcenie może być prowadzone, na przykład, w popularnym wśród uczniów i rodziców trzyletnim liceum ogólnokształcącym (liceum). Ostatnio pojawiają się istotne zmiany strukturalne w polskiej edukacji akademickiej w związku z wytycznymi Deklaracji Bolońskiej. Dlatego od roku 2007/2008, wszystkie uczelnie zapewniają dwustopniowy system kształcenia szkolenia przyszłych geografów. To stwarza nowe możliwości edukacyjne i zawodowe dla studentów (i potencjalnych studentów), w tym szerszego wyboru przedmiotów do studiowania nowych specjalności, jak również pomaga w podejmowaniu decyzji co do dalszej ścieżki edukacyjnej (dla studiów drugiego stopnia). Poszczególne wydziały i instytuty geografii niedawno dokonały istotnych zmian w swoich planach i programach nauczania.

II. część Innowacyjne metody w nauczaniu geografii – sposoby na aktywizację uczniów

Rozdział 5 Nauka geografii fizycznej poprzez doświadczenie

Geografia pełni niezastąpioną rolę wśród nauk przyrodniczych. Oprócz swojej interdyscyplinarności ma także bliskie powiązania ze wszystkimi przekrojowymi tematami. W związku z tym jest to odpowiedni przedmiot do stosowania procedur metodycznych IBL (ang. inquiry-based learning) czyli nauki poprzez doświadczanie i dociekanie. Nauczanie geografii z wykorzystaniem metod IBL wzbudza zainteresowanie uczniów ich otoczeniem, przyrodą i krajobrazem, w którym żyją. To rozwija ich umiejętności oraz zdolności zastosowania wiedzy teoretycznej w praktyce, a więc takich zdolności i umiejętności, których czeskim uczniom nadal brakuje. Jednocześnie rozdział Zajęcia terenowe w nauczaniu geografii, praktyka i zastosowanie, w ramowym programie nauczania na poziomie podstawowym jest realizowany przy pomocy nauczania problemowego. Obecnie jedno z głównych pytań badawczych dotyczących IBL w praktyce to: jak osiągnąć IBL w klasie? Przedstawiona analiza stara się udzielić odpowiedzi na to pytanie, głównie z czeskiego punktu widzenia. Czytelnik zapozna się z różnym podejściem, metodami i możliwościami korzystania z IBL w czeskich szkołach średnich. W tym samym czasie przedstawione są przykłady zagadnień badawczych i studiów przypadku szczególnie z Czech, Polski, Słowacji (i innych krajów), które mogą stanowić odpowiednią inspiracją dla nauczycieli. Studia przypadków będą koncentrować się głównie na przykładowych metodach nauczania z wykorzystaniem IBL w geografii fizycznej. Głównym powodem jest niewielkie zainteresowanie czeskich uczniów w obszarze nauk przyrodniczych i słaba motywacja do angażowania się w tą dziedzinę nauki. Opracowanie to ma na celu dostarczenie inspirujących pomysłów i sposobów na poprawę nauczania geografii fizycznej przez IBL i wzbudzenie zainteresowanie uczniów.

Rozdział 6 Edukacja dla rozwoju globalnego: podejście czeskie do szerzenia globalnej świadomości

Zakotwiczenie Edukacji dla rozwoju globalnego w praktyce szkolnej Europy Centralnej oraz w dokumentach szkolnych jest nadal w fazie początkowej. Charakterystyczny rozwój społeczny tych krajów jest główną przyczyną wolniejszej realizacji głównych tematów Edukacji dla rozwoju globalnego w nauczaniu. Po aksamitnej rewolucji w 1989 roku, nastąpiła transformacja z reżimu totalitarnego do demokratycznego społeczeństwa. Jednakże zmiany kulturowe i wartości etyczne nie nadążają za zmianami w obszarach politycznych i gospodarczych. Ludzie stają się zbyt apatyczni by dostrzegać różnice społeczne, biedę, problemy grup mniejszościowych (ludności romskiej, imigrantów), problemy środowiskowe, itp. W takim klimacie społecznym trudno jest zatem zacząć uczyć dzieci myślenia o różnych wydarzeniach w skali i ujęciu globalnym. W praktyce szkolnej, nauczyciele często zadają pytanie, jak "uchwycić" globalne tematy. Metody implementacji Edukacji dla rozwoju globalnego w czeskiej praktyce edukacyjnej są stale w fazie początkowej. Artykuł ten zatem najpierw sprawdza czym jest Edukacja dla rozwój globalnego, co jest jej głównym celem i dlaczego jest ona konieczna, aby potem wykorzystać ją w praktyce pedagogicznej. Wspomniany zostanie także rozwój i zastosowanie metodologii Edukacji dla rozwoju globalnego w Republice Czeskiej. Autorzy tego opracowania przedstawiają przykłady dobrych praktyk, a także studia przypadków, głównie z Czech, aby pokazać, że wzbudzanie zainteresowanie dzieci imprezami oraz inicjatywami w innych częściach świata może być bardzo użytecznym narzędziem do wspierania aktywnego uczenia się i właściwego dostrzegania także drażliwych tematów w ich otoczeniu. Niniejszy materiał zbada również możliwość aplikacji Edukacji dla rozwoju globalnego na różnych szczeblach systemu czeskiej edukacji. Jednak główna uwaga zostanie poświęcona studiom przypadków z wykorzystaniem metod w edukacji dla globalnego rozwoju w szkołach średnich.

Rozdział 7 „Studia przypadku w edukacji geograficznej jako przykład „efektywnego nauczania"

Studia przypadku przedstawiają odpowiednią formę i sposób dostarczania uczniom rozwiązań w rzeczywistych sytuacjach z otoczenia, w którym żyją. Jest to tzw. "efektywne nauczanie" i jest ono tak zaprojektowane, aby pomóc uczniom i studentom w radzeniu sobie z trudnościami codziennego życia poprzez nauczanie-uczenie się geografii. Ta metoda nie jest tak dobrze znana i stosowana w Republice Czeskiej jak poza jej granicami, gdzie funkcjonuje pod nazwą "powerful knowledge" lub "powerful teaching". Część wprowadzająca w tym rozdziale poświęca dużo miejsca na samo zrozumienie idei "efektywnego nauczania" i odnotowuje, czym różni się ono od IBL, od nauczania z wykorzystaniem projektu, od nauczania problemowego i od konstruktywistycznego podejścia do nauki. Opisane jest studium przypadku, gdzie wiedza z czeskiej edukacji geograficznej jest używana w procesie decyzyjnym, w którym studenci rozważają opcje i skutki budowy ośrodka narciarskiego w Brnie (w Czechach). Złożą oni później swoje wnioski do oceny rady gminy.

Rozdział 8 Nauczanie geografii z wykorzystaniem projektu

Nauczanie z wykorzystaniem projektu jest filozofią edukacji, używaną od niepamiętnych czasów. Bardzo często jest mylony z nauczanie projektowym. Jednak o atrakcyjności nauczania z wykorzystaniem projektu nie stanowią same prace projektowe, jak wielu ludzi wierzy, ale stymulowanie wewnętrznych zainteresowań uczniów. Podczas gdy większość metod nauczania stara się bardziej motywować uczniów, metoda nauczania z wykorzystaniem projektu opiera się na procesie uczenia naturalnego i działa z własną wewnętrzną motywacją uczniów i rozwijaniem ich hobby. Podstawową cechą nauczania z wykorzystaniem projektu jest partnerstwo między nauczycielem i uczniem. W nauczaniu z wykorzystaniem projektu wszystkie działania związane są z uczniem, a wszystkie wybory i decyzje związane z jego potrzebami. Jest to nauka skoncentrowana na uczniu. Stosowanie zasad nauczania z wykorzystaniem projektu w nauczaniu-uczeniu się geografii prowadzi do znacznie skuteczniejszej nauki, ułatwia nabycie odpowiedniej wiedzy lub umiejętności, aby zaoszczędzić czas zarówno dla ucznia i nauczyciela. Ale przede wszystkim pozwala na powrót do satysfakcji z uczenia się przez ucznia, a także do odczuwania satysfakcji z nauczania przez nauczyciela. Poniższa publikacja prezentuje podejście nauczania z wykorzystaniem projektu jako narzędzia do nabycia niezbędnej wiedzy i umiejętności. Celem tego rozdziału jest przedstawienie różnych metod i strategii do realizacji nauczania z wykorzystaniem projektu w ramach czeskiej edukacji geograficznej. W studiach przypadków zwrócono uwagę na typowe błędy nauczycieli w realizacji nauczania z wykorzystanie projektu. Końcowe podsumowanie próbuje uchwycić główne różnice i podejścia do stosowania tej strategii nauczania w edukacji geograficznej w Czechach w kontekście międzynarodowej praktyki szkolnej (głównie w krajach Europy Środkowej).

III. część Technologie informacyjne w nauczaniu geografii

Rozdział 9 Zastosowanie technologii informacyjno-komunikacyjnej i zasobów internetu w kształceniu przyrodniczym

Aktualnie uważa się, że technologie informacyjne stanowią zasadniczą część istotnych narzędzi nauczania, tj. takich narzędzi, których nauczyciel i uczeń używają, aby osiągnąć wskazane cele edukacyjne. Wykorzystanie mediów wizualnych, słuchowych i audiowizualnych stało się konieczne we współczesnej praktyce edukacyjnej. Dotyczy to zarówno modernizacji metod kształcenia, jak i utrzymania zainteresowania studentów, którzy są bardziej biegli w korzystaniu z tych technologii. Geografia – przedmiot poruszający problem zmieniającego się świata, nawiązując do przyczyn i skutków globalnych, regionalnych i lokalnych zmian, zarówno w środowisku naturalnym lub kulturowym, stanowi jeden z głównych obszarów do wykorzystywania nowoczesnych technologii w nauczaniu.

Rozdział 10 Technologie informacyjne w nauczaniu geografii – z perspektywy nauczyciela

Niniejszy rozdział przedstawia różne opcje i narzędzia do nauczania geografii w chmurze (np. Google Disk, YouTube, formularze, wyszukiwanie zaawansowane i łatwe publikowanie informacji, współpraca online w nauczaniu, etc.). W tym samym czasie prezentuje również praktyczny przykład wykorzystania technologii mobilnych, sieci społecznościowych oraz infografiki w nauczaniu geografii. Wprowadza się również narzędzie o nazwie ArcGIS online do tworzenia map i aplikacji z nimi związanych, które to narzędzie jest bardzo pomocne w inicjowaniu nauczania interaktywnego.

Rozdział 11 Rozwój zdolności kartograficznych i identyfikacji wizualnej obrazów – analiza (czytanie) obrazów, map oraz fotografii satelitarnych

Jako wskazują obecne trendy, zdjęcia satelitarne i lotnicze są niezwykle przydatne narzędzia, które mogą być zintegrowane w nauczaniu geografii. Zrealizowane badania przedstawia wkład w promowania wykorzystania obrazów satelitarnych i lotniczych w edukacji. Uczestnicy badania (uczniowie czeskiej szkoły w wieku 11, 15 i 19 lat) zrobili zadania przestrzenne na zdjęciach i mapach różnych typów. Przeanalizowano różnice wyników przetwarzania zadań na różnych rodzajów podłoży. Wyniki były oceniane według struktury wieku i płci badanych. Podczas uczniowie i studenci prezentowali swoje subiektywne opinie na temat trudności w czytaniu map i zdjęć różnych typów i na temat ich osobistych preferencji i obrazów map i zdjęć jako źródeł pozyskiwania i wyszukiwania informacji.

IV. część Bieżące tematy polityczne, regionalne i społeczne w geografii oraz ich zastosowanie w dydaktyce

Rozdział 12 Nauczanie geografii regionalnej i jej miejsce we współczesnym społeczeństwie: praca z informacją

Rozdział koncentruje się na podziale świata na główne regiony geograficzne. Na początku omawiane są paradygmaty geograficzne i proces globalizacji. Przedstawione są różne koncepcje proponowane zarówno przez geografów, jak i ekonomistów, lingwistów, etnologów i ekspertów w analizach politycznych. Omawiane i oceniane są różne podejścia wobec głównych regionów geograficznych. Wiele z nich zostało pokazanych na mapach. Ostatni rozdział dotyczy analizy "map mentalnych", w których pokazano, jak studenci z trzech różnych krajów postrzegają ewentualny podział planety na główne regiony geograficzne. Wyniki te następnie porównywano z powszechnie stosowanymi podziałami regionów świata. Wielu uczniów szkół średnich, a nawet studenci nie byli w stanie poradzić sobie z tym zadaniem. Uczniowie nie potrafili wypełnić prawidłowo ankiety w sposób pełny. Około 40% czeskich studentów zwróciło je w takim stanie, że prawidłowy rodzaj ewaluacji nie był niemożliwy. Jest to dość zaskakujące, zwłaszcza gdy weźmie się pod uwagę, że główne regiony świata musiały być przecież przedstawiane na kursach geografii w szkołach średnich oraz na uczelniach wyższych.

Rozdział 13 Nowa geografia polityczna

Aktualnie, po przemianach społeczno-gospodarczych na świecie, stosunki polityczne, etniczne, religijne, społeczne i ekonomiczne weszły w nową fazę rozwoju, starając się odkryć nowe formy równowagi. Struktury polityczne świata są w trakcie zmian w różnych wymiarach przestrzennych - z globalnych poprzez makro, regionalne do lokalnych. Jesteśmy świadkami procesów prowadzących zarówno do demokratyzacji systemów politycznych, jak przeciwnych trendów. Te procesy polityczne są bardzo silnie uzależnione od etnicznych i religijnych konfliktów, czy tendencji nacjonalistycznych - od pozytywnej emancypacji do negatywnej ksenofobii. W zakresie edukacji, wadą odzwierciedlającą zmiany struktur politycznych jest ich tempo, które nie może znaleźć odzwierciedlenia w podręcznikach, co wymaga stałej aktualizacji ze strony nauczycieli. Ważne jest, aby znaleźć optymalny poziom polityczno-geograficznej wiedzy dla każdej klasy w szkole, aby uniknąć konieczności zwykłego zapamiętywanie faktów. Stosunek do struktur politycznych jest najbardziej dotknięty przez subiektywne opinie nauczycieli, którzy przekazują je swoim uczniom. Dlatego duży nacisk należy położyć na szkolenia nauczycieli w kontekście nowoczesnej edukacji geografii politycznej i geografii kultury. W poniższym artykule zostały odzwierciedlone polityczne i gospodarcze zmiany w niektórych krajach Europy Środkowej i ich postrzeganie przez uczniów w szkołach średnich.

Rozdział 14 Zmiany w zachowaniu demograficznym w drugiej połowie 20 wieku i na początku 21 wieku: Możliwe wykorzystanie wyników w praktyce dydaktycznej

Zmiany zachowań demograficznych otrzymały wiele uwagi ze strony różnych dyscyplin naukowych oraz praktyk społecznych. Od lat 70-tych XX wieku, zmiany w zachowaniach demograficznych ludności zostały odzwierciedlone w krajach Europy Zachodniej i Północnej, później zidentyfikowano je w Europie Środkowej, Wschodniej i Południowej. Przejawiają się one silnie w następujących obszarach:

1. W obszarze zachowań reprodukcyjnych – zauważany jest spadek liczby urodzeń, płodności i reprodukcji.
2. W tym samym czasie ma miejsce proces starzenia się ludności – liczba i odsetek osób starszych rośnie. Duże zmiany mają miejsce w obszarze zmian modelu rodziny. Artykuł zawiera analizę zmian zachowań demograficznych ludności europejskiej, ze szczególnym uwzględnieniem populacji Czechów, Słowaków i innych populacji środkowoeuropejskiech. Tematyka ta może być ciekawa i atrakcyjna dla dzieci już w szkołach podstawowych. Celem niniejszej publikacji jest wsparcie nauczycieli, aby nauczanie odpowiednich tematów na lekcjach geografii było bardziej atrakcyjne. Aby zachęcić uczniów do aktywnego udziału w badaniach, prowadząc ich do rozpoznania, ocenienia i przedstawienia ich wiedzy demograficznej.

MIX
Papier aus verantwortungsvollen Quellen
Paper from responsible sources
FSC® C105338

If you have any concerns about our products,
you can contact us on
ProductSafety@springernature.com

In case Publisher is established outside the EU,
the EU authorized representative is:
Springer Nature Customer Service Center GmbH
Europaplatz 3, 69115 Heidelberg, Germany

Printed by Libri Plureos GmbH
in Hamburg, Germany